Molecular Analysis
and
Genome Discovery

Molecular Analysis
and
Genome Discovery

Edited by

Ralph Rapley

University of Hertfordshire, UK

and

Stuart Harbron

The Enzyme Technology Consultancy, UK

JOHN WILEY & SONS, LTD

Other Wiley Editorial Offices

John Wiley & Sons Inc.,
111 River Street, Hoboken, NJ 07030, USA

Jossey-Bass, 989 Market Street, San Francisco, CA 94103-1741, USA

Wiley-VCH Verlag GmbH, Boschstr. 12, D-69469 Weinheim, Germany

John Wiley & Sons Australia Ltd,
33 Park Road, Milton, Queensland 4064, Australia

John Wiley & Sons (Asia) Pte Ltd,
2 Clementi Loop #02-01, Jin Xing Distripark, Singapore 129809

John Wiley & Sons Canada Ltd,
22 Worcester Road, Etobicoke, Ontario, Canada M9W 1L1

Wiley also publishes its books in a variety of electronic formats. Some content that appears in print may
not be available in electronic books.

Library of Congress Cataloging-in-Publication Data

Molecular analysis and genome discovery/edited by Ralph Rapley and Stuart Harbron.
      p.; cm.
   Includes bibliographical references and index.
   ISBN 0-471-49847-5 (hardback : alk. paper) – ISBN 0-471-49919-6 (pbk. : alk. paper)
 1. Molecular diagnosis. 2. Genomics. 3. Proteomics. 4. Pharmacogenomics. 5. Polymerase chain reaction.
6. DNA microarrays.
   [DNLM: 1. Pharmacogenetics–methods. 2. Biological Markers. 3. Drug Design. 4. Genetic Techniques.
5. Pharmaceutical Preparations–metabolism. 6. Quantitative Structure-Activity Relationship. QV 38 M717
2004] I. Rapley, Ralph. II. Harbron, Stuart.
   RB43. 7. M595 2004
   615'. 7–dc22

                                                                                          2004001166

British Library Cataloguing in Publication Data

A catalogue record for this book is available from the British Library

ISBN 0 471 49847 5 hardback
      0 471 49919 6 paperback

Typeset in 10.5 on 13pt by Kolam Information Services, Pvt. Ltd, Pondicherry, India.
Printed and bound in Great Britain by TJ International, Padstow, Cornwall
This book is printed on acid-free paper responsibly manufactured from sustainable forestry in which
at least two trees are planted for each one used for paper production.

Contents

Molecular Analysis and Genome Discovery edited by Ralph Rapley and Stuart Harbron
© 2004 John Wiley & Sons, Ltd   ISBN 0 471 49847 5 (cased)   ISBN 0 471 49919 6 (pbk)

Preface

There can be no doubt that the face of diagnostics and drug discovery has changed beyond recognition over the last decade. Advances in the techniques of molecular analysis, and the ever-increasing use of automation, especially high throughput approaches, have paved the way for new, rapid and more reliable diagnostic tests. Completion of the Human Genome project has, in addition, been a startling achievement that can only accelerate the discovery of new genes and biomarkers for disease processes.

The genomics revolution promises to replace current prescription medicines with a new class of much more potent and efficacious medicines. The success of this revolution, the process of converting the claims of personalized medicine into an armamentarium specific for each patient, is dependent on continuing advances in molecular analysis and genome discovery.

This book aims to bring together these two aspects and show how the rapid advances in molecular analysis are bringing new tools and approaches to genome discovery. In combination, these provide the impetus for the development of new diagnostic techniques.

Following a historical overview of pharmacogenetics and pharmacogenomics by Werner Kalow, the first part of the book has detailed explanations of the latest techniques from experts in the field, copiously illustrated with examples from their own laboratories.

Chapters from Jörg Dötsch and Elaine Lyon present alternative approaches to quantitative and real-time PCR techniques and how they may be used in tumour and mutation detection.

Ivo Gut covers Genotyping and SNP analysis, and issues such as improving the economy of this approach are explored. Emerging techniques and the promise of high throughput SNP genotyping methods are presented.

Paal Andersen and Lars Larsen provide a comprehensive overview of mutation detection and screening. Methods for mutation scanning are also discussed.

Pyrosequencing is an exciting new approach to genomic analysis, and its principles and applications are explained and illustrated by Elahe Elahi and Mostafa Ronaghi.

Development of DNA-based microarrays have been a key area over the last decade, both from a research point of view and commercially. Their fabrication

Molecular Analysis and Genome Discovery edited by Ralph Rapley and Stuart Harbron
© 2004 John Wiley & Sons, Ltd   ISBN 0 471 49847 5 (cased)   ISBN 0 471 49919 6 (pbk)

and use are expertly covered in chapters from Magdalena Gabig-Ciminska and Andrzej Ciminski, and from Janette Burgess; a chapter from Jon Terrett's group focuses on the complementary chip-based proteomic technology.

Ciara O'Sullivan *et al.* discuss the powerful approach enabled by aptamer technology. Uniquely she relates how aptamers may be selected for use in both proteomic *and* therapeutic applications.

Kin-Ying To describes how a number of these molecular techniques may be applied to an analysis of differential gene expression.

The remaining chapters focus on the latest approaches to the identification of new target compounds, and Roberto Solari provides an excellent overview of the multi-stage drug discovery process.

Huang and Jong provide an interesting insight into potentially powerful approaches to analysing global features of microorganisms as they relate to the elucidation of new antibacterial agents.

The discovery of new target compounds based on an analysis of known structural motifs and elements is covered by Kan *et al.* and David Winkler, while Hudes, Menon and Golemis contribute a chapter investigating the impact of drugs on protein–protein interactions.

In compiling *Molecular Analysis and Genome Discovery* we have sought to combine both current and emerging approaches to analysing DNA, proteins and genomes, whilst not losing sight of the challenging outcomes required for successful drug discovery. In achieving this aim we have benefited from calling on the expertise of a distinguished and creative group of contributing authors; we have thoroughly enjoyed working with them.

Ralph Rapley
Stuart Harbron

Contributors

Andersen, Paal Skytt
Department of Clinical Biochemistry, Statens Serum Institut, Artillerivej 5, DK-2300 Copenhagen, Denmark

Baldrich Rubio, Eva
Department of Chemical Engineering, Universitat Rovira i Virgili, Avinguda Paisos Catalans 26, 43007 Tarragona, Spain

Barry, Richard
Oxford Glycosciences, Abingdon Science Park, Abingdon, UK

Burgess, Janette K.
Respiratory Research Group, Department of Pharmacology, Bosch Building, D05, University of Sydney, Sydney, NSW, Australia

Chen-Chen Kan
Keck Graduate Institute of Applied Life Sciences, 535 Watson Dr., Claremont, CA, USA, and Eidogen, 1055 East Colorado Blvd, Suite 550, Pasadena, CA 81106, USA

Ciminski, Andrzej
Department of Electronics, Telecommunications and Informatics, Gdansk University of Technology, 80–952 Gdansk Wrzeszcz, Poland

Debe, Derek A.
Eidogen, 1055 East Colorado Blvd, Suite 550, Pasadena, CA 81106, USA

Dötsch, Jörg
Department of Pediatrics, University of Erlangen-Nürnberg, Germany

Elahi, Elahe
Faculty of Science, University of Tehran, Tehran, Iran

Molecular Analysis and Genome Discovery edited by Ralph Rapley and Stuart Harbron
© 2004 John Wiley & Sons, Ltd   ISBN 0 471 49847 5 (cased)   ISBN 0 471 49919 6 (pbk)

Gabig-Ciminska, Magdalena
Department of Biotechnology, Royal Institute of Technology (KTH), S-10691
Stockholm, Sweden, and Laboratory of Molecular Biology (affiliated with the
University of Gdansk), Institute of Biochemistry and Biophysics, Polish Academy
of Sciences, Kladki 24, 80–822 Gdansk, Poland

Golemis, Erica A.
Division of General Science, Fox Chase Cancer Center, 8801 Burholme Ave.,
Philadelphia, PA 19111

Gut, Ivo Glynne
Centre National de Génotypage, Bâtiment G2, 2 rue Gaston Crémieux, CP 5721,
91057 Evry Cedex, France

Hambly, Kevin
Eidogen, 1055 East Colorado Blvd, Suite 550, Pasadena, CA 81106, USA

Homs, Mònica Campàs i
Department of Chemical Engineering, Universitat Rovira i Virgili, Avinguda
Paisos Catalans 26, 43007 Tarragona, Spain

Hudes, Gary
Division of Medical Science, Fox Chase Cancer Center, 8801 Burholme Ave.,
Philadelphia, PA 19111, USA

Jong, Ambrose
Children's Hospital, Los Angeles, and the University of Southern California, Los
Angeles, CA 90027, USA

Kalow, Werner
Faculty of Medicine, University of Toronto, Department of Pharmacology,
1 King's College Circle, Toronto, M5S 1A8, Canada

Kin-Ying To
Institute of BioAgricultural Sciences, Academia Sinica, Taipei 115, Taiwan

Larsen, Lars Allan
Wilhelm Johansen Centre for Functional Genome Research, Department of
Medical Genetics, IMBG, University of Copenhagen, Blegdamsvej 3, DK-2200
Copenhagen, Denmark

Lyon, Elaine
ARUP Laboratories, Department of Pathology, University of Utah, 500 Chipeta Way, Salt Lake City, UT 84108, USA

Menon, Sanjay
Morphochem Inc., Monmouth Junction, NJ 08852, USA

O'Sullivan, Ciara K.
Department of Chemical Engineering. Universitat Rovira i Virgili, Avinguda Paisos Catalans 26, 43007 Tarragona, Spain, and Institució Catalana de Recerca i Estudis Avançats, Passeig Lluis Companys 23, 08010 Barcelona, Spain

Rascher, Wolfgang
Department of Pediatrics, University of Erlangen-Nürnberg, Germany

Ronaghani, Mostafa
Stanford Genome Technology Center, Stanford University, 855 California Ave., Palo Alto, CA 9434, USA

Schoof, Ellen
Department of Pediatrics, University of Erlangen-Nürnberg, Germany

Sheng-He Huang
Children's Hospital Los Angeles, and the University of Southern California, Los Angeles, CA 90027, USA

Solari, Roberto
Apax Partners Ltd, 15 Portland Place, London W1B 1PT

Soloviev, Mikhail
Oxford Glycosciences, Abingdon Science Park, Abingdon, UK

Terrett, Jon
Oxford Glycosciences, Abingdon Science Park, Abingdon, UK

Winkler, David A.
CSIRO Molecular Science, Private Bag 10, Clayton South MDC, Clayton 3169, Australia

1

Pharmacogenetics and Pharmacogenomics: An Overview

W. Kalow

Pharmacogenomics has recently arisen from the well established science of pharmacogenetics; in fact, both names are sometimes mixed or interchanged by students of abnormal drug effects. The current overview will therefore start with a discussion of pharmacogenetics, its origin and its coverage; this will be followed by a description of the new aspects which pharmacogenomics brings into the study of interaction between genes and the drugs given to patients.

The origins of pharmacogenetics

Pharmacogenetics emerged from the combination of the old sciences of genetics and pharmacology. Sir Archibald Garrod (1931) anticipated the occurrence of individual differences in human reaction to drugs and environmental chemicals in his book *Inborn Factors in Diseases*. J.B.S. Haldane (1949) studied biochemical individuality and also predicted the occurrence of unusual reactions to drugs. By that time, a few genetic differences in drug response had been seen.

Snyder (1932) had described the inheritance of a deficient ability of some people to taste phenylthiocarbamide (PTC), fundamentally a pharmacologic observation. Sawin and Glick (1943) noticed a genetic lack of atropine esterase in some rabbits, a deficiency that restricted their consumption of belladonna-containing plants. During World War II, the generally safe antimalarial drug primaquine

Molecular Analysis and Genome Discovery edited by Ralph Rapley and Stuart Harbron
© 2004 John Wiley & Sons, Ltd ISBN 0 471 49847 5 (cased) ISBN 0 471 49919 6 (pbk)

caused haemolysis in many American soldiers, but only in soldiers of African descent; later, the event was shown to be due to deficiency of the enzyme glucose-6-phosphate dehydrogenase (G6PD), a deficiency frequent in Africans, where the mutant tended to cumulate because it protected its carriers from malaria (Beutler 1993). The new exciting tuberculosis-fighting drug isoniazid was seen (Hughes *et al.* 1954) to have neurological effects in some people, those later shown to have a familial deficiency of its destroying enzyme, N-acetyltransferase. Also in the 1950s, the new muscle-paralysing drug succinylcholine, used during anaesthesia, killed some patients while most had no trouble; the cause was found to be a mutation which inactivated plasma cholinesterase, the succinylcholine-destroying enzyme (Kalow 1956).

Several physicians and scientists felt stimulated by these reports. A committee of the American Medical Association invited the geneticist Motulsky (1957) to write a paper entitled 'Drug Reactions, Enzymes, and Biochemical Reactions' for their journal. The geneticist Vogel (1959) in Germany coined the word 'Pharmacogenetics' in a paper describing 'Modern Problems of Human Genetics'. I was in the process of summarizing all pertinent findings in a book (Kalow 1962). Pharmacogenetics was now an established entity.

The initial progress of pharmacogenetics

As time went by, more pharmacogenetic discoveries were made. Denborough (Denborough and Lovell 1960; Denborough *et al.* 1962) reported the occurrence of excessively high body temperatures (hyperpyrexia), rigidity and death among various family members in response to general anaesthetics; this condition is now called malignant hyperthermia. Dundee *et al.* (1962) emphasized the paralytic effect of barbiturates in cases of hepatic porphyria. Aebi *et al.* (1961) observed in Switzerland cases with a genetic lack of catalase activity, cases as seen before only in Japan. Von Wartburg *et al.* (1965) reported genetic variation of alcohol dehydrogenase. Further pharmacogenetic cases were reported in the following years, but, in addition, some of the older reports were extended. For instance, Kalow *et al.* (1970) showed that malignant hyperthermia represented a biochemical defect in skeletal muscle that caused the muscle to respond abnormally to caffeine, thereby making *in vitro* tests and predictions of this condition possible.

All these cases represented relatively rare observations and therefore did not dramatically affect the practice of medicine. This changed with the report on familial failures of the oxidation of debrisoquine, a sympatholytic, antihypertensive drug. Dr R.L. Smith (1986), a medical investigator in England, took the drug personally for experimental purposes, but suffered a marked and extensive hypotensive episode. He therefore started to investigate the metabolism of the drug, found the metabolic defect, and reported the findings in an excellent publication (Mahgoub *et al.* 1977). A short time later, Eichelbaum *et al.* (1979) in Germany reported that some people could not metabolize sparteine, an anti-arrhythmic drug. Subsequently, other investigators (Bertilsson *et al.* 1980; Inaba *et al.* 1980) found that the same liver enzyme metabolized both debrisoquine and sparteine. All measurements were based on the drug/metabolite ratio in urine.

The medical interest in this topic was stimulated by several facts. Oxidation is an important and widespread metabolism in human and animal bodies. The defect was not rare but obviously affected quite a few people; that is, it was a polymorphism, a frequently occurring genetic alteration. The enzyme was a cytochrome P450 (CYP2D6), an important class of drug-metabolizing enzymes. It was soon found that the enzyme metabolized many drugs, presently thought to be as many as 20 per cent of all clinically used drugs. Medline cites currently almost 2000 publications dealing with CYP2D6.

The usual clinical effect of the deficiency is an over-response, or toxic response, to any drug that is not metabolized by CYP2D6 (Meyer 2001). However, if a drug needs metabolic activation, an under-response, or lack of response, can also occur. For instance, the debrisoquine-metabolizing enzyme is also the one which converts codeine into morphine. If the enzyme lacks, this conversion does not take place, and codeine is not a pain-killing analgesic (Dayer *et al.* 1988).

Molecular genetic methods enlarged pharmacogenetics

As we can see, pharmacogenetics was in the beginning entirely concerned with functional differences in drug response or drug metabolism, that is, *phenotype* differences which could be directly observed or measured in family members. Pharmacogenetics became much enriched when new methods, described in the following chapters, allowed a direct look at the DNA of mutant genes, thereby determining *genotype*.

The use of the methods of molecular genetics disclosed many kinds of mutations (Stockley and Ray 2001), revealed as changes of nucleic acids. The most frequent variants are SNPs, *S*ingle *N*ucleotide *P*olymorphisms, indicating simple exchanges of two nucleotides; many genes carry several, often many SNPs (Grant and Phillips 2001 and see Gut; Chapter 4). For instance, about 130 SNPs have been described for G6PD, the enzyme glucose-6-phosphate dehydrogenase (http://www.bioinf.org.uk/g6pd/). However, some SNPs do not alter the amino acid of the derived protein, and therefore do not affect protein structure and function; they are called *silent*. Some mutations may change reaction rates, substrate binding, sometimes affecting only selected substrates. There are also other complexities. For instance, absence of an enzyme activity could be caused by any one of four kinds of mutations: frame shift, splicing defect, strip codon or gene deletion. In addition, genetic tests sometimes revealed gene duplication or even multiplication, thereby greatly decreasing the action of dependent drugs. Table 1.1 lists the number of alleles among members of the drug-metabolizing

Table 1.1 Cytochrome P450 variations

CYP Designations	No.	
	Alleles	Nucleotide Changes
1A1	9	14
1A2	14	16
1B1	20	16
2A6	18	29
2B6	16	13
2C8	7	19
2C12	12	6
2C19	11	20
2D6	77	249
2E1	13	17
2J7	7	6
3A4	17	11
3A5	12	6
3A7	5	10
4B1	4	7
5A1	12	11
Sum 16	**254**	**450**

From: Home page of the Human Cytochrome Allele Nomenclature Committee. Oct. 2002 (*http://www.imm.ki.se/CYPalleles/*).

P450 cytochromes. Note that some alleles have many nucleotide changes, and they are often defined by different nucleotide change compositions.

For many years, any attempt to search for the presence of a metabolic defect had required the application of a test drug to the subject. The ability to look directly for a gene alteration made such procedures unnecessary. It meant less bother to the patient, simplified the search, allowed an increase in the number of people to be tested, and thus improved the assessment of its frequency.

Drug receptors

A dramatic effect of the introduction of molecular methodology was the ability directly to investigate the structure and function of drug receptors (Weber 1997). Prior to that time, virtually all of pharmacogenetics pertained to alterations of drug metabolism, which could be determined by using traditional chemical methods to measure drug or metabolite concentrations. Drug receptors are located in brain or other tissues and are the elements to which a drug must bind in order to deliver its message. To study the structure of the receptors located in cells or on their surface is difficult, but when done allows an identification or reconstruction of its gene. Since all body cells carry all genes, one can look for the now recognizable receptor genes in blood cells, and thus look for mutations of such a gene in blood leukocytes. Weber (2001) described variants of many receptors of pharmacogenetic interest, for instance, glucocorticoid, androgen, oestrogen, retinoid, vitamin D, arylhydrocarbon (AH), ryanodine receptors. By 1998, there were almost 3000 human receptor polymorphisms. Thus, the study of genetic variation of drug receptors has become an important aspect of pharmacogenetics.

There is still a lower frequency of receptor variants than of variants of drug-metabolizing enzymes. This could have methodological reasons: metabolic variants could be determined traditionally with chemical methods while the identification of receptor variants requires molecular methodology. However, the frequency difference might also indicate Darwinian selection: receptors are vital components of our internal messenger systems and thus always have physiological functions; their variation may have pathological consequences, while the mutation of many drug-metabolizing enzymes is very often immaterial in the absence of drugs.

Interethnic Pharmacogenetics

Several interethnic differences in drug response were known early, as for instance the above-named primaquine haemolysis (Beutler 1993) of the Second World War, recognizable as a phenotypic difference. When structural enzyme variants were chemically and functionally established and their inheritance shown by family studies, interethnic differences were reported as being genetic (Kalow 1982), although only the introduction of molecular genetic methodology is able to provide the definite proof that interethnic pharmacological differences are genetic.

The reason is that there are many environmental factors which may alter the metabolism of a drug and which may differ between human populations, such as kinds of food, level of nutrition, climate, status of health and probably some customs. Within a population, family studies can prove the genetic nature of a drug response variation, but a difference between populations cannot thus shown to be genetic. Even if a drug response is genetically controlled in both of two populations, its difference between the populations can still have environmental causes. Prior to the arrival of DNA tests, some observed interethnic differences were called 'transcultural' (Blackwell 1976; Chien 1978). Some investigators did not want to be considered racist. In short, a firm proof that any interethnic difference of a drug response is genetic requires DNA tests. There are now many such proofs (Kalow *et al.* 2001).

Reports of genetic differences between human races have sometimes caused emotional debate (Schwartz 2001). Thus, the validity of customary racial divisions has been questioned by some authors who do not believe in the biological utility of these divisions. For instance, Kurzban *et al.* (2001) stated that the division of humanity into races is a mental construct, based on our wish to categorize *us* versus *them*. By contrast, Cavalli-Sforza *et al.* (1994) have tested 48 variable genes in 42 geographically separated populations. They determined the similarities and the differences, and then calculated genetic distances between populations; they concluded that there are nine population 'clusters', consisting of one kind of African, two of Caucasians, three of Asians, plus Amerinds, Pacific Islanders and New Guineans-Australians. Africans were most different from all other populations. Since mutation rates tend to be constant over long time scales, this suggests that the separation of Africans from all other human populations is very old. This fits to the archaeological concept that *Homo sapiens* spread out of Africa about 100 000 years ago stepwise to populate the earth (Bräuer 1989). If a population migrates, it carries its genes along and is recognizable as a separate entity, unless it mixes with another population. In short, our ethnic or racial divisions are based on genetic facts as well as on time and geography.

Population differences of drug effects are well established because they are readily measurable. Drugs, their metabolites, and most of their effects can be quantified and thus determined with precision. Except for measurements of body dimensions, this is in sharp contrast to many other human features; Stephen Jay Gould's book (1981) entitled *The Mismeasure of Man* well illustrates the problem. Thus, the abilities to determine via DNA which genes are variable, and to measure precisely the magnitude of expressed variation, have turned pharmacogenetics into a valuable component of human genetics.

Interethnic pharmacogenetics has become a clinically important science. Of 42 drug-metabolizing enzymes which are known to be genetically variable, 27 (i.e. 64%) have also shown a population difference (Kalow 2001); since population comparisons have never been made for most of the rest, the true percentage must be higher.

In many cases, the difference consisted simply of a different frequency of a given variation. For instance, the genetic lack of CYP2D6 activity has an approximately 7% occurrence in Europeans but only about 1% in both Chinese and Africans. However, the kinds of variants also often differ (Kalow 2001): a study of 11 nucleotide changes in the gene of cytochrome CYP2D6, in many subjects of European, Japanese, Chinese and African populations, indicated that only one of these (G-4268-C) occurred in all populations. This suggests that this mutation occurred before humans left Africa. Of the 11 nucleotide changes, eight were found when the data from different European countries were combined. There were five mutations among Chinese, three among Japanese and four among Africans. The data confirm that the time of occurrence of a mutation determines many population differences.

Pharmacogenetics: a summary

Pharmacogenetics initially meant a delineation of monogenic person-to-person differences in drug response or metabolism. The introduction of molecular genetic methodologies allowed genetic studies of drug receptors, and it proved the genetic nature of many pharmacological differences between ethnic groups. Because drug effects and drug metabolites tend to be clearly measurable entities, pharmacogenetics is a valuable component of the science of human genetics.

Multifactorial variation

All examples quoted so far represent pharmacological consequences of one or other mutation of a single gene. In other words, we dealt with monogenic variations. However, modern pharmacogenetics also encompasses multifactorial variations, that is, differences of drug response which cannot be explained by one or other mutation, but only by a combination of effects of several or many mutant genes, usually interacting with environmental factors. A first indication of this truth appeared 75 years ago (Trevan 1927).

The fact that all individuals, be they humans or animals, tend to respond differently to any given drug was established by Trevan's (1927) work. He introduced the concept and term LD_{50}, the dose of a drug that killed 50% of a group of animals; soon, the concept was broadened by the new term ED50, the *effective dose* for 50% of the subjects receiving the drug. Thus, the always-present person-to-person differences of drug response were established three-quarters of a century ago. For a long time, nobody asked whether or not genetic factors contributed to these differences. It therefore was an important step when twin studies showed that many pharmacological variations had a genetic cause, even when any well-established Mendelian inheritance was absent (Vesell 1989).

It is good to remember that even a simple and single alteration of a drug response or a drug's fate may have numerous potentially complicated causes. As an example, let me enumerate some factors that may change the rate of metabolism of a drug:

1. There may be a genetic change of the enzyme.

2. The gene may be of normal structure, but it may be poorly expressed.

3. The enzyme protein may be degraded too fast.

4. Hormones, foods, or other drugs may inhibit the enzyme.

5. The drug may not readily reach the intracellular enzyme because of poor blood flow, too much binding to albumin, or because of a genetic failure of a drug transporter.

Thus, even a simple event can have many, often interacting causes. The existence of monogenic alterations – and thereby the existence of classical pharmacogenetics – is almost surprising.

Data sets of multifactorial pharmacogenetics, or non-Mendelian pharmacogenetics, are usually reported in terms of mean or average, while the data distribution is graphically illustrated by smooth, bell-shaped (Gaussian) curves. These are the mathematical descriptions of most pharmacological events, but they also raise several questions.

As stated above, multifactorial variation usually represents an interplay of genetic and environmental factors. Sometimes, one may wish to know whether a variation is overwhelmingly determined by one or the other. The traditional way to answer this question is the use of twin studies (Vesell 1989) which compare the differences between pairs of identical and of fraternal twins; if the differences between identical are much smaller than between fraternal pairs, the difference is heritable and thus mostly genetic. For many human features, twin studies are the only means to test the nature of variation. However, as shown in Table 1.2, this is not so for drug response variants: a drug effect comes and goes, and therefore can be repeated in a given person. It is thus possible to give a drug two or more times to each subject in a group of perhaps 20 people, and then compare mathematically the magnitudes of the within-person and the between-person differences. If the drug is given at the same dose, at the same time of day, after an appropriate interval, the result is an indication of the genetic component of the variation of the drug response (Kalow *et al.* 1998, 1999; Ozdemir *et al* 2000). Since recruiting twins for a study is often difficult, this method is comparatively easy. Sometimes, bioequivalence studies can be evaluated by this method, thereby providing additional information at relatively low additional cost. If the pharmaceutical industry tests a new drug in many volunteers, some of these may receive the drug twice, again providing new information.

The normal distribution of a drug response does not usually raise much interest except when there is a difference between two populations. For example, codeine

Table 1.2 Repeated drug administration

Give a person a drug twice
    a few days apart under identical conditions,
    response differences will be mostly environmental.

Compare drug effects between people,
    response differences will be genetic and environmental.

Calculate standard deviations of within-person differences as SD_w and of between-person differences as SD_b.

Square SD_W and Sd_b to become SD_W^2 and Sd_b^2.
    $Sd_b^2 - SD_W^2 / SD_b^2 = r_{GC} =$ Genetic component of the response variation.

glucuronidation is normally distributed with significally different means in groups of Swedish and Chinese people (Yue and Säwe *1992*). Such differences between population averages may be epidemiologically significant even if the difference is relatively small in comparison to the scatter of responses within the populations (Kalow 1992; Kalow and Bertilsson 1994). Consider the fact that people with abnormal drug reactions represent the edge, not the mean, of a distribution curve. If the means or averages of the distribution curves differ, the edges of the curves are also likely to differ. It thus can happen that the percentage of people badly affected by a drug differ significantly between the populations, although the difference between the averages seems to have no biological significance.

Multifactorial variation can affect either or both pharmacokinetic or pharmacodynamic events, which all represent complex processes.

Pharmacokinetic changes mean that there is an abnormal drug concentration at the intended site of action or elsewhere in the body. This can happen via a failure of drug absorption, a change of drug metabolism or via changes of kidney function or bile secretion. There may be too much, or too little binding of a drug to albumin or other tissues. Drug transport may be abnormal.

Pharmacodynamic changes usually mean that the receptor, the site of drug attachment, may not respond with proper action. The receptor may be unable to bind the drug as expected, perhaps because the receptor is structurally changed; such a change may cause a drug to bind to a wrong receptor. The receptor may be occupied by another drug or chemical, displaying drug–drug interaction. The receptor may bind the drug but not properly submit his message downstream.

New problems and opportunities

The science of genetics is rapidly changing as the ability to determine human and other genomes makes new comparisons possible. Since a knowledge of the genes themselves has only restricted utility for medicine, work is focusing on the study of gene products, proteins and their functions, using the science of bioinformatics to interpret the data. At the same time, many genetic concepts are undergoing revision due to two realizations. First, gene function can be much affected by differences of gene expression. Secondly, investigations of the multiple genetic causes of common diseases promise the development of new drugs and of the emergence of personalized medicine.

Variation of gene expression

Gene expression is principally measured by the amount of RNA (see Chapters 2 and 3) which is formed by the DNA of the gene; RNA formation determines the protein production, the building stones of our bodies which also control our functions. The technical ability to measure gene expression is based on the creation of gene arrays (Kaufman 2000; Madden *et al.* 2001 see Chapters 7, and 8). Differences of gene expression create the differences between the cells of our organs, and cause many differences between the sexes.

The realization that gene expression can change is affecting the science of genetics by limiting the importance of the search for gene mutations. A functional difference of the gene of a drug-metabolizing enzyme does not necessarily indicate the presence of a mutation but could be caused by an altered gene expression.

Furthermore, the ability to measure gene expression is opening a new means to study gene function (see Chapter 9). By comparing gene expression between a healthy and a diseased tissue gene alterations may be determined which cause or are altered by the disease (Bals and Jany 2001). A similar comparison of tissues prior to and after drug exposure can tell us which genes are directly or indirectly affected by the drug (Steiner and Anderson 2000). Thus, gene expression studies promise to extend the science of pharmacology.

For instance, cigarette smoke contains chemicals that bind to the Ah receptor (Nebert *et al.* 1991), thereby increasing the expression of a gene which controls the formation of the cytochrome CYP1A2; the net effect for the patient is an increased metabolism and thereby reduced action of drugs metabolized by it. For example, the ratio of caffeine metabolites which indicates the activity of cytochrome CYP1A2 is significantly larger in smokers than in non-smokers (Kalow and Tang 1991). Rae *et al.* (2001) found that rifampin did not increase the mRNA of CYP2C18, CYP2E1 and CYP2D6, but it increased 3.7-fold that of CYP2C9, 6.5 times that of CYP2C8, but 55.1 times that of CYP3A4. Another example of a drug effect on gene expression is drug addiction; it results from changes of gene expression in the brain (Torres and Horowitz 1999). This accounts for the long duration of addictions which persist even if the addicting drug is no longer measurable in the body (Nestler 2001).

It is a very important fact that gene expression can be affected by drugs or toxicants. This can imply that pharmacogenetics is turned on its head. Pharmacogenetics taught us that genes affect drug action; now we know that drugs may affect gene function. In fact, we often do not know whether a drug affects a protein directly, or whether it acts by changing expression of the protein's gene. In addition, many hormones also change gene expression, and so do some pathological factors. Furthermore, gene expression can be caused by gene–gene interaction.

Epigenetic change may also affect gene expression (Wolffe and Matzke 1999), for example, by methylation of the cytosine in genes, a process that inactivates them, this is often but not always a temporary inactivation (Petronis 2001).

Epigenetics, as part of the science of gene expression, also serves physiologically to create genetic differences between the cells of different organs, or between the two sexes.

Genes, drugs and diseases

Genetic diseases like cystic fibrosis or Wilson's disease are rare, because the mutations which cause them represent a biological disadvantage, and thus tend to be eliminated in any population by evolutionary forces. By contrast, all common diseases can be caused or enhanced by numerous genes, usually inter-acting with environmental factors. For example, at least seven genes are able to contribute to Alzheimer's disease (Lehmann *et al.* 2001), often with some partici-pation of aluminium exposure (Flaten 2001). At least 20 independent gene loci influence susceptibility to asthma (Ober 2001). This kind of knowledge is cur-rently stimulating the search for new drugs which may affect such causative genes or their products. However, not all the genes capable of causing the disease are usually doing so in all patients; only one or a few such genes are likely to be responsible for the illness of a particular patient. When knowing which gene is acting in that patient, the drug can be administered that was designed to counter-act the effect of that gene, and thereby eliminate the disease in that patient. This is the hope held out by personalized medicine.

This concept fits well with the clinical experience, that some patients with the same disease and treated with the same drug will benefit from it, while others do not. The trouble is that a physician's diagnosis of a disease is ordinarily based on the patient's symptoms and histories but not on the cause of the disease. Knowing the gene which contributes to the disease in a given patient means some knowledge of the disease's cause. The occurrence of drug response differ-ences between people on the basis of disease-carrying genes is a medical experi-ence which lies outside traditional pharmacogenetics.

The procedures to test the causes of common diseases require the investigation of many genes, more genes and their variants than one can hope readily to pinpoint by traditional genetic methods. This is where genomics comes in as a new science, based on technologies which allow a simultaneous look at numerous genes and their mutations (see Chapter 9). Since we know from pharmacogenetics that most drug responses are controlled or influenced by genes, and since the creation of personalized medicine requires many studies involving genes and drugs, the use of genomic methods will be practical, this is pharmacogenomics (Kalow *et al.* 2001).

One can summarize the situation with the statement that pharmacogenetics was mostly concerned with drug safety, while pharmacogenomics will be aimed in addition to promote drug efficacy.

Summary and conclusions

The biological existence of pharmacogenetic events was perceived in the early decades of the 20th century by some geneticists, and some pertinent observations were reported at about the same time by a few pharmacologists; thus the established sciences of genetics and of pharmacology became combined. Some scientific reviews established pharmacogenetics as a new entity in the 1950s.

Following a number of new observations, the discovery of the metabolic deficiency of debrisoquine metabolism in some families, resulted in medical attention to pharmacogenetics: it was a deficiency of drug oxidation, occurring with relative frequency in different subjects, and is now known to affect a substantial percentage of all drugs. A couple of thousand scientific publications are dealing with this deficiency.

The science of pharmacogenetics changed with the introduction of the methods of molecular genetics. Many drug-metabolizing enzymes were shown to carry several mutations. It become possible to look for genetic variation of drug receptors. The genetic nature of many interethnic differences in drug metabolism or response was proven.

Traditional pharmacogenetics dealt entirely with the effects on drug response of single gene mutations. Further observations indicated the occurrence of multifactorial variation of drug responses, that is variation caused by simultaneous alterations of numerous genes. Twin studies proved the heritability, and thereby the pharmacogenetic nature, of many such alterations, but twin studies can be replaced by assaying repeated drug administrations.

The realization that gene expression is variable, and that it may be affected by drugs, brings a totally new aspect to pharmacogenetics: it is not only that genes determine a drug response, but drugs may determine gene function.

The realization that all common diseases may be contributed to by many genes raises hopes for the identification of new drug targets. Since not all possibly contributing genes are present in all patients, we have an explanation for interindividual difference in the response to a given drug, an explanation which lies outside the lessons tought by traditional pharmacogenetics. It is hoped that by knowing a patient's genes, the right drug can be selected for a given patient, thereby creating personalized medicine. The genetic work required is very extensive but will be helped by the use of genomic methods; thus, pharmacogenomics enters.

References

Aebi, H., Heiniger, J.P., Butler, R. and Hassig, A. (1961) Two cases of acatalasia in Switzerland. *Experientia* **17**: 466.

Bals, R. and Jany B. (2001) Identification of disease genes by expression profiling. *Eur Respir J* **18**: 882–889.

Bertilsson, L., Dengler, H.J., Eichelbaum, M. and Schulz, H-U. (1980) Pharmacogenetic covariation of defective N-oxidation of sparteine and 4-hydroxylation of debrisoquine. *Eur J Clin Pharmacol* **17**: 153–155.

Beutler, E. (1993) Study of glucose-6-phosphate dehydrogenase: History and molecular biology. *Am J Hematol* **42**: 53–58.

Blackwell, B. (1976) Culture, morbidity, and the effects of drugs. *Clin Pharmacol Ther* **19**: 668–674.

Bräuer, G. (1989) The evolution of modern humans: A comparison of the African and non-African evidence, in Mellars, P. and Stringer, C. (eds) *The Human Revolution: Behavioural and Biological Perspectives on the Origins of Modern Humans*, Princeton NJ, Princeton University Press, pp. 123–154.

Cavalli-Sforza, L.L., Manozzi, P. and Piazza, A. (1994) *The History and Geography of Human Genes*, Princeton, NJ, Princeton University Press.

Chien, C. (1978) Transcultural psychopharmacology. Summary of the symposium *Psychopharmacol Bull* **14**: 89–91.

Dayer, P., Desmeules, J., Leemann, T. and Striberni, R. (1988) Bioactivation of the narcotic drug codeine in human liver is mediated by the polymorphic monooxygenase catalyzing debrisoquine 4-hydroxylation. *Biochem Biophys Res Comm* **152**: 411–416.

Denborough, M.A. and Lovell, R.R.H. (1960) Anaesthetic deaths in a family. *Lancet* **ii**: 45.

Denborough, M.A., Forster, J.F.A., Lovell, R.R.H., Maplestone, P.A. and Villiers, J.D. (1962) Anaesthetic deaths in a family. *Br J Anaesth* **34**: 395–396.

Dundee, J.W., McClerry, W.N.C. and McLoughlin, G. (1962) The hazard of thiopental anaesthesia in porphyria. *Anesth Analg* **41**: 567–574.

Eichelbaum, M., Spannbrucker, N., Steinke, B. and Dengler, H.J. (1979) Defective N-oxidation of sparteine in man: A new pharmacogenetic defect. *Eur J Clin Pharmacol* **16**: 183–187.

Flaten, T.P. (2001) Aluminium as a risk factor in Alzheimer's disease, with emphasis on drinking water. *Brain Res Bull* **55**: 187–196.

Garrod, A.E. (1931) *Inborn Factors in Disease: An Essay*, Oxford University Press, (Abstract).

Gould, S.J. (1981) *The Mismeasure of Man*, New York, London, Norton & Co.

Grant, D.M. and Phillips, M.S. (2001) Technologies for the analysis of single-nucleotide polymorphisms, in Kalow, W., Meyer, U.A. and Tyndale, R.F. (eds) *An Overview, Pharmacogenomics*, New York, Basel, Marcel Dekker, Inc., pp. 183–190.

Haldane, J.B.S. (1949) Disease and evolution. *La Ricerca Scientifica* **19**: 68–75.

Hughes, H.B., Biehl, J.P., Jones, A.P. and Schmidt, L.H. (1954) Metabolism of isoniazid in man as related to the occurrence of peripheral neuritis. *Am Rev Tuberculosis* **70**: 266–273.

Inaba, T., Otton, S.V. and Kalow, W. (1980) Deficient metabolism of debrisoquine and sparteine. *Clin Pharmacol Ther* **27**: 547–549.

Kalow, W. (1956) Familial incidence of low pseudocholinesterase level. *Lancet* **ii**: 576–577.

Kalow, W. (1962) *Pharmacogenetics: Heredity and the Response to Drugs*, Saunders Comp. Philadelphia, London.

Kalow, W. (1982) Ethnic differences in drug metabolism. *Clin Pharmacokinet* **7**: 373–400.

Kalow, W. (1992) Pharmacoanthropology and the genetics of drug metabolism, in Kalow, W. (Ed.) *Pharmacogenetics of Drug Metabolism*, Internatl. Encycl. Pharmacol. Therap. Sect. 137, New York, Pergamon Press, Inc, pp. 865–877.

Kalow, W. (2001) 'Interethnic differences in drug response,' in Kalow, W., Meyer, U.A. and Tyndale, R. (eds) *Pharmacogenomics*, New York, Marcel Dekker Inc., pp. 109–134.

Kalow, W. and Bertilsson, L. (1994) Interethnic factors affecting drug response. *Adv Drug Res* **25**: 1–59.

Kalow, W. and Tang. B-K. (1991) Caffeine as a metabolic probe: Exploration of the enzyme-inducing effect of cigarette smoking. *Clin Pharmacol Ther* **49**: 44–48.

Kalow, W., Britt, B.A., Terreau, M.E. and Haist, C. (1970) Metabolic error of muscle metabolism after recovery from malignant hyperthermia. *Lancet* **ii**: 895–898.

Kalow, W., Tang, B-K. and Endrenyi. L. (1988) Hypothesis: Comparisons of inter- and intra-individual variations can substitute for twin studies in drug research. *Pharmacogenetics* **8**: 283–289.

Kalow, W., Ozdemir, V., Tang, B-K., Tothfalusi, L. and Endrenyi, L. (1999) The science of pharmacological variability: an essay. *Clin Pharmacol Ther* **66**: 445–447.

Kalow, W., Meyer, U.A. and Tyndale, R.F. (2001) *Pharmacogenomics. Drugs and the Pharmaceutical Sciences*, New York, Marcel Dekker Inc.

Kaufman, R.J. (2000) Overview of vector design for mammalian gene expression. *Mol Biotechnol* **16**: 151–160.

Kurzban, R., Tooby, J. and Cosmides. L. (2001) Can race be erased? Coalitional computation and social categorization. *Proc Natl Acad Sci USA* **98**: 15387–15392.

Lehmann, D.J., Williams, J., Mcbroom. J. and Smith, A.D. (2001) Using meta-analysis to explain the diversity of results in genetic studies of late-onset Alzheimer's disease and to identify high-risk subgroups. *Neuroscience* **108**: 541–554.

Madden, S.L., Wang, C. and Landes, G. (2001) Serial analysis of gene expression: transcriptional insights into functional biology, in Kalow, W., Meyer, U.A. and Tyndale, R.F. (eds), *Pharmacogenomics*, New York, Basel, Marcel Dekker, Inc., pp. 223–252.

Mahgoub, A., Dring, L.G., Idle, J.R., Lancaster, R. and Smith, R.L. (1977) Polymorphic hydroxylation of Debrisoquine in man. *Lancet* **ii**: 584–586.

Meyer, U.A. (2001) Clinical viewpoints, in Kalow, W., Meyer, U.A. and Tyndale, R.F. (eds), *Pharmacogenomics*, New York, Marcel Dekker, Inc, pp. 135–150.

Motulsky, A.G. (1957) Drug reactions, enzymes, and biochemical genetics. *JAMA* **165**: 835–837.

Nebert, D.W., Petersen, D.D. and Puga, A. (1991) Human AH locus polymorphism and cancer: Inducibility of CYP1A1 and other genes by combustion products and dioxin. *Pharmacogenetics* **1**: 68–78.

Nestler, E.J. (2001) Molecular basis of long-term plasticity underlying addiction. *Nat Rev Neurosci* **2**: 119–128.

Ober, C. (2001) Susceptibility genes in asthma and allergy. *Curr Allergy Asthma Rep* 1, pp. 174–179.

Ozdemir, V., Kalow, W., Tang, B-K., et al, (2000) Evaluation of the genetic component of variability in CYP3A4 activity: a repeated drug administration (RDA) method, *Pharmacogenetics*, **10**, pp.

Petronis, A. (2001) 'Human morbid genetics revisited: Relevance of epigenetics'. *Trends Genet* **17**: 142–146.

Rae, J.M., Johnson, M.D., Lippman, M.E. and Flockhart, D.A. (2001) Rifampin is a selective, pleiotropic inducer of drug metabolism genes in human hepatocytes: Studies with cDNA and oligonucleotide expression arrays. *J Pharmacol Exp Ther* **299**: 849–857.

Sawin, P.B. and Glick, D. (1943) Hydrolysis of atropine by esterase present in rabbit serum. *Proc Nat Acad Sci* **29**: 55–59.

Schwartz, R.S. (2001) Racial profiling in medical research. *New Engl J Med* **18**: 1392–1393.

Smith, R.L. (1986) Introduction: Human genetic variations in oxidative drug metabolism. *Xenobiotica* 16: 361–365.

Snyder, L.H. (1932) Studies in human inheritance, IX. The inheritance of taste sensitivity in man. *Ohio J Sci* **32**: 436–440.

Steiner, S. and Anderson, N.L. (2000) Expression profiling in toxicology – potentials and limitations. *Toxicol Lett* **112–113**: 467–471.

Stockley, T.L. and Ray, P.N. (2001) Molecular diagnostics and development of biotechnology-based diagnostics, in Kalow, W., Meyer, U.A. and Tyndale, R.F. (eds) *Drugs and the Pharmaceutical Sciences*, Pharmacogenomics, Marcel Dekker Inc., New York, Vol. 113, pp. 169–181.

Torres, G. and Horowitz, J.M. (1999) Drugs of abuse and brain gene expression. *Psychosom Med* **5**: 630–635.

Trevan, J.W. (1927) The error of determination of toxicity. *Royal Society of London, Proceedings 1 Series B* **101**: 483–514.

Vesell, E.S. (1989) Pharmacogenetic perspectives gained from twin and family studies. *Pharmac Ther* **41**: 535–552.

Vogel, F. (1959) Moderne Probleme der Humangenetik. *Ergebnisse der inneren medizin und kinderheilkunde* **12**: 65–126.

Von Wartburg, J.P., Papenberg, J. and Aebi, H. (1965) An atypical human alcohol dehydrogenase. *Can J Biochem* **43**: 889–898.

Weber, W.W. (1997) *Pharmacogenetics*, New York, Oxford University Press, Inc.

Weber, W.W. (2001) Pharmacogenetics – receptors, in Kalow, W., Meyer, U.A. and Tyndale, R.F. (eds) *Pharmacogenomics*, New York, Basel, Marcel Dekker, Inc, pp. 51–80.

Wolffe, A.P. and Matzke, M.A. (1999) Epigenetics: regulation through repression. *Science* **15**: 481–486.

Yue, Q. and Säwe, J. (1992) Interindividual and interethnic differences in codeine metabolism, in Kalow, W. (Ed.), *Pharmacogenetics of Drug Metabolism*, Int Encycl Pharmacol Therap Sect 137, New York, Oxford, Pergamon Press, pp. 721–727.

2

Quantitative TaqMan Real-time PCR: Diagnostic and Scientific Applications

Jörg Dötsch, Ellen Schoof and **Wolfgang Rascher**

Summary

This chapter summarizes the present data on the diagnostic and scientific applications of TaqMan real-time PCR, a fluorescence-based method for quantification of gene expression and gene copy number. Particular emphasis is spent on precision and accuracy and the comparison of TaqMan real-time PCR with more traditional methods for competitive quantitative PCR. Apart from TaqMan real-time PCR, a number of similar approaches have been validated such as molecular beacons, hybridization probes and DNA-binding dyes. Pitfalls of TaqMan PCR include improper use (lack of standardization), the need for an experienced technician, and the danger of false-negative results.

Provided a proper assessment for each new marker is made, this method has been shown to be an easy, fast and reliable alternative to the more traditional methods for the determination of gene amplification and gene expression. The combination with techniques such as single cell picking might be a future target of the application of real-time PCR.

Molecular Analysis and Genome Discovery edited by Ralph Rapley and Stuart Harbron
© 2004 John Wiley & Sons, Ltd   ISBN 0 471 49847 5 (cased)   ISBN 0 471 49919 6 (pbk)

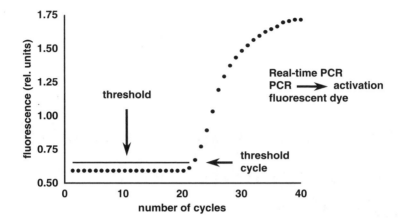

Figure 2.1 Sketch of the principle of TaqMan PCR for the quantification of gene expression. By measuring the amplicon concentration in the early exponential phase of the PCR reaction, the exhaustion of reagents is avoided. (Modified according to Dötsch *et al.* 2001)

Introduction

The invention of real-time PCR has revolutionized the quantification of gene expression and DNA copy number measurements. However, after the first documentation of real-time PCR in 1993 (Higuchi *et al.* 1993) it took several years for this method to become a mainstream tool. PCR generates DNA copies in an exponential way. As soon as resources are exhausted, however, the so-called plateau phase of PCR reaction is reached, making quantification very unreliable. Therefore, quantification appears most reliable in the early exponential phase of PCR, i.e. in a 'real-time' fashion. To ensure measurements in this phase of the PCR cycle, real-time PCR measures as soon as the threshold of detection is definitely reached. The cycle of PCR at which this occurs is then named the threshold cycle (Ginzinger 2002) (Figure 2.1). It is the objective of this chapter to describe the possibilities of TaqMan real-time PCR for mRNA and DNA quantification and to discuss pitfalls and alternatives.

Principles of TaqMan real-time PCR

The use of the TaqMan reaction has been described in a number of original and review articles (Dötsch *et al.* 1999, 2001; Orlando *et al.* 1998). This approach makes use of the 5' exonuclease activity of the DNA polymerase (AmpliTaq Gold). Briefly, within the amplicon defined by a gene specific PCR primer pair an oligonucleotide probe labelled with two fluorescent dyes is created, and designated as a TaqMan probe. As long as the probe is intact, the emission of the reporter dye (i.e. 6-carboxy-fluorescein, FAM) at the 5′-end is quenched by the second fluorescence dye (6-carboxy-tetramethyl-rhodamine, TAMRA) at the 3′-end. During the extension phase of PCR, the polymerase cleaves the Taq-Man probe resulting in a release of reporter dye. The increasing amount of reporter dye emission is detected by an automated sequence detector combined with a special software (ABI Prism 7700 Sequence Detection System, Perkin-Elmer, Foster City, CA, USA). The algorithm normalizes the reporter signal (Rn) to a passive reference. Next, the algorithm multiplies the standard deviation of the background Rn in the first few cycles (in most PCR systems cycles 3–15, respectively) by a default factor of 10 to determine a threshold. The cycle at which this baseline level is exceeded is defined as the threshold cycle (Ct) (Figure 2.1). Ct has a linear relation with the logarithm of the initial template copy number. Its absolute value additionally depends on the efficiency of both DNA amplification and cleavage of the TaqMan probe. The Ct values of the samples are interpolated to an external reference curve constructed by plotting the relative or absolute amounts of a serial dilution of a known template vs. the corresponding Ct values.

The oligonucleotides of each target of interest can be designed by the Primer Express software (Perkin-Elmer) using uniform selection parameters which allow the application of standard cycle conditions.

Reliability and validation of TaqMan real-time PCR

The use of TaqMan real-time PCR for the quantification of gene expression has been shown to be at least as reliable as the application of other quantitative PCR techniques like competitive PCR (Förster 1994; Dötsch *et al.* 2000). Whereas the expression of highly expressed genes like the house-keeping gene glyceraldehyde-3-phosphate is well correlated between the two methods, for the determination of genes with lower expression, like the neuropeptide Y, TaqMan PCR is much more sensitive than competitive PCR (Dötsch *et al.* 2000) (Figure 2.2). In addition, the spectrum of linear measurements for real-time PCR is in the range of 10^6, in contrast to 10^2 in competitive RT-PCR. Finally, a considerably higher number of samples per day can be measured by real-time PCR (up to 400 measurements). In comparison to the Northern blot assessment, only a minimal fraction of mRNA is necessary to quantify gene expression by real-time PCR (Heid *et al.* 1996, Gibson *et al.* 1996).

RNA may be extracted using standard techniques like commercial RNA isolation kits (e.g. RNAzol-B isolation kit, WAK-Chemie Medical GmbH, Bad Homburg, Germany) or conventional phenol-chloroform extraction for DNA (Raggi *et al.* 1999).

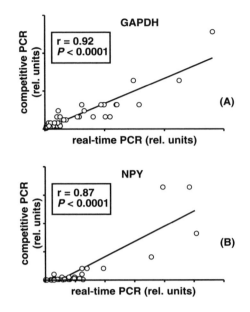

Figure 2.2 Relation of gene expression as assessed by competitive quantitative and TaqMan real-time PCR. (**A**) mRNA expression of GAPDH. r = 0.92, $P < 0.001$. (**B**) mRNA expression of NPY. $R = 0.87$, $P < 0.001$ (Adapted from Dötsch *et al.* 2000)

Applications for TaqMan real-time PCR

A number of applications for TaqMan real-time PCR have been introduced in the last number of years, the most important being the quantification of gene expression. Some of the most important applications and potential applications will be discussed below.

Quantification of gene expression

Quantification of gene expression has been facilitated to a considerable degree by the use of real-time techniques such as TaqMan PCR. This technique has practically replaced less sensitive and more time-consuming methods such as Northern blot or RNAse protection assay. Quantitative competitive PCR (Förster 1994) is both less effective and less sensitive (Dötsch *et al.* 2000).

Crucial aspects in the measurements of gene expression by real-time PCR are preparation of RNA, especially in samples that only contain small amounts of the specific mRNA, such as single cell picking or micro dissection (see Section 7). On the other hand, quantification may be difficult. In general, house-keeping genes may be applied, e.g. by using duplex approaches. Alternatively, external standards may be considered (Klein 2002).

In most cases of mRNA quantification, gene expression has to be related to house-keeping genes that are expressed relatively stably throughout the cells. In studies performed on gene expression in neuroblastomas so far, three different house-keeping genes have been assessed: the more traditional genes glyceralde-hyde-3-phosphate (GAPDH) (Dötsch *et al.* 2000) and β-actin (Raggi *et al.* 1999) and the neuronal marker protein gene product 9.5 (PGP9.5) (Dötsch *et al.* 2000). One major general difficulty in the use of house-keeping genes for the determination of mRNA transcript ratios is the possibility that their expression might also be altered by co-expression of pseudogenes and environmental changes, for example by hypoxia in the case of GAPDH (Bustin 2000). Pseudogenes may be eliminated using primer combinations that are intron spanning. However, un-identified influences cannot be dealt with as easily. Therefore, at least two house-keeping genes have been used for quantification of mRNA expression. However, this aspect clearly needs further evaluation. In our group, a number of primers and TaqMan probes have been used for house-keeping gene amplification (Dötsch *et al.* 2000; Schoof *et al.* 2001) (Table 2.1).

Quantification of gene copy number, determination of minimal residual disease and allelic discrimination in malignant tumours

MYCN detection by TaqMan PCR in neuroblastoma tissue

DNA copy number, MYCN amplification in neuroblastomas, is of great potential interest with regard to the prognosis of disease. In fact, in clinical practice

Table 2.1 Primers and TaqMan probes used for the quantification of human house-keeping gene expression

β-ACTIN	probe	CCAGCCATGTACGTTGCTATCCAGGC
	forward	GCGAGAAGATGACCCAGGATC
	reverse	CCAGTGGTACGGCCAGAGG
GAPDH	probe	CCTCAACTACATGGTTTACATGTTCCAATATGATTCCAC
	forward	GCCATCAATGACCCCTTCATT
	reverse	TTGACGGTGCCATGGAATTT
PBGD	probe	CTTCGCTGCATCGCTGAAAGGGC
	forward	TGTGCTGCACGATCCCG
	reverse	ACACTGCAGCCTCCTTCCAG
HPRT	probe	CGCAGCCCTGGCGTCGTGATTA
	forward	CCGGCTCCGTTATGGC
	reverse	GGTCATAACCTGGTTCATCATCA
β2MG	probe	TGATGCTGCTTACATGTCTCGATCCCA
	forward	TGACTTTGTCACAGCCCAAGATA
	reverse	CCAAATGCGGCATCTTC

Abbreviations: GAPDH: glyceraldehyd-3-phosphate dehydrogenase, PBGD: porphobilinogen deaminase, HPRT: hypoxanthine-guanine-phosphoribosyl-transferase, β2MG: β2-microglobulin (Dötsch *et al.* 2000; Schoof *et al.* 2001, 2003; Trollmann *et al.* 2003).

MYCN gene expression correlates with both, advanced disease stage (Brodeur et al. 1984) and rapid tumour progression (Seeger et al. 1985).

Several methods have been used for the detection of MYCN mainly based on Southern or dot blot (Pession et al. 1997; de Cremoux et al. 1997), on quantitative PCR (Sestini et al. 1991), and on fluorescent in situ hybridization techniques (Shapiro et al. 1993). The use of most PCR methods is restricted, however, by the fact that end-point measurements are used for quantification. Therefore, Raggi and co-workers (1999) introduced a TaqMan real-time base method for the determination of MYCN amplification in neuroblastomas. The authors demonstrate a precise assay with an inter-assay coefficient of variation of 13 per cent and an intra-assay coefficient of variation of 11 per cent. The threshold cycle for the detection of MYCN correlates in an inverse linear way with the logarithm of the input of genomic DNA molecules. There is a good linear relationship between the MYCN amplification measured by TaqMan real-time PCR and competitive PCR. Using Kaplan–Meier survival curves the authors showed that the amplification of MYCN as assessed by TaqMan real-time PCR is closely linked to cumulative survival as this had already been demonstrated with several other techniques for the quantification of MYCN amplification.

Minimal residual disease

An other important aspect of real-time PCR in the field of oncology is the detection of minimal residual disease that is 100–1000-fold more sensitive than traditional methods. As few as five copies may be detected in one reaction (Klein 2002) but the maximal input of DNA during sample preparation is the limiting step. This limitation must therefore be considered when looking for minimal residual disease in malignant diseases such as childhood or adulthood acute lymphoblastic leukaemia (Kwan et al. 2000; Cilloni et al. 2002).

Allelic discrimination and haploinsufficiency

Various methods have been used to study tumour cytogenetic aberrations in malignant tumours for clinical decision making. Initially, conventional cytogenetic techniques were applied. Karyotyping by conventional cytogenetics, however, depends on dividing cells and successful evaluations are often hampered by inferior metaphase quality. Subsequently, Restriction Fragment Length Polymorphisms (RFLPs), PCR-based microsatellite and fluorescence in-situ hybridization (FISH) analyses have been applied to overcome the restrictions of conventional cytogenetics. However, RFLP- and PCR-based microsatellite analyses depend on informative loci in the region of interest and the need for normal reference DNA of the respective patient, whereas FISH analyses are sometimes hampered by inferior tissue quality and hybridization probe availability. The latest technique used for routine detection of cytogenetic aberrations is comparative genomic hybridization CGH (Brinkschmidt 1997). CGH has proved to be consecutively applicable to tumour specimens for the detection of most aberrations known from conventional cytogenetics, but deletions of smaller DNA regions may be undetectable. Thus, this technique should be used to pre-screen tumour samples for gross cytogenetic aberrations and be supplemented by a TaqMan PCR-based approach to detect loss or gain of DNA on the single gene level.

Pathogene detection and quantification

Real-time PCR may be of great benefit in the detection and quantification of pathogens such as viruses, bacteria and fungi, whether for environmental detection (Brunk and Avanis-Aghajani 2002) or in infected patients.

Several difficulties, however, may occur (Klein 2002): in contrast to traditional methods of microbiology, vital and dead microorganisms are both detected. Secondly, most infectious organisms are characterized by a high mutation rate which may influence the estimation of viral or bacterial load dramatically (Klein *et al.* 2001). Finally, quantification necessitates the use of reliable standards. This may be achieved by using duplex or multiplex assays (Gruber *et al.* 2001). To assure permanent quality and to compare results international standardization will be required.

Limitations and pitfalls in the use of TaqMan real-time PCR

The use of TaqMan PCR may be particularly difficult if gene expression at a low level is to be quantified. One major pitfall in this context is the accidental determination of genomic DNA when RT-PCR is intended. There are various approaches to meet this problem. If possible primer combinations that are intron spanning (Ginzinger 2002) should be selected. If there is no possibility to select intron spanning primers, RNA samples may be pretreated with DNAse. However, this measure should not be chosen routinely and may also be deleterious if only small amounts of RNA are present that are partially destroyed as well (Bustin 2002).

One other difficulty is the high degree of technical expertise that is required to achieve as low a variation coefficient and as sensitive a measurement as possible. It could be shown that the degree of technical expertise may alter the gene level that is measured by up to 1000-fold (Bustin 2002).

Since real-time PCR is highly sensitive the risk of having interference with minor contamination is quite considerable. On the other hand, the risk of false-negative results must not be underestimated since post-hoc PCR steps are not visible to the degree that is provided by the more traditional methods for gene quantification (Klein 2002).

Alternative real-time PCR methods

There are other methods for real-time PCR not relying on exonuclease cleavage of a specific probe to generate a fluorescence signal. One of them, the LightCycler System (Roche Molecular), makes use of so called 'hybridization probes' (Wittwer *et al.* 1997; Giulietti *et al.* 2001 and see Chapter 3). Like exonuclease probes, hybridization probes are used in addition to the PCR primers, but unlike the first, hybridization probes combine two different fluorescent labels to allow resonance energy transfer. One of them is activated by external light. When both probes bind very closely at the DNA molecules generated by the PCR amplification process, the emitted light from the first dye activates the fluorescent dye of the second probe. This second dye emits light with a longer wavelength which is measured every cycle. Thus, the fluorescence intensity is directly correlated to the extent of probe hybridization and, subsequently, directly related to the amount of PCR product. A major advantage of the LightCycler is the very short time of a PCR run. However, a higher number of samples at one time can be analysed using the TaqMan system and, at least theoretically, might be more accurate as there is an internal reference dye to monitor minor variations in sample preparation.

Apart from exonuclease and hybridization methods for real-time PCR there are other options, including hairpin probes, hairpin primers and intercalating fluorescent dyes. Hairpin probes, also known as molecular beacons, contain reverse complement sequences at both ends, binding together while the rest of the strand remains single stranded, creating a panhandle-like structure. In addition, there are fluorescent dyes at both ends of the molecule, a reporter and a quencher similar to the TaqMan probes. In the panhandle-like conformation there is no fluorescence as the fluorescent reporter at one end and the quencher at the other end of the probe are very close to one another (Tyagi and Kramer 1996). As the central part of a molecular beacon consists of a target specific sequence, both ends are separated from each other when this part of the molecule is bound to the PCR product and a fluorescence signal can be emitted from the reporter dye. Hairpin primers, also named 'Amplifluor primers' (Intergen), are similar to molecular beacons, but fluorescence is generated as they become incorporated into the double-stranded PCR product during amplification. Another very simple technique to monitor the generation of PCR product in a real-time fashion is the use of intercalating dyes, such as EtBr and SYBR Green I, which do not bind to single-stranded DNA but to the double-stranded PCR product (Zubritsky 1999;

Giuletti *et al.* 2001). However, hairpin primers and intercalating dyes do not offer the high specificity of the probe-based techniques and a positive signal might even be generated by primer dimers.

Future developments

From a diagnostic point of view, one interesting aspect for the future might be the use of semi-automated or automated real-time devices for the assessment of gene expression and amplification, given the relatively ease of this approach. For diseases such as neuroblastoma real-time PCR might help to quantify more prognostic markers like the nerve growth factor receptor (TRKA gene) (Naka-gawara *et al.* 1993), the expression of genes involved in multidrug resistance (MDR1 and MRP) (Chan *et al.* 1991; Norris *et al.* 1996), and genes related to tumour invasion and metastasis (nm23 and CD44) (Leone *et al.* 1993; Favrot *et al.* 1993). There are first reports on the use of multiplex real-time PCR, a development that is certainly going to facilitate diagnostic procedure in the next 5 years (Hahn *et al.* 2000; von Ahsen *et al.* 2000).

From a research point of view, real-time PCR might facilitate the identification of new prognostic markers, since a large number of samples can be processed in a relatively short time (Norris *et al.* 2000; Kawamoto *et al.* 2000). Another future application of TaqMan PCR is the confirmation of results obtained by cDNA micro-arrays, which will be abundantly used for cDNA screening. Of particular interest might be the chance to determine gene expression in very few cells using single cell picking (Fink *et al.* 1998; Von der Hardt *et al.* 2003). Using this approach, not only minimal involvement of tumour cells can be visualized but also non-homogeneous distributions in malignant tumour might be monitored with respect to essential prognostic markers. It is of importance that in situ gene expression after laser capture microdissection may not only be performed in frozen sections but also from formalin-fixed and paraffin-embedded biopsies (Lehmann and Kreipe 2001).

Conclusions

TaqMan real-time PCR provides a reliable new technology for the quantification of gene expression. However, a number of preconditions have to be met for each new marker. First, the system parameters reflecting amplification efficiency (slope) and linearity should match the minimal requirements. This should include the calculation of the intra- and inter-assay coefficient of variation with regard to the threshold cycle at which the amplification signal is detected. Secondly, the assay itself should be carefully evaluated with regard to a linear relationship between threshold cycle and the logarithm of a serial dilution of a reference sample. Thirdly, real-time PCR results should be compared with a second, independent method for quantification like quantitative competitive PCR. Finally, it should be assessed whether the results obtained with regard to clinical outcome represent the experiences obtained with other methods for gene detection.

Apart from a high degree of precision, practical advantages of real-time PCR are easy handling, rapid measurements and a broad linear range for the measurements. Whereas competitive PCR only allows for the determination of a few samples in one assay, TaqMan real-time PCR provides the opportunity to measure more than 80 samples at one time in a 96-well plate together with the control reactions needed.

References

Brinkschmidt, C., Christiansen, H., Terpe, H.J., Simon, R., *et al.* (1997) Comparative genomic hybridization (CGH) analysis of neuroblastomas – an important methodological approach in pediatric tumour pathology. *J Pathol* **181**: 394–400.

Brodeur, G.M., Seeger, R.C., Schwab, M., Varmus, H.E. and Bishop, J.M. Amplification of N-myc in untreated human neuroblastomas correlates with advanced disease stage. *Science* **224**: 1121–1124.

Brunk, C.F., Li, J. and Avaniss-Aghajani, E. (2002) Analysis of specific bacteria from environmental samples using a quantitative polymerase chain reaction. *Curr Issues Mol Biol* **4**: 13–18.

Bustin, S.A. (2000) Absolute quantification of mRNA using real-time reverse transcription polymerase chain reaction assays. *J Mol Endocrinol* **25**: 169–193.

Bustin, S.A. (2000) Quantification of mRNA using real-time reverse transcription PCR (RT-PCR): trends and problems. *J Mol Endocrinol* **29**: 23–39.

Chan, H.S., Haddad, G., Thorner, P.S., DeBoer, G., *et al.* (1991) P-glycoprotein expression as a predictor of the outcome of therapy for neuroblastoma. *N Engl J Med* **325**: 1608–1614.

Cilloni, D., Gottardi, E., De Micheli, D., Serra, A., *et al.* (2002) Quantitative assessment of WT1 expression by real-time quantitative PCR may be a useful tool for monitoring minimal residual disease in acute leukemia patients. *Leukemia* **16**: 2115–2121.

de Cremoux, P., Thioux, M., Peter, M., Vielh, P., *et al.* (1997) Polymerase chain reaction compared with dot blotting for the determination of N-myc gene amplification in neuroblastoma. *Int J Cancer* **72**: 518–521.

Dötsch, J., Nüsken, K.D., Knerr, I., Kirschbaum, M., *et al.* (1999) Leptin and neuropeptide Y gene expression in human placenta: ontogeny and evidence for similarities to hypothalamic regulation. *J Clin Endocrinol Metab* **84**: 2755–2758.

Dötsch, J., Harmjanz, A., Christiansen, H., Hänze, J., *et al.* (2000) Gene expression of neuronal nitric oxide synthase and adrenomedullin in human neuroblastoma using real-time PCR. *Int J Cancer* **88**: 172–175.

Dötsch, J., Repp, R., Rascher, W. and Christiansen, H. (2001) Diagnostic and scientific applications of TaqMan real-time PCR in neuroblastomas. *Expert Rev Mol Diagn* **1**: 233–238.

Favrot, M.C., Combaret, V. and Lasset, C. (1993) CD44–a new prognostic marker for neuroblastoma. *N Engl J Med* **329**: 1965.

Fink, L., Seeger, W., Ermert, L., Hänze, J., *et al.* (1998) Real-time quantitative RT-PCR after laser-assisted cell picking. *Nat Med* **4**: 1329–1333.

Förster E. (1994) Rapid generation of internal standards for competitive PCR by low-stringency primer annealing. *Biotechniques* **16**: 1006–1008.

Gibson, U.E., Heid, C.A. and Williams, P.M. (1996) A novel method for real-time quantitative RT-PCR. *Genome Res* **6**: 995–1001 .

Ginzinger, D.G. (2002) Gene quantification using real-time quantitative PCR: an emerging technology hits the mainstream. *Exp Hematol* **30**: 503–512.

Giulietti, A., Overbergh, L., Valckx, D., Decallonne, B., *et al.* (2001) An overview of real-time quantitative PCR: applications to quantify cytokine gene expression. *Methods* **25**: 386–401.

Gruber, F., Falkner, F.G., Dorner, F. and Hammerle, T. (2001) Quantitation of viral DNA by real-time PCR applying duplex amplification, internal standardization, and two-color fluorescence detection. *Appl Environ Microbiol* **67**: 2837–2839.

Hahn, S., Zhong, X.Y., Burk, M.R., Troeger, C. and Holzgreve, W. (2000) Multiplex and real-time quantitative PCR on fetal DNA in maternal plasma. A comparison with fetal cells isolated from maternal blood. *Ann NY Acad Sci* **906**: 148–152.

Heid, C.A., Stevens, J., Livak, K.J. and Williams, P.M. (1996) Real-time quantitative PCR. *Genome Res* **6**: 986–994.

Higuchi, R., Fockler, C., Dollinger, G. and Watson, R. (1993) Kinetic PCR analysis: real-time monitoring of DNA amplification reactions. *Biotechnology (NY)* **11**: 1026–1030.

Kawamoto, T., Shishikura, T., Ohira, M., Takayasu, H., *et al.* (2000) Association between favorable neuroblastoma and high expression of the novel metalloproteinase gene, nbla3145/XCE, cloned by differential screening of the full-length-enriched oligo-capping neuroblastoma cDNA libraries. *Med Pediatr Oncol* **35**: 628–631.

Klein, D. (2002) Quantification using real-time PCR technology: applications and limitations. *Trends Mol Med* **8**: 257–260.

Klein, D., Leutenegger, C.M., Bahula, C., Gold, P., *et al.* (2001) Influence of preassay and sequence variations on viral load determination by a multiplex real-time reverse transcriptase-polymerase chain reaction for feline immunodeficiency virus. *J Acquir Immune Defic Syndr* **26**: 8–20.

Kwan, E., Norris, M.D., Zhu, L., Ferrara, D., *et al.* (2000) Simultaneous detection and quantification of minimal residual disease in childhood acute lymphoblastic leukaemia using real-time polymerase chain reaction. *Br J Haematol* **109**: 430–434.

Lehmann, U. and Kreipe, H. (2001) Real-time PCR analysis of DNA and RNA extracted from formalin-fixed and paraffin-embedded biopsies. *Methods* **2**: 409–418.

Leone, A., Seeger, R.C., Hong, C.M., Hu, Y.Y., *et al.* (1993) Evidence for nm23 RNA overexpression, DNA amplification and mutation in aggressive childhood neuroblastomas. *Oncogene* **8**: 855–865.

Nakagawara, A., Arima-Nakagawara, M., Scavarda, N.J., Azar, C.G., *et al.* (1993) Association between high levels of expression of the TRK gene and favorable outcome in human neuroblastoma. *N Engl J Med* **328**: 847–584.

Norris, M.D., Bordow, S.B., Marshall, G.M., Haber, P.S., *et al.* (1996) Expression of the gene for multidrug-resistance-associated protein and outcome in patients with neuroblastoma. *N Engl J Med* **334**: 231–238.

Norris, M.D., Burkhart, C.A., Marshall, G.M., Weiss, W.A. and Haber M. (2000) Expression of N-myc and MRP genes and their relationship to N-myc gene dosage and tumor formation in a murine neuroblastoma model. *Med Pediatr Oncol* **35**: 585–589.

Orlando, C., Pinzani, P., Pazzaggli, M. (1998) Developments in quantitative PCR. *Clin Chem Lab Med* **36**: 255–269.

Pession, A., Trere, D., Perri, P., Rondelli, R., *et al.* (1997) N-myc amplification and cell proliferation rate in human neuroblastoma. *J Pathol* **183**: 339–344.

Raggi, C.C., Bagnoni, M.L., Tonini, G.P., Maggi, M., *et al.* (1999) Real-time quantitative PCR for the measurement of MYCN amplification in human neuroblastoma with the TaqMan detection system. *Clin Chem* **45**: 1918–1924.

Schoof, E., Girstl, M., Frobenius, W., Kirschbaum, M., *et al.* (2001) Decreased gene expression of 11beta-hydroxysteroid dehydrogenase type 2 and 15-hydroxyprostaglandin dehydrogenase in human placenta of patients with preeclampsia. *J Clin Endocrinol Metab* **86**: 1313–1317.

Seeger, R.C., Brodeur, G.M., Sather, H., Dalton, A., *et al.* (1985) Association of multiple copies of the N-myc oncogene with rapid progression of neuroblastomas. *N Engl J Med* **313**: 1111–1116.

Sestini, R., Orlando, C., Zentilin, L., Lami, D., *et al.* (1994) Gene amplification for c-erbB-2, c-myc, epidermal growth factor receptor, int-2, and N-myc measured by quantitative PCR with a multiple competitor template. *Clin Chem* **41**: 826–832.

Shapiro, D.N., Valentine, M.B., Rowe, S.T., Sinclair, A.E., *et al.* (1993) Detection of N-myc gene amplification by fluorescence in situ hybridization. Diagnostic utility for neuroblastoma. *Am J Pathol* **142**: 1339–1346.

Tyagi, S. and Kramer, R. (1996) Molecular beacons: probes that fluorescence upon hybridization. *Nat Biotechnol* **14**: 303–308.

von Ahsen, N., Oellerich, M. and Schutz, E. (2000) A method for homogeneous color-compensated genotyping of factor V (G1691A) and methylenetetrahydrofolate reductase (C677T) mutations using real-time multiplex fluorescence PCR. *Clin Biochem* **33**: 535–539.

Von der Hardt, K., Kandler, M.A., Fink, L., Schoof, E., *et al.* (2003) Laser-assisted microdissection and real-time PCR detect anti-inflammatory effect of perfluorocarbon. *Am J Physiol Lung Cell Mol Physiol* **285**: 55–62.

Wittwer, C., Herrmann, M., Moss, A., and Rasmussen, R. (1997) BioTechniques. Continuous fluorescence monitoring of rapid cycle DNA amplification. **22**: 130–138.

Zubritsky, E. (1999) Widespread interest in gene quantification and high-throughput assays are putting quantitative PCR back in the spotlight. *Anal Chem* **71**: 191A–195A.

3
Hybridization Probes for Real-time PCR

Elaine Lyon

Introduction

Molecular techniques have traditionally used two methods to visualize DNA: DNA binding dyes such as ethidium bromide or, to improve sequence specificity, hybridization probes. Originally, hybridization probes were radioactively labelled with ^{32}P and used for Southern analysis (Southern 1975) or allele specific hybridization assays (*in situ* dot blots; Wu *et al*. 1989). Probes are now labelled with fluorescent dyes and are the basis for new technologies such as real-time PCR or micro-arrays. For real-time PCR applications, probes do not interfere with the amplification reaction and are added directly to the PCR, eliminating post-PCR manipulation for detection. This chapter will discuss the use of fluorescent labelled probes for real-time PCR. It will focus on two applications, quantification and mutation detection. Examples will be given using the LightCycler (Roche Molecular Biochemicals, Mannheim, Germany), although other instruments are commercially available.

Probe chemistry

Numerous probe and primer chemistries have been described for real-time PCR. 'Primers' refer to oligonucleotides that are extended during PCR to form amplicon. If primers are labelled, the product itself is also labelled. 'Probes' are

Molecular Analysis and Genome Discovery edited by Ralph Rapley and Stuart Harbron
© 2004 John Wiley & Sons, Ltd   ISBN 0 471 49847 5 (cased)   ISBN 0 471 49919 6 (pbk)

oligonucleotides that are blocked from extending during PCR and are used only for product detection. Both labelled primers and labelled probes can be used for real-time PCR. Amplicon detection by fluorescence labelled primers or probes utilizes two basic mechanisms: release from quenching or fluorescence resonance energy transfer (FRET). Release from quenching is the principle behind TaqMan chemistry where a single probe is labelled with two different fluorophores that are in close proximity (Lee *et al.* 1993; Kreuzer *et al.* 2000; see Chapter 2). In this dual-labelled, single probe system one fluorophore acts as a quencher and prevents excitation of the other. During the amplification process, probes hybridize to the increasing number of amplicons. As Taq polymerase extends the bases of the forming product, the probes are degraded by the 5' exonuclease activity of the polymerase, releasing the fluorophore into solution. Without the quenching effect, the other fluorophore is excited and emits its energy as fluorescent signal. These probes are more correctly referred to as hydrolysis probes, since it is in their degradation that fluorescent signal is detected.

Another probe system, molecular beacons, applies the same quenching mechanism, but not by hydrolysis (Tyagi and Kramer 1996; Giesendorf *et al.* 1998). Instead, the probes are designed with a hairpin structure introduced into the probe. When in solution, the hairpin structure brings the fluorophores in close proximity. As the probe hybridizes to the product, the hairpin is interrupted, and the fluorophores are separated physically. Upon hybridization, the quenching is released and fluorescence is emitted.

A similar mechanism is used for labelled primers. Scorpion primers utilize either a stem-loop structure like molecular beacons, or a 'duplex' format (Thelwell *et al.* 2000; Solinas *et al.* 2001). The duplex format uses a separate oligonucleotide complementary to the 5' sequence of the primer. This oligonucleotide serves the same purpose as the stem-loop structure in quenching fluorescence from the labelled primer.

A third design utilizing fluorescent quenching is a single labelled probe (Crockett and Wittwer, 2001). The fluorescent label is positioned at the end of a probe close to a guanine in the complementary target sequence. As the probe is bound, the fluorophore is brought next to the guanine, which has an inherent quenching ability. As PCR product accumulates, more of the probe's fluorophore is quenched, showing a loss of fluorescent signal, rather than a fluorescent gain due to release of quenching.

Fluorescence resonance energy transfer

Fluorescent energy resonance transfer (FRET) has been used in a variety of biological applications, but has recently been adapted to molecular detection (Wittwer *et al.* 1997; Lay and Wittwer 1997). In FRET, a donor fluorophore is excited at a specific wavelength. When a second fluorophore (acceptor) is in close proximity, the energy from the donor fluorophore is transferred non-radiatively to the acceptor fluorophore. The transfer of energy excites the acceptor fluorophore, which then emits fluorescence at a longer wavelength. Fluorescence is detected by the instrument. Figure 3.1 shows the principle of FRET using hybridization probes.

The LightCycler traditionally has been capable of detecting two dyes simultaneously (Bernard *et al*. 1999; Phillips *et al*. 2000) but newer versions can detect up to six dyes. Two common acceptor fluorophores are LCRed 640 and LCRed 705. The same donor fluorophore (usually fluorescein) is capable of exciting both the LCRed 640 and 705 dyes. However, the emission spectrum of these two dyes is at different wavelengths. A mathematical algorithm to compensate for bleedthrough between channels allows simultaneous detection (Bernard *et al*. 1999; Bagwell and Adams 1993). These dyes are incorporated into either end of the probe or primer, or on an internal base (to allow primer extension). Probes labeled on the 3′ end are blocked from acting as a primer, and therefor can be included with the PCR. Probes internally labelled or labelled on the 5′ end require a block such as a phosphate group on the 3′ end to prevent extension. Although these fluorophores can be attached at any of these positions, usually fluorescein is attached at the 3′ end and LCRed 640 or 705 dyes are labelled on the 5′ end.

Hybridization probe design

One of the first LightCycler applications used a primer labelled fluorescently on a base internal to the 3′ end to allow extension from the primer (Lay and Wittwer 1997). A single probe is designed complementary to the extended amplicon and labelled on its 3′ end. As the probe hybridizes to the extended strand, fluorescent resonance energy transfer occurs between the probe and labelled amplicon.

The most common FRET design is a dual probe system, with each probe labelled with a single fluorophore (Figure 3.1; Lyon 2001). FRET occurs only when both probes are hybridized to the PCR product. If one of the probes dissociates from the target, fluorescence signal is lost. Probes are 15–40 bases in length and spaced so that fluorophores are 1–3 bases apart upon probe hybridization. For quantitative applications, both probes can be designed with similar Tm's. For mutation detection, probes are designed with different Tm's. One probe covers the mutation or base alteration of interest and is referred to as a 'reporter' or 'mutation' probe. The other probe is designed to remain hybridized at higher temperatures and is considered an 'anchor probe'. The higher Tm of the anchor probe is accomplished by increasing base length or GC content.

Reporter probes are typically designed to be 15–25 bases. The mutation position is centred within the probe, although mutations at nearly any position within the region of the probe will be detected. End positions however, destabilize the probe to a lesser extent. Depending on the sample's genotype, the probe and

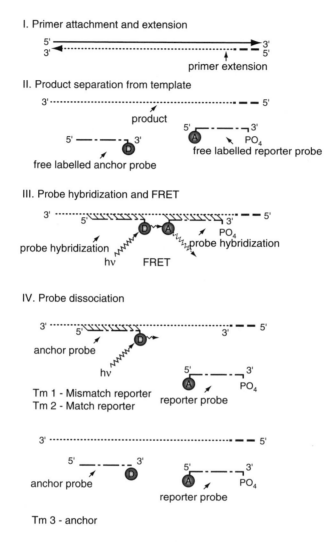

Figure 3.1 Dual probe FRET design. The most common design for hybridization probes is a dual probe system. I. Primers extend to form the target amplicon during PCR. II. Upon denaturation, single-stranded products are formed. III. As the temperature is lowered, the probes hybridize with the single-stranded complementary sequence, resulting in FRET and fluorescence signal. IV. As the temperature is increased, the reporter probe with a mismatch dissociates first (Tm 1), followed by dissociation of the perfectly matched probe (Tm 2). As reporter probes dissociate, fluorescence signal is lost. Finally, the anchor probe dissociates (Tm 3)

target will be either perfectly matched or form a mismatch upon hybridization. The optimal Tm of the perfectly matched reporter probe is about 10°C lower than the anchor probe. The Tm difference between anchor and reporter allows a mutation to exist under the anchor probe without dissociating from the target below the Tm of the reporter probe. A mismatch decreases the Tm of the reporter probe 3–12°C from the perfectly matched reporter probe.

Factors that affect probe stabilization are probe length (the shorter the probe, the greater the destabilization of a single mismatch), the GC content, the base change and position, the nearest neighbours of the base change and the number of base alterations within the region of the probe. All of these factors can be used to optimize an assay. For example, a G–T mismatch is more stable than a C–T or C–A mismatch (Allawi and SantaLucia 1998a, 1998b). Probes can be designed to match either the sense or the anti-sense strand of the target nucleic acid depending on the desired temperature shift of the unstable configuration. Instability of specific mismatches is predicted by nearest neighbour thermodynamics (Allawi and SantaLucia 1998b; SantaLucia et al. 1996; Peyret et al. 1999). Computer programs based on nearest neighbour calculations are available to predict the Tm of a perfect match and mismatch oligonucleotide (Schütz and von Ahsen 1999; Schüz et al. 2000).

A minor groove binding protein improves probe stability, and allows shorter probes and better temperature discrimination between wild-type and mutant sequences (De Kok et al. 2002). Shorter probes lessen the chance of other mutations – either expected or unexpected – interfering with the assay. This is important when sequence discrimination is difficult, for example with viral or bacterial genomes of closely related species or sub-species with limited unique sequences for probe design.

FRET applications

Real-time quantification

PCR has an exponential phase with amplicon theoretically doubling every cycle, followed by a plateau phase. The plateau phase may be due to exhaustion of reagents such as primers or enzyme. Alternatively, the plateau phase may be result of build-up of PCR products self-hybridizing and competing with primer annealing. Whatever the reason, nucleic acid differing in initial concentration may plateau at the same end concentration, making end-point analysis unreliable. Before real-time technology, quantitative PCR relied on labour-intensive analysis by manually removing replicate samples at different cycles. Competitive PCR used artificial nucleic acids such as PCR amplicons or plasmids designed to have similar, yet distinguishable sequences as the target. Similar target and competitor sequences ensure close PCR efficiencies between the target and competitor. A common competitor design used small insertions (i.e. 20 base pairs) to differentiate between target and competitor (Diviacco et al. 1992; Sestini et al. 1995).

Real-time PCR revolutionized quantitative PCR. Real-time quantification was originally done with a DNA binding dye such as ethidium bromide or Syber

A.

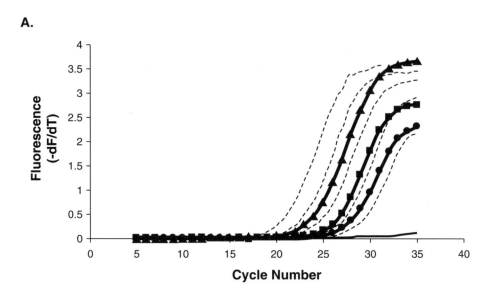

B.

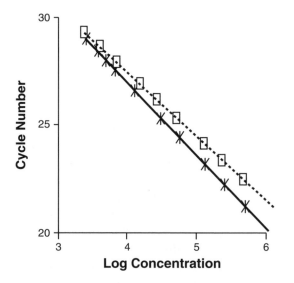

Figure 3.2 Amplification curves of HER2/neu cell lines. A. Three cell lines with differing HER2/neu copies are amplified by real-time PCR and hybridization probes. MRC-5, non-amplified (●), T47D, low-level amplification (■), and SKBR3, high-level amplification (▲) are shown. A standard curve is shown with dashed lines. B. Standard curves for HER2/neu (☆) and beta globin (⊟) are shown with Cp's plotted against the log of the concentration. Differing slopes indicate differing efficiencies between reactions

Green (Wittwer *et al.* 1997). The same application is possible using hybridization probes (Lassmann *et al.* 2002) to increase specificity for the target. Fluorescence acquisition must be at a temperature that allows probe hybridization to the product, often at the annealing stage. During amplification, fluorescence data is acquired once each cycle. As product accumulates and probes hybridize, fluorescence is detected above background. The cycle that signal is detected above background is the crossing point (Cp) and is used for quantification. The more copies of the initial template, the earlier the cycle that fluorescence will be detected. By monitoring product accumulation during the reaction, differences in initial quantities are distinguishable (Figure 3.2).

Absolute values can be obtained by real-time PCR. For absolute quantification, standards with known concentrations are tested along with unknown samples. The concentrations (log) of the standards are plotted against the Cp. The Cps of the samples are compared with the standard curve to determine the initial nucleic acid copy number of the sample. Absolute quantification requires accurate concentration of standards.

Relative quantification is another use of real-time PCR. Relative quantification detects changes in target quantity relative to a reference gene. Reference genes serve as controls for PCR amplification to ensure against PCR inhibitors in samples. Reference genes may also be used to normalize the amount of DNA or RNA. For accurate quantification, relative quantification requires equal or near-equal PCR efficiencies between the reference and target PCR. Ideally the efficiency of the PCR should be 2, with the number of target sequences doubling every cycle. Factors that affect efficiency are salt concentrations, low initial target copy numbers, primer sequences, concentrations and purity. Small differences in efficiencies between reactions become major differences after 30 cycles of PCR. PCR efficiencies should be tested with each new batch of primers by serial diluting standards and plotting the Cp versus the log of the initial template concentration. The slope of the line is the negative log of the efficiency. Differences in primer concentration or purity may require re-optimization of PCR to maintain equal efficiencies or mathematical correction to adjust for unequal efficiencies (Rasmussen 2001).

Hybridization probes can combine absolute and relative quantification in a single reaction. In this application, target amplicon is monitored with LCRed 640 probes, while the reference amplicon is detected with LCRed 705. Co-amplifying target and reference genes eliminate pipetting differences between separate reactions. Standards can also be co-amplified, using standard curves specific for both reference and target. By comparing each reaction to its own standard curve, PCR efficiency differences between target and reference reactions will not affect calculations. After absolute quantities of target and reference are determined, a target copy number can be normalized to its reference copy number.

Real-time PCR and quantification has been used for diverse applications. Expression levels of mRNA have been quantified using RT-PCR. Some examples are CK20 (Lassmann *et al.* 2002), cytokine expression (Kuhne and Oschmann 2002). Levels of translocation transcripts such as bcr-abl can be quantified to monitor therapy (Bolufer *et al.* 2000). These applications quantify expression levels relative to a housekeeping gene – a gene expressed in the tissue at a constant level and not affected by the specific treatment or therapy. DNA gene amplification for genes such as HER2/*neu* or N-myc involved in tumour growth is another application (Millson *et al.* 2003). Real-time quantification has also been used extensively for infectious disease (Kosters *et al.* 2002; Stocher and Berg 2002).

Mutation detection

Hybridization probes and melting curve analysis have also been used for mutation detection. The first application described was for the Factor V Leiden mutation, a cause of inherited thrombophilia (Lay and Wittwer 1997). Since then, numerous reports describe melting curve assays for SNPs (Bernard *et al.* 1998b; Von Ahsen *et al.* 1999), dinucleotide or mononucleotide repeat tracts (Von Ahsen *et al.* 2000; Nauck *et al.* 1999), and small deletions (Gundry *et al.* 1999). The reporter probe is designed to be complementary to a sequence containing a mutation or polymorphism. If a mismatch exists within the region of the probe, the probe/target complex will be destabilized. After amplification PCR products are denatured, cooled, then slowly heated with a transition rate of 0.1–0.3°C/s, with continual fluorescence monitoring. As the temperature increases, the probe will dissociate first from the mismatched complex, followed by dissociation from the perfectly matched template (Figure 3.1). The anchor probe, with a higher Tm than the reporter probe, dissociates last. As the reporter probe melts, FRET is lost between the reporter and anchor, resulting in loss of fluorescence signal. Calculating the negative derivative of the curves allows easy visualization of alleles. Homozygous wild-type, heterozygous and homozygous mutant genotypes are easily distinguished by the derivative melting curves (Figure 3.3).

Although mutation analysis has also been described using TaqMan or other probe systems, these systems use a mutation specific and a wild-type specific probe, each labelled with a different fluorophore (Sanders 2000). A temperature is chosen to allow hybridization only with the perfectly matched probe. Signal is read for each probe separately, and a signal ratio of the two probes determines the wild-type, heterozygote or homozygous mutant genotypes. This procedure is similar to in situ dot hybridization using temperature stringencies to distinguish genotypes. Melting curve analysis uses a single reporter probe that distinguishes both mutant and wild type by a temperature gradient. Thus melting curve analysis can be considered a 'dynamic dot blot'.

Multiple mutations can be distinguished simultaneously using melting curve analysis. Reactions can be multiplexed by Tm or by fluorophore. To multiplex by Tm, reporter probes for the two loci are labelled with the same fluorophore and

A. Wild –type Probe

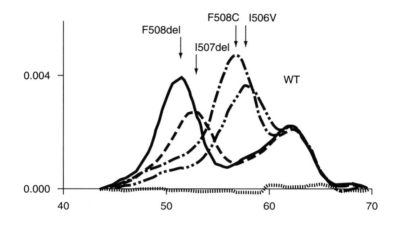

B. F508del Probe

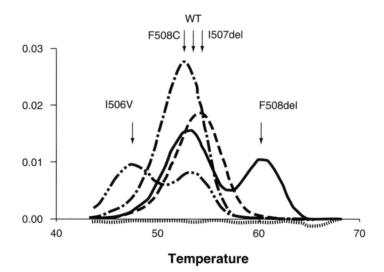

Temperature

Figure 3.3 Derivative melting curves for cystic fibrosis F508del locus. Heterozygous samples of F508del (—), F508C (– – –), I506V(– ⋅⋅ –), I507del(----), are shown using A. a wild-type specific probe and B. a mutation specific (F508del) probe. The wild-type specific probe separates normal alleles from the mutations. All mutations can be identified by Tm, although the I506V and F508C Tm's are similar. To confirm genotypes, a mutation specific probe can clearly differentiate between I506V and F508C. No template control (.........)

designed with different Tm's (Bernard *et al.* 1998a). A 25°C temperature range (45–70°C) allows separation of matches and mismatches for at least two loci. To multiplex by fluorophore, two probe pairs are used: a fluorescein – LCRed 640 pair and a fluorescein – LCRed705 pair. LCRed 640 is detected in the F2 channel, and the LCRed 705 in the F3 channel of the LightCycler (Bernard *et al.* 1999; Phillips *et al.* 2000). As for quantification applications, a colour compensation file compensates for bleed-through into different channels. By using a combination of Tm and fluorophore multiplexing (including a quenching affect), up to eight mutations have been multiplexed in a single reaction (Schütz *et al.* 2000).

Different mutations can also be detected by a single reporter probe and distinguished by Tm (Phillips *et al.* 2000; Herrmann *et al.* 2000; Elenitoba-Johnson *et al.* 2001). An example of this is the locus for the common cystic fibrosis mutation F508del, a three base-pair deletion that results in the deletion of the amino acid phenylalanine in the CFTR protein. Within nine bases, five mutations or polymorphisms, F508del, I507del, I507V, I506V and F508C exist. A wild-type probe distinguishes wild type from all mutations. A combination of a wild-type and mutation probe distinguishes all genotypes (Figure 3.3).

Mutations that have not been previously described may also be detected by melting curve analysis, showing curves with unexpected Tm's. Examples that have been reported include previously unknown mutations in Factor V (Lyon *et al.* 1998), Factor II (Warshawsky *et al.* 2002) and haemochromatosis (Phillips *et al.* 2000). These mutations may go undetected, give false-positive, false-negative or inconclusive results by other technologies such as restriction digest or fluorescent ratios.

Melting curve analysis can also quantify alleles using areas under the derivative melting curves. Several methods with different levels of control have been described. The first level of control co-amplifies a reference gene and distinguishes target and reference by Tm (Chui *et al.* 2001). Concerns of differing efficiencies of reactions exist with this design, similar to Cp analysis. Competitive PCR adds an additional level of control. The competitor in quantitative melting analysis is a single nucleotide change from the wild-type sequence (Ruiz-Ponte *et al.* 2000; Lyon *et al.* 2001b). By co-amplifying an artificial template with a single base difference, PCR efficiencies should be equal. Differing concentrations of competitor are added to the reaction and the competitor copy number resulting in equal allele areas between competitor and target is the copy number of the target. Using this method, absolute quantification is achieved. As with quantification by crossing thresholds, a reference gene and its competitor can be amplified in the same reaction and detected in a different fluorescent channel (Lyon *et al.* 2001a).

A third method for allele quantification uses naturally occurring SNPs as an internal control (Ruiz-Ponte *et al.* 2000). In this application, the alleles in a heterozygous sample are compared with each other. This method does not require an additional reference gene or competitor, since the heterozygous locus is its own internal control. This method is sensitive enough to detect a ratio of 2 : 1, making it a good application for detecting gene duplications. However, not all samples will have naturally occurring SNPs at a specific locus. To compensate, a panel of SNPs within the region will increase the probability that an informative SNP locus can be found for any given sample. Alternatively, a competitor can be added to samples homozygous for a particular locus.

Summary

Hybridization probes give sequence specificity for molecular detection. Fluorescently labelled probes allow real-time PCR detection for absolute or relative quantification. Several levels of control are available, namely, reference genes, standard curves or competitors. Melting curve analysis is a simple way to genotype known alleles. The strength of melting curve analysis is its ability to distinguish between multiple base changes within the region of the probe. This allows a single probe to detect multiple mutations, some of which will be previously undetected by other methods.

References

Allawi, H.T. and SantaLucia J. Jr. (1998a) Thermodynamics of internal C → T mismatches in DNA. *Nucleic Acids Res* **11**: 2694–2701.

Allawi, H.T. and SantaLucia, J. Jr. (1998b) Nearest neighbor thermodynamic parameters for internal G.A mismatches in DNA. *Biochemistry* **8**: 2170–2179.

Bagwell, C.B. and Adams, E.G. (1993) Fluorescence spectral overlap compensation for any number of flow cytometry parameters. *Ann NY Acad Sci* **677**: 167–184.

Bernard, P.S., Ajioka, R.S., Kushner, J.P. and Wittwer, C.T. (1998a) Homogenous multiplex genotyping of hemochromatosis mutations with fluorescent hybridization probes. Am J Pathol **153**: 1055–1061.

Bernard, P.S., Lay, M.J. and Wittwer, C.T. (1998b) Integrated amplification and detection of the C677T point mutation in the methylenetetrahydrofolate reductase gene by fluorescence resonance energy transfer and probe melting curves. *Anal Biochem* **255**: 101–107.

Bernard, P.S., Pritham, G.H., and Wittwer, C.T. (1999) Color multiplexing hybridization probes using the apolipoprotein E locus as a model system for genotyping. *Anal Biochem* **273**: 221–228.

Bolufer, P., Sanz, G.F., Barragan, E., Sanz, M.A., *et al.* (2000) Rapid quantitative detection of BCR-ABL transcripts in chronic myeloid leukemia patients by real-time reverse transcriptase polymerase-chain reaction using fluorescently labeled probes. *Haematologica* **85**: 1248–1254.

Chui, R.W., Murphy, M.F., Fidler, C., Wainscoat, J.S. and Lo, Y.M. (2001) Technical optimization of RhD zygosity determination by real-time quantitative polymerase chain reaction: implication for fetal RhD status determination by maternal plasma. *Ann NY Acad Sci* **945**: 156–160.

Crockett, A.O. and Wittwer, C.T. (2001) Fluorescein-labeled oligonucleotides for real-time PCR: using the inherent quenching of deoxyguanosine nucleotides. *Anal Biochem* **290**: 89–97.

De Kok, J.B., Wiegerinck, E.T., Giesendorf, B.A., and Swinkels, D.W. (2002) Rapid genotyping of single nucleotide polymorphisms using novel minor groove binding DNA oligonucleotides (MGB probes). *Hum Mutat* **19**: 554–559.

Diviacco, S., Norio, P., Zentilin, L., Menzo, S., *et al.* (1992) A novel procedure for quantitative polymerase chain reaction by coamplification of competitive templates. *Gene* **122**: 313–320.

Elenitoba-Johnson, K.S.J., Bohling, S.D., Wittwer, C.T. and King, T.C. (2001) Multiplex PCR by multicolor fluorimetry and fluorescence melting curve analysis. *Nat Med* **7**: 249–253.

Giesendorf, B.A., Vet, J.A., Tyagi, S., Mensink, E.J., *et al.* (1998) Molecular beacons: a new approach for semiautomated mutation analysis. *Clin Chem* **44**: 482–486.

Gundry, C.N., Bernard, P.S., Herrmann, M.G., Reed, G.H. and Wittwer, C.T. (1999) Rapid F508del and F508C assay using fluorescent hybridization probes. *Genet Test* **3**: 365–370.

Herrmann, M.G., Dobrowolski, S.F., and Wittwer, C.T. (2000) Rapid beta-globin genotyping by multiplexing probe melting temperature and color. *Clin Chem* **46**: 425–428.

Kosters, K., Reischl, U., Schmetz, J., Riffelmann, M. and Wirsing von Konig, C.H. (2002) Real-time LightCycler PCR for detection and discrimination of *Bordetella pertussis* and *Bordetella parapertussis*. *J Clin Microbiol* **40**: 1719–1722.

Kreuzer, K.A., Bohn, A., Lass, U., Peers, U.R., and Schmidt, C.A. (2000) Influence of DNA polymerases on quantitative PCR results using TaqMan probe format in the LightCycler instrument. *Mol Cell Probes* **14**: 57–60.

Kuhne, B.S. and Oschmann, P. (2002) Quantitative real-time RT-PCR using hybridization probes and imported standard curves for cytokine gene expression analysis. *Biotechniques* **33**: 1080–1082.

Lassmann, S., Bauer, M., Soong, R., Schregimann, J., *et al.* (2002) Quantification of CK20 gene and protein expression in colorectal cancer by RT-PCR and immunohistochemistry reveals inter-and intratumour heterogeneity. *J Pathol* **198**: 198–206.

Lay, M.J. and Wittwer, C.T. (1997) Real-time fluorescence genotyping of factor V Leiden during rapid-cycle PCR. *Clin Chem* **43**: 2262–2267.

Lee, L.G., Connell, C.R. and Bloch, W. (1993) Allelic discrimination by nick translation PCR with fluorogenic probes. *Nucl Acids Res* **21**: 3761–3766.

Lyon, E. (2001) Mutation detection using fluorescent hybridization probes and melting curve analysis. *Exp Rev Mol Diagn* **1**: 17–26.

Lyon, E., Millson, A., Phan, T. and Wittwer C.T. (1998) Detection and identification of base alterations within the region of factor V Leiden by fluorescent melting curves. *Molec Diag* **3**: 203–210.

Lyon, E., Millson, A. and Suli, A. (2001a) HER2/*neu* gene amplification quantified by PCR and melting peak analysis using a single base alteration competitor as an internal standard, in S. Meuer, C. Wittwer and K. Nakagawara (eds) *Rapid Cycle Real Time PCR-Methods and Application*, Springer-Verlag, Heidelberg, Germany: 207–217.

Lyon, E., Millson, A., Lowery, M.C., Woods, R. and Wittwer, C.T. (2001b) Quantification of HER2/neu gene amplification by competitive PCR using fluorescent melting curve analysis. *Clin Chem* **47**: 844–851.

Millson, A., Suli, A., Hartung, L., Kunitake, S., Bennett, A., Nordberg, M.C.L., Hanna, W., Wittwer, C.T., Seth, A. and Lyon, E. (2003) Comparison of two quantitative PCR methods for detecting HER2/neu amplification. *J Mol Diagn* **5**: 184–190.

Nauck, M., Wieland, H. and Marz, W. (1999) Rapid, homogenous genotyping of the 4G/5G polymorphism in the promoter region of the PAII gene by fluorescence resonance energy transfer and probe melting curves. *Clin Chem* **45**: 1141–1147.

Peyret, N., Seneviratne, P.A., Allawi, H.T. and SantaLucia, J. (1999) Nearest-neighbor thermodynamics and NMR of DNA sequences with internal A.A, C.C, G.G, and T.T mismatches. *Biochemistry* **12**: 3468–3477.

Phillips, M., Meadows, C.A., Huang, M.Y., Millson, A. and Lyon E. (2000) Simultaneous detection of C282Y and H63D hemochromatosis mutations by dual-color probes. *Mol Diag* **5**: 107–116.

Rasmussen, R. (2001) Quantification of the LightCycler, in S. Meuer, C. Wittwer and K. Nakagawara (Ed.), *Rapid Cycle Real Time PCR-Methods and Application*. Springer-Verlag, Heidelberg, Germany, pp. 207–217.

Ruiz-Ponte, C., Loidi, L., Vega, A., Carracedo, A. and Barros, F. (2000) Rapid real-time fluorescent PCR gene dosage test for the diagnosis of DNA duplications and deletions. *Clin Chem* **46**: 1574–1582.

Sanders, S.J. (2000) Factor V Leiden genotyping using real-time fluorescent polymerase chain reaction. *Mol Cell Probes* **14**: 249–253.

SantaLucia, J. Jr., Allawi, H.T. and Seneviratne, P.A. (1996) Improved nearest-neighbor parameters for predicting DNA duplex stability. *Biochemistry* **19**: 3555–3562.

Schütz, E. and von Ahsen, N. (1999) Spreadsheet software for thermodynamic melting point prediction of oligonucleotide hybridization with and without mismatches. *Biotechniques* **27**: 1218–1222, 1224.

Schütz, E., von Ahsen, N. and Oellerich, M. (2000) Genotyping of eight thiopurine methyltransferase mutations: three-color multiplexing, 'Two-Color/Shared' anchor, and fluorescence-quenching hybridization probe assays based on thermodynamic nearest-neighbor probe design. *Clin Chem* **46**: 1728–1737.

Sestini, R., Orlando, C., Zentilin, L., Donatella, L., *et al.* (1995) Gene amplification for cerbB-2, c-myc, epidermal growth factor receptor, int-2, and N-myc measured by quantitative PCR with a multiple competitor template. *Clin Chem* **41**: 826–832.

Solinas, A., Brown, L.J., McKeen, C., Mellor, J.M., *et al.* (2001) Duplex scorpion primers in SNP analysis and FRET applications. *Nucleic Acids Res* **29**: E96.

Southern, E.M. (1975) Detection of specific sequences among DNA fragments separated by gel electrophoresis. *J Mol Biol* **98**: 503–517.

Stocher, M. and Berg, J. (2002) Normalized quantification of human cytomegalovirus DNA by competitive real-time PCR on the LightCycler Instrument. *J Clin Microbiol* **40**: 4547–4553.

Thelwell, N., Millington, S., Solinas, A., Booth, J. and Brown, T. (2000) Mode of action and application of scorpion primers to mutation detection. *Nucleic Acids Res* **28**: 3752–3761.

Tyagi, S. and Kramer, F.R. (1996) Molecular beacons-probes that fluoresce upon hybridization. *Nat Biotechnol* **14**: 303–308.

Von Ahsen, N., Oellerich, M., Armstrong, V.W. and Schütz E. (1999) Application of a thermodynamic nearest-neighbor model to estimate nucleic acid stability and optimize probe design. Prediction of melting points of multiple mutations of apolipoprotein

B-3500 and Factor V with a hybridization probe genotyping assay on the LightCycler. *Clin Chem* **45**: 2094–2101.

Von Ahsen, N., Oellerich, M. and Schutz, E. (2000) DNA base bulge vs unmatched end formation in probe-based diagnostic insertion/deletion genotyping: genotyping the UGT1A1 (TA)(n) polymorphism by real-time fluorescence PCR. *Clin Chem* **46**: 1939–1945.

Warshawsky, I., Hren, C., Sercia, L., Shadrach, B., *et al.* (2002) Detection of a novel point mutation of the prothrombin gene at position 20209. *Diagn Mol Path* **11**: 152–156.

Wittwer, C.T., Herrmann, M.G., Moss, A.A. and Rasmussen, R.P. (1997) Continuous fluorescent monitoring of rapid cycle DNA amplification. *Biotechniques* **22**: 130–138.

Wu, D.Y., Nozari, G., Schold, M., Conner, B.J. and Wallace, R.B. (1989) Direct analysis of single nucleotide variation in human DNA and RNA using in situ dot hybridization. *DNA* **8**: 135–142.

4

An Overview of Genotyping and Single Nucleotide Polymorphisms (SNP)

Ivo Glynne Gut

Introduction

Genotyping, the measurement of genetic variation has many applications: disease gene localization and identification of disease causing variants of genes, quantitative trait loci (QTL) mapping, pharmacogenetics, identity testing based on genetic fingerprinting, just to mention the major ones. Genotyping is used in genetic studies: in humans, animals, plants and other species. Classical strategies for identifying disease genes rely on localizing a region harbouring a disease-associated gene by genome scan. Two approaches can be taken: the linkage study, where related individuals are analysed, and the association study where individuals are solely selected on the basis of being affected by a phenotype or not. For a linkage study, related individuals with affected and non-affected members are genotyped with 400 microsatellite markers that are fairly evenly-spaced throughout the genome (there is a commercially available panel of microsatellite markers from Applied Biosystems; www.appliedbiosystems.com). A genomic region where affected individuals have the same genotype and which is different from the non-affected individuals has an increased probability of harbouring a disease-causing variant of a gene. A major drawback is that with only 400 microsatellite markers many genes fall into the interval between any two neighbouring markers.

Molecular Analysis and Genome Discovery by Ralph Rapley and Stuart Harbron
© 2004 John Wiley & Sons, Ltd   ISBN 0 471 49847 5 (cased)   ISBN 0 471 49919 6 (pbk)

To home in on the right gene the region has to be saturated with additional markers. In the human genome one microsatellite marker is found roughly every one million bases, so this leaves on average eight to ten genes between any two microsatellite markers. Finally, the candidate genes have to be sequenced. This strategy has successfully been applied to the identification of genes responsible for many monogenic disorders. However, for pathologies with potentially many genes throughout the genome implicated, this strategy loses power. In order to maintain power the number of individuals entered into a study would need to be increased as well as the numbers of markers that are tested. Excessively large families affected by a single complex pathology are extremely rare. This forces researchers to move from a linkage to an association study (Cardon and Bell 2001).

Microsatellite markers (also called short tandem repeats – STRs) are stretches of a repetitive sequence motif of a few bases. The standard microsatellite markers for linkage studies have a two base CA sequence repeat motif. Alleles of a microsatellite carry a different number of repetitions of the sequence motif and thus a large number of different alleles are possible. The wide variety of possible alleles has made microsatellites popular with geneticists as this provides ample statistical power.

There are some technical challenges associated with the analysis of microsatellites. They are analysed by sizing PCR products on automated sequencers. PCR of microsatellites results in stutter products, due to slippage of the template strand, which gives rise to parasite peaks in the electropherograms. Particularly in cases where two alleles are close together, automated interpretation is very difficult due to problems of deconvolution. Consequently, substantial manual labour is involved in their interpretation. The fact that microsatellites are largely intergenic in the best case puts them into complete linkage disequilibrium with a disease-causing variant of a gene – never is an allele of a microsatellite the causative variant. SNPs in contrast can be intragenic and can thus represent disease-causing variants of genes.

Genotyping of Single Nucleotide Polymorphisms

Single Nucleotide Polymorphisms (SNPs, Figure 4.1) are changes in a single base at a specific position in the genome, in most cases with only two alleles (Brookes 1999). SNPs are found at a frequency of about one every 1000 bases in humans (Kruglyak 1997). By definition the rarer allele should be more abundant than 1 per cent in the general population. The relative simplicity of methods for SNP genotyping and the abundance of SNPs in the human genome have made them very popular in recent years. Pilot studies using SNPs as markers for genotyping have been carried out (Wang *et al.* 1998). Yet, there still is some debate about the usefulness of SNP markers compared with microsatellite markers for linkage studies and how many SNP markers will have to be analysed for association studies

ACGTGTCAGTCTTAGCTACGGATTACGT
TGCACAGTCAGAATCGATGCCTAATGCA

ACGTGTCAGTCTTCGCTACGGATTACGT
TGCACAGTCAGAAGCGATGCCTAATGCA

Figure 4.1 A single nucleotide polymorphism (SNP)

to be meaningful (Kruglyak 1999; Weiss and Terwilliger 2000). It has been reported that the human genome is structured in blocks of complete linkage disequilibrium that coincides with ancient ancestral recombination (Reich *et al.* 2001; Patil *et al.* 2001). Using high-frequency SNPs usually only a very limited number of haplotypes are detected and depending on the region of the genome LD can extend more than 100 Kb. Recently, a large-scale project – the HapMap project – was announced (Couzin 2002). The objective is to provide a selection of SNP markers that tag haplotype blocks in order to reduce the number of genotypes that have to be carried out for a genome-wide association study (Weiner and Hudson 2002).

In the last few years there has been a flurry of activity in the arena of SNP genotyping technology development. SNP genotyping methods are constantly being improved, perfected, integrated and new methods are still emerging to satisfy the needs of genomics and epidemiology.

Many SNP genotyping technologies have reached maturity in recent years and have been integrated into large-scale genotyping operations. No single SNP genotyping method fulfils the requirements of every study that might be undertaken. The choice of method depends on the scale and the scientific question a project is trying to answer. A project might require genotyping of a limited number of SNP markers in a large population or the analysis of a large number of SNP markers in one individual. Flexibility in choice of SNP markers and DNA to be genotyped or the possibility to quantify precisely an allele frequency in pooled DNA samples might be issues.

SNP genotyping methods are very diverse (Syvänen 2001). Broadly, each method can be separated into two elements, the first of which is a method for interrogating a SNP. This is a sequence of molecular biological, physical and chemical procedures for the distinction of the alleles of a SNP (Gut 2001; Figure 4.2). The second element is the actual analysis or measurement of the allele-specific products, which can be an array reader, a mass spectrometer, a plate reader, a gel separator/reader system, or other. Often very different methods share elements, for example, reading out a fluorescent tag in a plate reader (SNP genotyping methods with fluorescent detection were reviewed by Landegren *et al.* 1998), or the method of generating allele-specific products (e.g. primer extension, reviewed by Syvänen 1999), which can be analysed in many different formats.

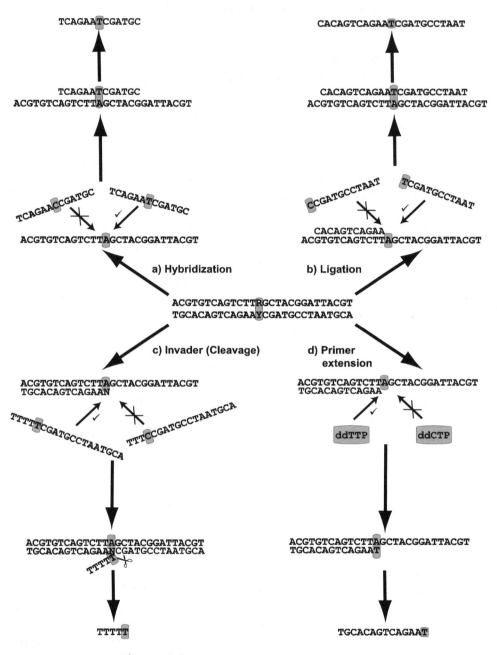

Figure 4.2   Methods for interrogating SNPs

Methods for interrogating SNPs (Table 4.1)

Hybridization

Alleles can be distinguished by hybridizing complementary oligonucleotide sequences to a target sequence (see Chapter 3). The stringency of hybridization is a physically controlled process. As the two alleles of a SNP are very similar in sequence significant cross-talk can occur. Allele-specific hybridization with two differently labelled oligonucleotides, one for each of the alleles, is generally used in conjunction with stringent washing to remove incompletely complementary probes. The stringency can be increased by the choice of appropriate buffers and temperatures, or by using modified oligonucleotides like peptide nucleic acids (PNAs) (Egholm *et al.* 1993) or locked nucleic acids (LNAs) (Ørum *et al.* 1999). Alternatively, molecular beacons – allele-specific oligonucleotide probes – use a fluorescent dye/quencher system for visualization (Tyagi *et al.* 1996; 1998). Molecular beacons are added to a PCR and hybridize to their complementary target sequence in the PCR product. In the case of the presence of an allele a dye and quencher molecules at either end of the molecular beacon are separated, which results in the emission of fluorescence. Multiple molecular beacons can be used simultaneously (Marras *et al.* 1999).

Dynamic allele-specific hybridization (DASH) relies on the different thermal stability of fully complementary probes compared with probes with one mismatch. Templates are immobilized and probes hybridized. The separation of the duplex is monitored in real-time (Howell *et al.* 1999). An improved version (DASH-2) was recently presented. DASH-2 makes use of induced fluorescence resonance energy transfer (iFRET, Howell *et al.* 2002), which allows simultaneous monitoring of multiple SNPs in a single process step and allows repeated measurement on the same sample (Jobs *et al.* 2003).

Another hybridization-based SNP genotyping method uses electrostringency on gel-arrays (Edman *et al.* 1997; Sosnowski *et al.* 1997). The hybridization is supported by applying an electric field, which results in a dramatic decrease of hybridization and wash times, while accuracy increases (www.nanogen.com). These devices available from Nanogen are suitable for genotyping on the order of 100 SNPs in an individual DNA samples. As these are well-packaged devices they seem suitable for DNA diagnostic applications at point of care.

Enzymatic methods

Adding an enzymatic step to distinguish alleles helps increase fidelity. A hybridization event is followed by the intervention of an enzyme. Many different

Table 4.1 A selection of SNP genotyping methods and their features

Method	Allele distinction/ Detection	TC	HT	Strength	Reference
Microarray					
GeneChip	H-2	yes	yes	sophisticated software	Wang et a., 1998, www.affymetrix.com
Tag array	P-2	yes	yes	detector for a high degree of multiplexing	www.affymetrix.com
APEX	P-2	yes	yes	data quality	Shumaker et al., 1996; Pastinen et al., 1997
OLA	O-2	yes	–	high multiplexing	Khanna et al., 1999
EF microarray	P-2	yes	–	flexibility, stringency	Edman et al., 1997; Sosnowski et al., 1997
Mass Spectrometry					
PNA	H-3	yes	–	simple principle, rapid data accumulation	Ross, Lee & Belgrader, 1997; Griffin, Tang & Smith, 1997
Masscode	H-3 (mass tag)	yes	yes	very high-throughput	www.qiagen.com
Mass tags	H-3 (mass tag)	yes	–	high-throughput	Shchepinov et al., 1999
PROBE	P-3	yes	–	accurate, rapid data accumulation	Braun, Little & Köster, 1997; Little et al., 1997
MassArray	P-3	yes	yes	accurate, high-throughput, complete system	www.sequenom.com
PinPoint	P-3	yes	–	high quality data, rapid data accumulation	Haff and Smirnov, 1997; Ross et al., 1998
GOOD	P-3	yes	yes	accurate, easy handling, no purification	Sauer et al., 2000 a and b
VSET	P-3	yes	–	accurate	Li et al, 1999
Invader	N-3	no	–	no PCR, isothermal	Griffin et al., 1999
Gel					
ARMS	H-1	yes	–	easy access	Newton et al., 1989
RFLP	N-1	yes	–	easy access, often used as reference	Botstein et al., 1980
MADGE	N-1/PCR-gel	yes	–	easy access, low set-up cost	Day and Humphries, 1994
SNaPshot™	P-1	yes	yes	multiplexing	www.appliedbiosystems.com
OLA	O-1	yes	–	high plex factor for allele generation	Baron et al., 1996
Padlock	O-1	no	–	no PCR	Nielsson et al., 1994; Landegren et al., 1996

(Continued)

Table 4.1 (*Continued*)

Method	Allele distinction/ Detection	TC	HT	Strength	Reference
Plate Reader					
Invader	N-4	no	yes	end point, no PCR, isothermal	DeFrancesco, 1998; Mein et al., 2000
SNP-IT	P-4	yes	yes	end point, integrated system for SNPstream UHT	www.orchidbio.com
OLA	O-4	yes	–	end point, easy format	Barany, 1991; Samiotaki et al., 1994
FP-TDI	P-4 (fluorescence polarization)	yes	yes	end point, homogeneous assay	Chen et al., 1997; 1999
5'-Nuclease (TaqMan™)	H-4 (FRET)	yes	yes	real-time or end point, homogeneous assay	Holland et al., 1991; Livak et al., 1995 www.appliedbiosystems.com
Molecular beacons	H-4 (FRET)	yes	–	real-time, homogeneous assay	Tyagi et al., 1996; 1998
kinetic PCR	H-4	yes	–	real-time, homogeneous assay	Germer, Holland & Higuschi, 2000
Amplifluor™	H-4	yes	–	real-time or end point, homogenous assay	Myakishev et al., 2001 www.serologicals.com
DASH/DASH-2	H-4 (iFRET)	yes	yes	low cost	Howell et al., 1999; Jobs et al., 2003
Flow Cytometer					
Coded spheres	O-2	yes	–	flexibility, high-throughput	Iannone et al., 2000
Coded spheres	P-2	yes	yes	flexibility, high-throughput	Cai et al., 2000; Chen et al., 2000
Sequencing					
Sequencing	P-1	yes	–	complete sequence information	www.appliedbiosystems.com
Pyrosequencing		yes	yes	quantitation	Ronaghi et al., 1996; 1999
Emerging Methods					
Illumina	allele-specific primer extension and ligation		yes	high degree of multiplexing	www.illumina.com
Parallele	single-base primer extension and ligation	yes	yes	high degree of multiplexing, detection on tag array	www.p-gene.com
SNPlex	OLA and PCR	yes	yes	multiplexing	www.appliedbiosystems.com

enzymes, like DNA polymerases, DNA ligases, or sequence- or structure-specific nucleases can be applied. Most enzymatic methods allow the generation of both allele products of a SNP in a single reaction.

5′-nuclease assay

This assay is also called the TaqMan™ assay (www.appliedbiosystems.com). Strictly speaking it is a hybridization-based assay, similar to molecular beacons, except that the probe is degraded enzymatically during PCR (see Chapter 2). A labelled oligonucleotide probe complementary to an internal sequence of a target DNA is added to a PCR (Holland *et al.* 1991, Livak *et al.* 1995; Kalinina *et al.* 1997). The nucleotide probe carries a fluorescent dye and a fluorescence quencher molecule. Successful hybridization of the oligonucleotide probe due to matching with one allele of the SNP results in its degradation by the 5′- to 3′-nuclease activity of the employed DNA polymerase whereby the fluorescent dye and quencher are separated, which promotes fluorescence. Since the inclusion of minor groove binders into the probes in 2001 the TaqMan™ assay has become a very reliable and high-throughput SNP genotyping method.

ARMS and kinetic PCR

The formation of a product in a PCR is dependent on complementary primers. For ARMS (amplification refractory mutation system, Newton *et al.* 1989), kinetic PCR (Germer *et al.* 2000), and the Amplifluor™ assay (Myakishev *et al.* 2001; www.serologicals.com) one of the two PCR primers ends 3′ with the correct complementary base to one allele of a SNP. Thus the formation or the rate of formation of a PCR product depends on the allele present in the template. One reaction for each allele is run in parallel. Successful generation of a PCR product is either analysed on a gel (ARMS) or in real-time by the increased fluorescence of dyes intercalating into the forming PCR products. For the Amplifluor™ assay (Myakishev *et al.* 2001) each one of the allele-specific primers has a 5′-tag sequence with a fluorescent dye/quencher system included. Successful PCR results in the separation of the quencher from the fluorescent dye, which promotes fluorescence. Unfortunately, with most DNA polymerases the 'wrong' allele is also amplified, but at a later thermal cycle. Thus endpoint measurement can be deceptive. A DNA polymerase with very high allele-specificity was used for molecular haplotyping (Tost *et al.* 2002). Usually, allele-specific PCR is extremely difficult to optimize for DNA fragments larger than 200 bases. In the case of molecular haplotyping allele-specific PCR products of 4 Kb were achieved without much optimization.

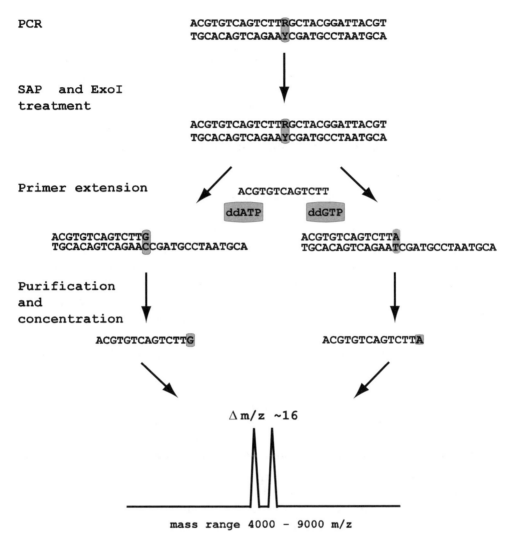

Figure 4.3 Example of SNP genotyping by primer extension with detection by MALDI mass spectrometry. Primer extension arrays require the preparation of template DNA by PCR, usually followed by at least partial clean-up prior to the primer extension reaction.

Primer extension (Figure 4.3)

Primer extension is a stable and reliable way of distinguishing alleles of a SNP. For primer extension an oligonucleotide is hybridized next to a SNP. Nucleotides are added by a DNA polymerase generating allele-specific products (Syvänen 1999).

Oligonucleotide ligation (OLA)

For OLA, two oligonucleotides adjacent to each other are ligated enzymatically when the bases next to the ligation position are fully complementary to the template strand by a DNA ligase (Barany 1991; Samiotaki *et al.* 1994; Baron *et al.* 1996). Padlock is a variant of OLA, in which one oligonucleotide is circularized by ligation (Nilsson *et al.* 1994, 1997; Landegren *et al.* 1996). Allele-specific products can be visualized by separation on gels or in plate reader formats.

Restriction Fragment Length Polymorphism (RFLP)

RFLP is one of the most commonly used and oldest formats for SNP genotyping in a standard laboratory (Botstein *et al.* 1980) that far predates the coining of the term SNP. PCR products are digested with restriction endonucleases that are specifically chosen for the base change at the position of the SNP, resulting in a restriction cut for one allele but not the other. Fragment patterns are used for allele assignment after gel separation. Due to the limited number of restriction enzymes, the complex patterns that may result, and gel separation, it is a very labour intensive method and does not lend itself to automation. The cost of RFLP depends on the required restriction endonuclease.

All the assays described above require PCR amplification of a stretch of genomic DNA prior to the allele-interrogation. In the past PCR has often been criticized, initially for being temperamental, then mainly for being expensive. However, it remains one of the most important tools in molecular biology. With respect to SNP genotyping, none of the large-scale studies carried out on the current generation of SNP genotyping methods manages to work without PCR. The Invader assay marks a departure from using PCR. It can be carried out without PCR at the expense of sacrificing copious amounts of genomic DNA for every SNP genotype. The Invader assay was applied for the largest association study carried out so far (Ozaki *et al.* 2002). In this instance multiplex PCR amplification was used to decrease the amount of genomic DNA consumed in each reaction. A previously published study using the Invader assay also used PCR to conserve DNA resources (Mein *et al.* 2000).

Flap endonuclease or invasive cleavage

The Invader assay makes use of flap endonucleases (FEN) for the discrimination between alleles of SNPs (Harrington and Lieber 1994). An invader oligonucleotide and a signal oligonucleotide with a 5′-overhang (flap) over the invader oligonucleotide are hybridized to a target sequence. Only in the case of a perfect match of the signal oligonucleotide with the target sequence is the flap cleaved off.

For the Invader squared assay the cleaved off flap is used to drive a secondary, universal cleavage reaction thereby increasing the rate of signal generation (Ryan *et al.* 1999). Cleavage of the flap can be linked to a change in fluorescence for example with a fluorescence resonance energy transfer system. Alternatively, the cleaved off flap can also be analysed by mass spectrometry (Griffin *et al.* 1999). Processing for the Invader assay is isothermal. The Invader assay is marketed by Third Wave Technologies (www.twt.com).

DNA sequencing

DNA sequencing is a viable method for assigning alleles of polymorphisms. Automated fluorescence Sanger sequencing currently is not competitive for the kind of throughput that is being targeted for association studies. However, in highly polymorphic regions of the genome like the MHC, when many SNPs can be captured in single sequencing reaction, it does become competitive. A very imaginative protocol for sequencing a few tens of bases from a primer is pyrosequencing (Ronaghi *et al.* 1996, 1999; Ronaghi 2001; www.pyrosequencing.com; see Chapter 6). Pyrosequencing is based on sequencing by following DNA synthesis. The protocol uses a DNA polymerase, an ATP sulfurylase, a luciferase and an apyrase to visualize the successful incorporation of a nucleotide. The DNA polymerase in the presence of one of the four nucleotide-triphosphate makes a template directed extension of a primer, if the next base on the template DNA strand is complementary to the available nucleotide-triphosphate. Pyrophosphate is released by a successful extension, which in turn is used to drive a chemiluminescent reaction with the luciferase, which is monitored in real-time. Nucleotide-triphosphates are added in turn to establish sequence ladders.

Analysis formats

Over the past years diverse analysis formats have been devised. They include gels, microtitre plate fluorescent readers with integrated thermocyclers, oligonucleotide microarrays (DNA chips), coded spheres with reading in a flow cytometer or on a bead array, and mass spectrometers. Nearly all of the above-described methods for allele-distinction have been combined with all of these analysis formats (Figure 4.4).

Gel-based analysis

DNA fragments can be separated by size by electrophoretic migration through gels. The current state-of-the-art is separation in 'gel-filled' capillaries. Instru-

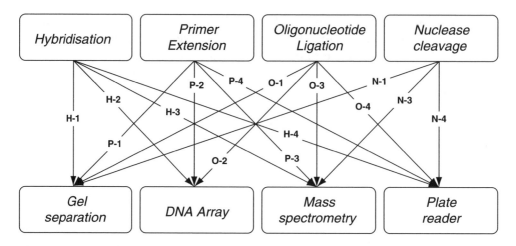

Figure 4.4 SNP genotyping methods separated into allele-preparation and analysis. Most SNP genotyping methods are a combination of one allele-preparation method with an analysis technique. Connection between allele-preparation methods and detection methods correspond to the abbreviations in Table 4.1

mentation with 96- or even 384-capillaries is commercially available. A SNP genotyping method with gel-based analysis is SNaPshot™ (www.appliedbiosystems.com). Primer extension is carried out in solution with four differently fluorescence-labelled ddNTPs. Allele calling is done by assigning colours at given positions in an electropherogram to an allele. This method is well suited to a multiplexing degree of about 10–15. Different SNPs can be spaced apart from each other in the electropherograms by adding bases on the 5′-end of extension primers so that individual zones for each of the SNPs are created. Alternatively individual SNPs can be injected into a capillary one-by-one without regenerating the capillary. A precursor of this method with separation on slab gels was shown by Pastinen *et al.* (1996). The oligonucleotide ligation assay with analysis on gels has also been carried out (Grossman *et al.* 1994; Day *et al.* 1995). Advantages of capillary systems over slab gel systems include the potential for 24-h unsupervised operation, the elimination of cumbersome gel pouring and loading, and that no lane tracking is required. A facile and quite efficient method using gel separation is microtitre array diagonal gel electrophoresis (MADGE, Day and Humphries, 1994). SNP genotyping methods using it are usually based on the generation of allele-specific products by ARMS or RFLP. On a horizontal gel system, 384 reactions are analysed simultaneously. A system can be acquired with very moderate capital investment. Melt-MADGE, a version that allows applying a temperature gradient during gel migration, was also shown. Melt-MADGE opens the path for further applications apart from SNP genotyping (Day *et al.* 1999) such as SNP discovery and mutation detection.

Fluorescent-reader based analysis

Several genotyping technologies use microtitre plate fluorescent readout systems for detection. Oligonucleotide ligation with oligonucleotides covalently linked to fluorescent chromophores is a sensitive analysis method (Barany 1991; Samiotaki *et al.* 1994). Template-directed dye-terminator incorporation (TDI) is a primer extension procedure adapted to a fluorescent reader system (Chen *et al.* 1997, Chen and Kwok 1997). Drawbacks of these systems are that unreacted fluorescent reagents have to be removed prior to reading, PCR products used as templates for the reactions have to be prepared beforehand, and the probing oligonucleotides have to be immobilized on a solid surface. Fluorescence polarization (FP) allows non-immobilized processing with real-time analysis without the separation of reacted and unreacted fluorescent dyes due to the change in polarization of the fluorescence after primer extension (FP-TDI, Chen *et al.* 1999). A fluorescence polarization reader is available from Perkin-Elmer as well as the FP-TDI technology (www.perkinelmer.com). Homogeneous assays, for which all reagents (DNA template, primers, DNA polymerase, dNTPs and buffers) are mixed in a tube or a well of a microtitre plate such as the TaqMan™ assay, molecular beacons, kinetic PCR, and the Amplifluor™ assay can be executed with similar detection devices using integrated thermocycling systems. In homogeneous methods the appearance of an allele of a SNP is monitored optically in real-time without sampling of the reaction. A number of suppliers offer instruments to carry out these assays (e.g. Applied Biosystems, BioRad, Stratagen, and MJ Research). A great advantage of some of these assays, like TaqMan™ or Amplifluor™ is that they can also be monitored by end-point measurement. Using an automated version of reader with a plate-feeder robot – like the ABI 7900HT – a single plate can be scanned in a couple of minutes. Throughput of several hundred thousand SNPs genotyped per day can be achieved this way. Applied Biosystems has prepared 200 000 SNPs throughout the human genome in TaqMan™ assays (Assay-On-Demand™), for which reagents are commercially available. However, localizing the proposed SNPs in the genome without access to the Celera sequence database can be difficult.

Array-based analysis

For oligonucleotide microarrays (DNA chips) a series of oligonucleotides (features) is chemically attached to a solid surface, usually a glass slide (Shalon *et al.* 1996; Ramsay 1998; see Chapters 7 and 8). The position of a specific oligonucleotide on the solid surface is used as the identifier. In excess of 10 000 different oligonucleotides can be arrayed on 1 cm^2 of the solid surface. The preparation of microarrays on a solid surface is achieved by one of two methods, either by spotting oligonucleotides or DNA (Cheung *et al.* 1999), or by in-situ synthesis

(Pease *et al.* 1994; Lipshutz *et al.* 1999). Hybridizing an unknown sequence to a known sequence on the DNA chip identifies a complementary sequence element in the unknown sequence. Microarrays have found their application mainly in expression analysis (Brown and Botstein 1999; Debouck and Goodfellow 1999). In principle there are three different formats that are used for SNP genotyping in an oligonucleotide microarray format:

1 Hybridization arrays: This format uses allele-specific hybridization. Oligonu-cleotides covering the complementary sequence of the two alleles of a SNP are on specific positions of the array. Fluorescently labelled PCR products containing the SNP sequences to be queried are hybridized to the array. In most cases it is not sufficient to represent only one sequence for each allele, rather a series of oligonucleotides that walk over each variant of the SNP are used (www.affyme-trix.com). This is referred to as tiling (Wang *et al.* 1998; Mei *et al.* 2000). The hybridization pattern of all oligonucleotides spanning the SNP is used to evalu-ate positive and negative signals. Despite redundancy, part of the assays fail to distinguish between heterozygous and homozygous samples. This approach was applied similarly by Patil *et al.* (2001) to resequence the entire chromosome 21 (www.perlegen.com). A practical difficulty is that the entire chromosome had to be amplified by PCR prior to hybridization, which is very labour intensive. The amount of information obtained this way is impressive.

2 Arrays with enzymatic processing: Arrayed primer extension (APEX) is a format, in which oligonucleotides corresponding to sequences immediately upstream of a SNP are arrayed on a solid surface (Shumaker *et al.* 1996; Pastinen *et al.* 1997). PCR products containing the SNP sequences are hybridized to the arrayed oligonucleotides. Each oligonucleotide on the array acts as a primer for a primer extension reaction with a DNA polymerase and four differently fluorescently labelled dideoxynucleotides. The fluorescence emission of the incorporated nucleotide identifies the next base on the hybrid-ized template and thus the allele of the SNP. In contrast to hybridization arrays, only one sequence per SNP is required on the array. The specificity of the genotyping is markedly increased due to the use of enzymatic discrimin-ation rather than differentiation by hybridization, yet enzymatic processing in the proximity of surfaces is not unproblematic. It was applied to resequence the p53 gene (Tonisson *et al.* 2002). An interesting new variant of this was pre-sented by Pastinen *et al.* (2000). In this approach the arrayed primers are allele-specific (the 3′-terminal base is complementary to one of the alleles of the SNP; two oligonucleotides are required for each SNP) and are extended with multiple fluorescent deoxynucleotides. More than one fluorescent chromophore is in-corporated for each matching allele, increasing the detection efficiency. An array-based detection of allele-specific products generated by OLA was used by Khanna *et al.* (1999).

3 Tag arrays: an array approach that has become very popular recently is the tag array (Fan *et al.* 2000). The features on the array represent tag oligonucleotides that are solely chosen based on their maximum sequence difference. The complementary tag sequences are included in the oligonucleotides for allele-distinction and are used to capture allele-specific products on complementarity. A completely integrated system of this kind is realized in Orchids SNPstream UHT system (www.orchid.com), which uses SNP-IT single-base primer extension for allele-discrimination.

The direct use of genomic DNA on a microarray without PCR amplification is not possible due to the high complexity and low concentration of target sequence. Thus all oligonucleotide array SNP genotyping approaches require the template DNA to be amplified by PCR. To match the highly parallel analysis qualities of the microarrays, either highly multiplex PCRs have to be established or multiple PCR products have to be pooled. Multiplex PCR is difficult to control and the risk of cross-talk increases. PCR has a bias towards smaller products, which means that only very small PCR products can be generated in highly multiplex PCR. Multiplexes beyond eight with small fragment size PCR products (< 200 bases) require a significant optimization effort.

Fluorescent emission from the features is recorded with an array scanner and used for allele calling. The majority of commercial array scanners are capable of detecting two different fluorescent colours (Cy3 and Cy5). For SNP genotyping four colour scanning would be desirable so that all four bases could be scanned simultaneously. Asper Biotech has commercialized a four-colour array scanner (www.asperbio.com).

Mass spectrometry-based analysis

Mass spectrometry, specifically MALDI mass spectrometry (matrix-assisted laser desorption/ionization time-of-flight mass spectrometry) has been demonstrated as an analysis tool for SNP genotyping (Little *et al.* 1997; Haff and Smirnov 1997; Li *et al.* 1999; Griffin *et al.* 1999; Sauer *et al.* 2000a, 2000b; for an exhaustive review see Tost and Gut 2002). Allele-specific products are deposited with a matrix on the metal surface of a target plate. The matrix and the analyte are desorbed into the gas phase with a short laser pulse. Analytes are ionized by collision and accelerated towards the detector. The time-of-flight of a product to the detector is directly related to its mass. The analysis is done serially (one sample after another) but at a very high rate. Mass spectrometers are capable of recording a single trace in far under a second. Apart from the speed and accuracy a major strength of mass spectrometric analysis is the number of available detection channels. Resolution of the current generation of mass spectrometers allows the distinction of base substitutions in the range of 1000–6000 Da (this corresponds to product sizes of 3–20 bases, the smallest mass difference for a base change thymine to adenine is 9 Da). In principle this would allow the analysis of hundreds of products in a single trace. However, for reasons of difficulty in multiplexing the sample preparation, only SNP genotyping multiplexes on the order of up to 12 SNPs have been shown (Ross *et al.* 1998). Sample consumption is very low, the data accumulation can be automatic, and absolute mass information allows for automatic allele calling with high confidence. The current generation of mass spectrometers is capable of recording close to 40 000 spectra per day. This puts the theoretical capacity for SNP genotyping of a single mass spectrometer in excess of 250 000 genotypes per day.

One of the major problems in the analysis of DNA by MALDI is the purity of the sample required for analysis. The purity has to be significantly higher than for other genotyping methods due to the sensitivity of the analysis to impurities. For the PROBE assay magnetic bead purification with a biotin/streptavidin binding system was applied (Braun *et al.* 1997; Little *et al.* 1997). An improved version with integrated automation is commercially available as the MassArray™ system (www.sequenom.com). The allele-preparation procedure is referred to as the MassExtend reaction. Haff and Smirnov (1997) demonstrated the PinPoint assay, for which purification with ZipTips (www.millipore.com) was applied. These are solid phase purification procedures that are not easy to automate. For magnetic beads batch to batch variability has to be dealt with, while ZipTips require frequent changing as they get blocked. These purification procedures can be quite costly. The GOOD Assay was recently presented (Sauer *et al.* 2000a, 2000b). Its main feature is that although the products are analysed by mass spectrometry, no purification is required due to a chemical modification strategy. This strategy does require additional reaction steps and modified oligonucleotides, but all reaction steps are simple additions of solutions into one reaction vial. For this reason this method is well suited for automated sample preparation. Another advantage of this preparation format is that the analysed products are in the size range of 4–5 bases, where the performance of MALDI mass spectrometers is optimal. For a protocol presented by Li *et al.* products are also in this favourable mass range, but a solid phase separation is applied (Li *et al.* 1999).

PNAs are a suitable substrate for mass spectrometric detection. SNP genotyping using MALDI mass spectrometry and hybridized PNAs has been demonstrated (Ross *et al.* 1997; Griffin *et al.* 1997), yet this approach has not been integrated into a high-throughput system.

Other mass spectrometric detection methods offer further possibilities for SNP genotyping. Mass tags that can be analysed on a number of different mass spectrometers have been shown (Shchepinov *et al.* 1999). A commercially available system applying this principle is MassCode (Kokoris *et al.* 2000; www.qiagen.com). A reporter entity (mass tag), similar to a fluorescent dye molecule, is attached as the identifier to an oligonucleotide complementary to one allele of a

SNP. For analysis the tags are cleaved from the oligonucleotides and analysed by mass spectrometry. In contrast to fluorescent dyes and their detection, mass tags and mass spectrometers offer a dramatically larger number of distinguishable entities. Using isotopically pure materials for the mass tags makes it possible to place one mass tag every few Da. This means that a fairly simple mass spectrometer can be used to distinguish thousands of mass tags at once.

How can SNP genotyping be made more economic? Miniaturization – Multiplexing

Miniaturization is widely considered the future of SNP genotyping in terms of decreasing cost and facilitating automation. However, it does require moving away from common laboratory equipment. There are efforts to place genotyping methods onto a microfluidic system (Kalinina et al. 1997; Anderson et al. 2000; Pang and Yeung 2000; Schmalzing et al. 2000). These solutions work along the principles for allele-distinction and detection presented above and there are limitations. DNA has to be added to a reaction with sufficient representation of the two alleles. About 1 ng of genomic human DNA corresponds to 300 copies of the genome. This is sufficient starting material for a genotyping experiment in a volume of a few μl. However, moving to nl volume PCR and reducing the amount of genomic DNA starting material proportionally has to result in allele-dropout.

SNP genotyping using fluorescence-coded beads avoids drawbacks of miniaturization (Cai et al. 2000; Iannone et al. 2000; Chen et al. 2000). These three publications describe the use of primer extension and oligonucleotide ligation for allele-discrimination with capture of the products on fluorescence-coded beads carrying complementary oligonucleotide sequences to tag sequences at the 5'-terminus of the extension or at one end of one of the ligation oligonucleotides. The fluorescent codes of beads and alleles information are read in a flow cytometer. The virtual reaction volume of a single SNP in a 10 μl reaction with 100 differently coded beads is only 100 nl. The emerging methods described below increase the degree of multiplexing in macroscopic volumes. Examples with 5000-plex reactions are shown.

Haplotyping

One of the major drawbacks of SNPs compared with microsatellites as genetic markers is the limited number of alleles they can present – usually two. A workaround is combining the information of several SNPs by determination of the phase – haplotypes. A method to carry out molecular haplotyping at high throughput has been presented. A multiplex primer extension SNP genotyping method with mass spectrometric detection was combined with allele-specific PCR (Tost *et al.* 2002).

DNA methylation analysis – analysis of methylation variable positions (MVPs)

DNA is made up of four nucleotides (C, T, G and A) and carries the 'blueprint' of an individuals phenotype. DNA is very static, while life is a dynamic process. Reversible methylation of cytosine in the sequence context CpG adds a dynamic component to DNA because it can act as a switch of transcription of a gene if the CpGs are in the promoter region of the gene (Novik *et al.* 2002). CpGs that can be methylated or not are termed methylation variable positions (MVPs). Measuring the degree of DNA methylation can be done by virtually any SNP genotyping method if a genomic DNA samples is converted with bisulfite. Bisulfite treatment of DNA results in conversion of non-methylated C into U, while methylated C remains unchanged. After bisulfite conversion and PCR amplification determination of the degree of methylation at a given MVP in the genomic DNA sample can be achieved by quantitating the degree of C and T at that position. The quantitative resolution of the genotyping method determines the accuracy of measurement that can be achieved. Accurate SNP quantitation can be achieved with a number of methods. However, one of the most accessible and easiest approaches for quantitating MVPs has to be pyrosequencing (Uhlmann *et al.* 2002; Tost *et al.* 2003; Chapter 6). Assay set-up is straightforward and the accuracy of quantitation is better than 2 per cent.

SNP genotyping and quantitation

SNP genotyping methods usually proceed along the axis of analysing a number of SNPs on an individual DNA sample. An axis that has also been explored is analysis of a large number of SNPs in pooled DNA samples, but this is of course at the expense of individual genotype information. Cases and controls are pooled individually and genotyped. The idea is that a SNP implicated in a pathology will show a deviation of the ratio of the two allele peaks in a pool of affected individuals compared with a pool of normal individuals. The number of DNA samples that can be pooled for one reaction depends on the resolution of the SNP genotyping method. As with MVP analysis, pyrosequencing provides excellent resolution and is well suited for pooled DNA SNP genotyping, also due to the ease of assay set-up (Rickert *et al.* 2002; Grub *et al.* 2002). Mass spectrometry has also been applied for quantitation of SNP genotyping (Ross *et al.* 2000; Buetow *et al.* 2001). Its resolution is slightly lower than pyrosequencing. On the other hand it can provide higher throughput. Further kinetic PCR is suitable for genotyping pooled DNA samples and is very effective (Germer *et al.* 2000).

Emerging methods

Recently, SNP genotyping methods have started emerging that dramatically increase the number of SNPs that are analysed in a single reaction. They move away from the principles outlined above. The main feature of three of these methods is that they invert the order of allele-distinction and amplification. Rather than amplifying genomic DNA (by PCR) to provide a sufficient amount of DNA template for allele query, the allele-distinction takes place at the level of genomic DNA by strategies with increased specificity. PCR is then used to provide a detectable amount of the allele-specific products using a universal primer system. Currently multiplexing with these methods exceeds 250-plex. Thus a 96-well plate can provide in excess of 25 000 genotypes. Manipulation of a single 96-well microtitre plate is easily achieved manually by a single technician without a need for extensive automation. On the other hand with automation it is feasible to record millions of SNP genotypes a day.

Companies spearhead these approaches. Illumina (www.illumina.com) uses coded-sphere arrays (Healey *et al.* 1997; Oliphant *et al.* 2002). The code on the spheres consists of an oligonucleotide tag capture sequence. The coded-spheres are lodged in solid phase wells at the end of a glass-fibre bundle. First the oligonucleotide tag of each coded sphere in its glass-fibre well is determined. The tags serve to capture the allele-identifying products. In order for the allele-specific reactions to remain allele-specific at the elevated degree of multiplexing multiple enzymatic methods are combined. The protocol of Illumina combines primer extension with ligation to a tag system. Allele-specific products are amplified by PCR and captured on the coded-sphere fibre array. 96/384 fibre-optic bundles are grouped to match standard microtitre plate formats. Each fibre-optic bundle contains 4000 fibres and is used to detect 1500-plex reactions.

Parallele is another company (www.p-gene.com) that has focused on increasing the multiplex factor for SNP genotyping. The molecular biological procedure that underlies the allele-distinction uses single-base primer extension in four individual reactions with each one of the four dNTPs, followed by ligation (Hardenbol *et al.* 2003). The combination of two allele-distinction methods achieves sufficient specificity required for a high level of multiplexing. To add further specificity this method operates with cyclization of oligonucleotides. The two ends of a single oligonucleotide are ligated in this procedure and thus a successful process results in a cyclic oligonucleotide. Thereafter non-cyclic probes are degraded by exonuclease digestion. This is followed by amplification of the residual, allele-specific probes and fluorescence marking. Each probe carries a tag sequence, which is used for capture on a tag array. The system is optimized for Affymetrix tag arrays that are currently available with a density of several thousand tags on a single array. According to Parallele 13 000-plex reactions are possible. A multiplex factor of 13 000 in a 20 μl reaction corresponds to a virtual volume of a single reaction of 1.5 nl. With a read-time of a few minutes per array millions of data points can be accumulated in a day.

Applied Biosystems has just introduced SNPlex (www.appliedbiosystems. com). Allele-distinction is achieved by oligonucleotide ligation followed by amplification with a universal PCR system. PCR products are captured by a biotin label in a streptavidin coated microtitre plate. Zip chutes are hybridized to the different Zip codes that are included in the individual SNP arrays of the multiplex. After washing, the hybridized zip chutes are loaded onto an ABI 3730 capillary sequencer and sized for allele calling. Currently a multiplex factor of 48 is offered.

Perlegen – a spin-off of Affymetrix – has adopted a brute force approach for re-sequencing entire genomes (www.perlegen.com). The entire sequence of a genome is synthesized on an array in short oligonucleotides. The DNA sequence of an individual is amplified by PCR and hybridized to the array. The absence of alleles of SNPs result in a decrease of hybridization signal. Effectively the sequence of an individual is compared with a reference sequence. Data derived for chromosome 21 by this approach was presented by Patil *et al*. (2001). Perlegen is currently resequencing an entire human genome in 2 weeks.

High-throughput SNP genotyping methods

A number of the methods described above have been integrated into high-throughput operations. Methods currently used at high-throughput are highlighted in Table 4.1. The first complete systems for the entire high-throughput SNP genotyping process, with automation and LIMS, are marketed as off-the-shelf products. Examples of this are systems of Orchid, Sequenom and Illumina.

The next generation

Essentially sequencing can be considered a genotyping method with high resolution. However, sequencing with the currently available methods is economically not competitive with genotyping methods. Beyond the emerging methods outlined above there are highly interesting methods for DNA sequencing appearing on the horizon. Most of these approaches currently still are very experimental, but promise to sequence an entire human genome in less than a day. The proponents are companies like Solexa (www.solexa.co.uk), 454 Corporation (www.454.com), US Genomics (www.usgenomics.com), Möbius Genomics, and Visigen (www. visigenbio.com). The approaches differ from the classical sequencing approaches and emerging genotyping methods mainly in that they rely on sequencing a single or small number of genomic DNA molecules. They do not use standard laboratory tools and methods like microtitre plates, PCR or PCR machines. In some examples the sequencing relies on directly monitoring a DNA polymerase incorporating dNTPs during DNA replication or pulling a single DNA molecule through a nanopore and measuring impedance changes. In other cases a multitude of single oligonucleotides are used to capture fragments of genomic DNA and then sequence 25 bases from these primers. Thereafter the sequence tags are aligned and compared with a reference sequence. Most of these innovative sequencing methods require extensive preparation of samples before the actual sequencing can start. However, once initiated they can produce vast amounts of DNA sequence data very rapidly (Leaman *et al.* 2003). The approach promises to go as far as accumulating the sequence of an entire human genome in a matter of a couple of hours. A major effort of these approaches concerns the DNA polymerases. Their performance has to be adapted to the application to reduce mis-incorporations as these would show up as polymorphisms. Further the demands on informatics will be extensive. In any case these are exciting prospects for genetics.

Conclusions

The field of genotyping technologies is still in great flux. There is no telling which method will emerge as the best. Some methods that were hailed as unbeatable a few years ago are no longer available. On the other hand the field seemed content when methods with one allele-discrimination step and processing of a few SNPs in a single reaction emerged and large-scale studies were initiated by automating this way of processing. Since then methods have emerged that combine multiple allele-discrimination procedures to increase the specificity, thus making it possible to advance on multiplex preparation. Considering that a few years ago people were struggling with multiplex reactions in the single digits, today multiplex factors of a few thousand are being achieved. This represents an increase of three orders of magnitude. Considering that the entire human genome probably has of the order of 10 million SNPs, it is another three orders of magnitude gain that is required. The prospect of obtaining whole genome sequences in a matter of a few hours is a great perspective for the future of variation analysis.

Acknowledgements

I would like to thank Jörg Tost for reading the manuscript and making valuable suggestions. Funding was provided by the French Ministry of Education, Research and Technology (MENRT).

References

Anderson, R.C., Su, X., Bogdan, G.J. and Fenton, J. (2000) A miniature integrated device for automated multistep genetic assays. *Nucl Acids Res* **28**: e60.

Barany, F. (1991) Genetic disease detection and DNA amplification using cloned thermostable ligase. *Proc Natl Acad Sci USA* **88**: 189–193.

Baron, H., Fung, S., Aydin, A., Bahring, S., *et al.* (1996) Oligonucleotide ligation assay (OLA) for the diagnosis of familial hypercholesterolemia. *Nat Biotechnol* **14**: 1279–1282.

Botstein, D., White, R.L., Skolnic, M. and Davis, R.W. (1980) Construction of a genetic linkage map in man using restriction fragment length polymorphisms. *Am J Hum Genet* **32**: 314–331.

Braun, A., Little, D.P. and Köster, H. (1997) Detecting CFTR gene mutations by using primer oligo base extension and mass spectrometry. *Clin Chem* **43**: 1151–1158.

Brookes, A.J. (1999) The essence of SNPs. *Gene* **234**: 177–186.

Brown, P.O. and Botstein, D. (1999) Exploring the new world of the genome with DNA microarrays. *Nat Genet Suppl* **21**: 33–37.

Buetow, K.H., Edmonson, M., MacDonald, R., Clifford, R., *et al.* (2001) High-throughput development and characterization of a genomewide collection of gene-based single nucleotide polymorphism markers by chip-based matrix-assisted laser desorption/ionization time-of-flight mass spectrometry. *Proc Natl Acad Sci USA* **98**: 581–584.

Cai, H., White, P.S., Torney, D., Deshpande, A., *et al.* (2000) Flow cytometry-base minisequencing: a new platform for high-throughput single-nucleotide polymorphism scoring. *Genomics* **66**: 135–143.

Cardon, L.R. and Bell, J.I. (2001) Association study designs for complex diseases. *Nat Rev Genet* **2**: 91–99.

Chen, J., Iannone, M.A., Li, M.-S., Taylor, D., *et al.* (2000) A microsphere-based assay for multiplexed single nucleotide polymorphism analysis using single base chain extension. *Genome Res* **10**: 549–557.

Chen, X. and Kwok, P.-Y. (1997) Template-directed dye-terminator incorporation (TDI) assay: a homogeneous DNA diagnostic method based on fluorescence resonance energy transfer. *Nucleic Acids Res* **25**: 347–353.

Chen, X., Zehnbauer, B., Girnke, A. and Kwok, P.-Y. (1997) Fluorescence energy transfer detection as a homogeneous DNA diagnostic method. *Proc Natl Acad Sci USA* **94**: 10756–10761.

Chen, X., Levine, L. and Kwok, P.-Y. (1999) Fluorescence polarization in homogeneous nucleic acid analysis. *Genome Res* **9**: 492–498.

Cheung, V.G., Morley, M., Aguilar, F., Massimi, A., *et al.* (1999) Making and reading microarrays. *Nat Genet Suppl* **21**: 15–19.

Couzin, J. (2002) Human Genome. HapMap launched with pledges of 100 M$. *Science* **298**: 941–942.

Day, D.J., Speiser, P.W., White, P.C. and Barany, F. (1995) Detection of steroid 21-hydroxylase alleles using gene-specific PCR and a multiplexed ligation detection reaction. *Genomics* **29**: 152–162.

Day, I.N.M. and Humphries, S.E. (1994) Electrophoresis for genotyping: microtitre array diagonal gel electrophoresis on horizontal polyacrylamide gels, hydrolink, or agarose. *Anal Biochem* **222**: 389–395.

Day, I.N., Spanakis, E., Chen, X. and O'Dell, S.D. (1999) Microplate array diagonal gel electrophoresis for mutation research in DNA banks. *Electrophoresis* **20**: 1250–1257

Debouck, C. and Goodfellow, P.N. (1999) DNA microarrays in drug discovery and development. *Nat Genet Suppl* **21**: 48–50.

Edman, C.F., Raymond, D.E., Wu, D.J., Tu, E., *et al.* (1997) Electric field directed nucleic acid hybridisation on microchips. *Nucleic Acids Res* **25**: 4907–4914.

Egholm, M., Buchardt, O., Christensen, L., Behrens, C., *et al.* (1993) PNA hybridizes to complementary oligonucleotides obeying the Watson-Crick hydrogen bonding rules. *Nature* **365**: 566–568.

Fan, J.-B., Chen, X., Halushka, M.K., Berno, A., *et al.* (2000) Parallel genotyping of human SNPs using generic high-density oligonucleotide tag arrays. *Genome Res* **10**: 853–860.

Germer, S., Holland, M.J. and Higuchi, R. (2000) High-throughput SNP allele-frequency determination in pooled DNA samples by kinetic PCR. *Genome Res* **10**: 258–266.

Griffin, T.J., Tang, W. and Smith, L.M. (1997) Genetic analysis by peptide nucleic acid affinity MALDI-TOF mass spectrometry. *Nat Biotechnol* **15**: 1368–1372.

Griffin, T.J., Hall, J.G., Prudent, J.R. and Smith, L.M. (1999) Direct genetic analysis by matrix-assisted laser desorption/ionization mass spectrometry, *Proc Natl Acad Sci USA* **96**: 6301–6306.

Grossman, P.D., Bloch, W., Brinson, E., Chang, C.C., *et al.* (1994) High-density multiplex detection of nucleic acid sequences: oligonucleotide ligation assay and sequence-coded separation. *Nucleic Acids Res* **22**: 4527–4534.

Gruber, J.D., Colligan, P.B. and Wolford, J.K. (2002) Estimation of single nucleotide polymorphism allele frequency in DNA pools by using pyrosequencing. *Hum Genet* **110**: 395–401.

Gut, I.G. (2001) Automation in genotyping of single nucleotide polymorphisms. *Human Mut* **17**: 475–492.

Haff, L. and Smirnov, I.P. (1997) Single-nucleotide polymorphism identification assays using a thermostable DNA polymerase and delayed extraction MALDI-TOF mass spectrometry. *Genome Res* **7**: 378–388.

Hardenbol, P., Baner, J., Jain, M., Nilsson, M., Namsataev, E.A., Karlin-Neumann, G.A., Fakhrai-Rad, H., Ronaghi, M., Willis, T.D., Landegren, U. and Davis, R.W. (2003) Multiplexed genotyping with sequence-tagged molecular inversion probes. *Nat Biotech* **21**: 673–678.

Harrington, J.J. and Lieber, M.R. (1994) Functional domains within FEN-1 and RAD2 define a family of structure specific endonucleases: implications for nucleotide excision repair. *Genes and Development* **8**: 1344–1355.

Healey, B.G., Matson, R.S. and Walt, D.R. (1997) Fiberoptic DNA sensor array capable of detecting point mutations. *Anal Chem* **251**: 270–279.

Holland, P.M., Abramson, R.D., Watson, R. and Gelfand, D.H. (1991) Detection of specific polymerase chain reaction product by utilizing the $5'3'$ exonuclease activity of Thermus aquaticus DNA polymerase. *Proc Natl Acad Sci USA* **88**: 7276–7280.

Howell, W.M., Jobs, M., Gyllensten, U. and Brookes, A.J. (1999) Dynamic allele-specific hybridization. A new method for scoring single nucleotide polymorphisms. *Nat Biotechnol* **17**: 17: 87–88.

Howell, W.M., Jobs, M. and Brookes, A.J. (2002) iFRET: an improved fluorescence system for DNA-melting analysis. *Genome Res* **12**: 1401–1407.

Iannone, M.A., Taylor, J.D., Chen, J., Li, M.-S., *et al.* (2000) Multiplexed single nucleotide polymorphism genotyping by oligonucleotide ligation and flow cytometry. *Cytometry* **39**: 131–140.

Jobs, M., Howell, W.M., Strömquist, L. and Brookes, A.J. (2003) DASH-2: Flexible,low-cost and high-throughput SNP genotyping by allele-specific hybridization on membrane arrays. *Genome Res* **13**: 916–924.

Kalinina, O., Lebedeva, I., Brown, J. and Silver, J. (1997) Nanoliter scale PCR with TaqMan detection. *Nucleic Acids Res* **25**: 1999–2004.

Khanna, M., Park, P., Zirvi, M., Cao, W., *et al.* (1999) Multiplex PCR/LDR for detection of K-ras mutations in primary colon tumors. *Oncogene* **18**: 27–38.

Kokoris, M., Dix, K., Moynihan, K., Mathis, J., (2000) High-throughput SNP genotyping with the masscode system. *Mol Diagnosis* **5**: 329–340.

Kruglyak, L. (1997) The use of a genetic map of biallelic markers in linkage studies. *Nat Genet* **17**: 21–24.

Kruglyak, L. (1999) Prospect for whole-genome linkage disequilibrium mapping of common disease genes. *Nat Genet* **22**: 139–144.

Landegren, U., Samiotaki, M., Nilsson, M., Malmgren, H. and Kwiatkowski, M. (1996) Detecting genes with ligases. *Methods: A Companion to Methods in Enzymology* **9**: 84–90.

Landegren, U., Nilson, M. and Kwok, P.-Y. (1998) Reading bits of genetic information: Methods for single-nucleotide polymorphism analysis. *Genome Res* **8**: 769–776.

Leaman, J.H., Lee, W.L., Tartaro, K.R., Lanza, J.R., Sarkis, G.J., deWinter, A.D., Berka, J. and Lohman, K.L. (2003) A massively parallel PicoTiterPlate based platform for discrete picoliter-scale polymerase chain reactions. *Electrophoresis* **24**: 3769–3777.

Li, J., Butler, J.M., Tan, Y., Lin, H. *et al.* (1999) Single nucleotide polymorphism determination using primer extension and time-of-flight mass spectrometry. *Electrophoresis* **20**: 1258–1265.

Lipshutz, R.J., Fodor, S.P.A., Gingeras, T.R. and Lockhart, D.J. (1999) High density synthetic oligonucleotide arrays. *Nat Genet Suppl* **21**: 20–24.

Little, D.P., Braun, A., Darnhofer-Demar, B., Frilling, A., *et al.* (1997) Detection of RET proto-oncogene codon 634 mutations using mass spectrometry. *J Mol Med* **75**: 745–750.

Livak, K., Marmaro, J. and Todd, J.A. (1995) Towards fully automated genome-wide polymorphism screening. *Nat Genet* **9**: 341–342.

Marras, S.A., Kramer, F.R. and Tyagi, S. (1999) Multiplex detection of single-nucleotide variations using molecular beacons. *Genet Anal* **14**: 151–156.

Mei, R., Galipeau, P.C., Prass, C., Berno, A., *et al.* (2000) Genome-wide detection of allelic imbalance using human SNPs and high-density DNA arrays. *Genome Res* **10**: 1126–1137.

Mein, C.A., Barrat, B.J., Dunn, M.G., Siegmund, T., *et al.* (2000) Evaluation of single nucleotide polymorphism typing with invader on PCR amplicons and its automation. *Genome Res* **10**: 330–343.

Myakishev, M.V., Khripin, Y., Hu, S. and Hamer, D.H. (2001) High-throughput SNP genotyping by allele-specific PCR with universal energy-transfer-labeled primers. *Genome Res* **11**: 163–169.

Newton, C.R., Graham, A., Heptinstall, L.E., Powell, S.J., *et al.* (1989) Analysis of any point mutation in DNA. The amplification refractory mutation system (ARMS). *Nucleic Acids Res* **17**: 2503–2516.

Nilsson, M., Malmgren, H., Samiotaki, M., Kwiatkowski, M., *et al.* (1994) Padlock probes: Circularizing oligonucleotides for localized DNA detection. *Science* **265**: 2085–2088.

Nilsson, M., Krejci, K., Koch, J., Kwiatkowski, M., *et al.* (1997) Padlock probes reveal single-nucleotide differences, parent of origin and in situ distribution of centromeric sequences in human chromosomes 13 and 21. *Nat Genet* **16**: 252–255.

Novik, K.L., Nimmrich, I., Genc, B., Maier, S. *et al.* (2002) Epigenomics: genome-wide study of methylation phenomena. *Curr Issues Mol Biol* **4**: 111–128.

Oliphant, A., Barker, D.L., Stuelpnagel, J.R. and Chee, M.S. (2002) BeadArray technology: enabling an accurate, cost-effective approach to high-throughput genotyping. *Biotechniques* (*June Suppl*): 56–61.

Ørum, H., Jakobsen, M.H., Koch, T., Vuust, J. and Borre, M.B. (1999) Detection of the factor V Leiden mutation by direct allele-specific hybridization of PCR amplicons to photoimmobilized locked nucleic acids. *Clin Chem* **45**: 1898–1905.

Ozaki, K., Ohnishi, Y., Iida, A., Sekine, A., *et al.* (2002) Functional SNPs in the lympho-toxin-alpha gene that are associated with susceptibility to myocardial infarction. *Nat Genet* **32**: 650–654.

Pang, H.-M. and Yeung, E.S. (2000) Automated one-step DNA sequencing based on nanoliter reaction volumes and capillary electrophoresis. *Nucleic Acids Res* **28**: e73.

Pastinen, T., Partanen, J. and Syvänen, A.-C. (1996) Multiplex, fluorescent, solid-phase minisequencing for efficient screening of DNA sequence variation. *Clin Chem* **42**: 1391–1397.

Pastinen, T., Kurg, A., Metspalu, A., Peltonen, L. and Syvänen, A.-C. (1997) Minisequencing: A specific tool for DNA analysis and diagnostics on oligonucleotide arrays. *Genome Res* **7**: 606–614.

Pastinen, T., Raito, M., Lindroos, K., Tainola, P., *et al.* (2000) A system for specific, high-throughput genotyping by allele-specific primer extension on microarrays. *Genome Res* **10**: 1031–1042.

Patil, N., Berno, A.J., Hinds, D.A., Berrett, W.A., *et al.* (2001) Blocks of limited halotype diversity revealed by high-resolution scanning of human chromosome 21. *Science* **294**: 1719–1723.

Pease, A.C., Solas, D., Sullivan, E.J., Cronin, M.T., *et al.* (1994) Light-generated oligo-nucleotide arrays for rapid DNA sequence analysis. *Proc Natl Acad Sci USA* **91**: 5022–5026.

Ramsay, G. (1998) DNA chips: state-of-the-art. *Nat Biotechnol* **16**: 40–44.

Reich, D.E., Cargill, M., Bolk, S., Ireland, J., *et al.* (2001) Linkage disequilibrium in the human genome. *Nature* **411**: 199–204.

Rickert, A.M., Premstaller, A., Gebhardt, C., and Oefner, P.J. (2002) Genotyping of Snps in a polyploid genome by pyrosequencing. *Biotechniques* **32**: 592–600.

Ronaghi, M., Karamohamed, S., Pettersson, B., Uhlén, M. and Nyren, P. (1996) Real-time DNA sequencing using detection of pyrophosphate release. *Anal Biochem* **242**: 84–89.

Ronaghi, M., Nygren, M., Lundeberg, J. and Nyren, P. (1999) Analyses of secondary structures in DNA by pyrosequencing. *Anal Biochem* **267**: 65–71.

Ronaghi, M. (2001) Pyrosequencing sheds light on DNA sequencing. *Genome Res* **11**: 3–11.

Ross, P.L., Lee, K. and Belgrader, P. (1997) Discrimination of single-nucleotide poly-morphisms in human DNA using peptide nucleic acid probes detected by MALDI-TOF mass spectrometry. *Anal Chem* **69**: 4197–4202.

Ross, P., Hall, L., Smirnov, I. and Haff, L. (1998) High level multiplex genotyping by MALDI-TOF mass spectrometry. *Nat Biotechnol* **16**: 1347–1351.

Ross, P., Hall, L. and Haff, L.A. (2000) Quantitative approach to single-nucleotide polymorphism analysis using MALDI-TOF mass spectrometry. *Biotechniques* **29**: 620–629.

Ryan, D., Nuccie, B. and Arvan, D. (1999) Non-PCR dependent detection of factor V Leiden mutation from genomic DNA using a homogeneous invader microtitre plate assay. *Mol Diagnosis* **4**: 135–144.

Samiotaki, M., Kwiatkowski, M., Parik, J. and Landegren, U. (1994) Dual-color detection of DNA sequence variants by ligase-mediated analysis. *Genomics* **20**: 238–242.

Sauer, S., Lechner, D., Berlin, K., Lehrach, H., *et al.* (2000a) A novel procedure for efficient genotyping of single nucleotide polymorphisms. *Nucleic Acids Res* **28**: e13.

Sauer, S., Lechner, D., Berlin, K., Plançon, C., *et al.* (2000b) Full flexibility genotyping of single nucleotide polymorphisms by the GOOD assay. *Nucleic Acids Res* **28**: e100.

Schmalzing, D., Belenky, A., Novotny, M.A., Koutny, L., *et al.* (2000) Microchip electrophoresis: a method for high-speed SNP detection. *Nucleic Acids Res* **28**: e43.

Shalon, D., Smith, S.J. and Brown, P.O. (1996) A DNA microarray system for analyzing complex DNA samples using two-color fluorescent probe hybridization. *Genome Res* **6**: 639–645.

Shchepinov, M.S., Chalk, R. and Southern, E.M. (1999) Trityl mass-tags for encoding in combinatorial oligonucleotide synthesis. *Nucleic Acids Symposium Series* **42**: 107–108.

Shumaker, J.M., Metspalu, A. and Caskey, C.T. (1996) Mutation detection by solid phase primer extension. *Hum Mut* **7**: 346–354.

Sosnowski, R.G., Tu, E., Butler, W.F., O'Connell, J.P. and Heller, M.J. (1997) Rapid determination of single base mismatch mutations in DNA hybrids by electric field control. *Proc Natl Acad Sci USA* **94**: 1119–1123.

Syvänen, A.-C. (1999) From gels to chips: 'Minisequencing' primer extension for analysis of point mutations and single nucleotide polymorphisms. *Hum Mut* **13**: 1–10.

Syvänen, A.-C. (2001) Accessing genetic variation: genotyping single nucleotide polymorphisms. *Nat Rev Genet* **2**: 930–942.

Tonisson, N., Zernant, J., Kurg, A., Pavel, H., *et al.* (2002) Evaluating the arrayed primer extension resequencing assay of TP53 tumor suppressor gene. *Proc Natl Acad Sci USA* **99**: 5503–5508.

Tost, J. and Gut, I.G. (2002) Genotyping single nucleotide polymorphisms by mass spectrometry. *Mass Spectrom Rev* **21**: 388–418.

Tost, J., Brandt, O., Boussicault, F., Derbala, D., *et al.* (2002) Molecular haplotyping at high throughput. *Nucleic Acids Res* **30**: e96.

Tost, J., Dunker, J. and Gut, I.G. (2003) analysis and quantification of multiple methylation variable positions in CpG islands by pyrosequencing. *Biotechniques* **35**: 152–156.

Tyagi, S. and Kramer, F.R. (1996) Molecular beacons: probes that fluoresce upon hybridization. *Nat Biotechnol* **14**: 303–308.

Tyagi, S., Bratu, D.P. and Kramer, F.R. (1998) Multicolor molecular beacons for allele discrimination. *Nature Biotechnol* **16**: 49–53.

Uhlmann, K., Brinckmann, A., Toliat, M.R., Ritter, H. and Nürnberg, P. (2002) Evaluation of a potential epigenetic biomarker by quantitative methyl-single nucleotide polymorphism analysis. *Electrophoresis* **23**: 4072–4079.

Wang, D.G., Fan, J.-B., Siao, C.-J., Berno, A., *et al.* (1998) Large-scale identification, mapping, and genotyping of single-nucleotide polymorphisms in the human genome. *Science* **280**: 1077–1082.

Weiner, M. and Hudson, T.J. (2002) Introduction to SNPs: discovery of markers for disease. *Biotechniques Suppl* **32**: 4–13.

Weiss, K.M. and Terwilliger, J.D. (2000) How many diseases does it take to map a gene with SNPs? *Nat Genet* **26**: 151–157.

5

High-throughput Mutation Screening

Paal Skytt Andersen and **Lars Allan Larsen**

Introduction

The rapid progress of the Human Genome Project has led to the discovery of more than a thousand genes associated with mendelian disorders (Online Mendelian Inheritance in Man 2002). Many more genes involved in disease will probably be revealed due to the increasing knowledge of gene functions in model organisms such as fly, worm and mouse. Furthermore, association studies using single nucleotide polymorphisms (SNPs) as high density markers is expected to be a powerful tool for identification of genes involved in complex disorders such as diabetes, schizophrenia and hypertension (Bader 2001).

Thus, the ability to detect specific mutations or search for unknown mutations in a large number of samples is becoming increasingly important for genetic diagnosis and for the process of discovery of novel disease genes. The latter may involve genotyping of thousands of SNPs in hundreds of samples and subsequently screening for mutations in a number of candidate genes in several disease loci. Although DNA sequencing has become a highly automated process, this technique is not always the best choice for large-scale mutation detection. Therefore, a number of methods for high-throughput mutation screening have emerged during the recent years. A successful outcome of a large mutation screening project is highly dependent on the choice of a suitable method for the specific project. Initially, it must be considered if the method is to be used for

Molecular Analysis and Genome Discovery edited by Ralph Rapley and Stuart Harbron
© 2004 John Wiley & Sons, Ltd   ISBN 0 471 49847 5 (cased)   ISBN 0 471 49919 6 (pbk)

Table 5.1   Things to be aware of when choosing the method for mutation screening

1. Is the method used for specific mutations or unknown mutations?

2. Cost:
Consumables cost
Instrumentation cost and versatility

3. Personnel:
Sample handling time (duration and hands-on time)
Data analysis (Is it time consuming and does it require highly experienced personnel?)

4. Assay strengths and weaknesses:
Throughput – does the method meet with your throughput requirements now and in a year?
Does sensitivity and specificity meet your requirements?
Is it a validated method, has the sensitivity and specificity been determined by independent
    labs?

scanning for unknown mutations or for detection of specific mutations in the sample material. From there on choosing the best method is not a simple task and the final choice is often based on a compromise between several parameters such as throughput, sensitivity and specificity of the methods, available human and economic resources and available instrumentation (Table 5.1). However, a prerequisite for a good choice is a general overview of the available methods. In this chapter, we will explain the principle behind the current methods, and we will discuss their advantages and drawbacks. The focus of the chapter is methods for detection of point mutations, thus methods for detection of large deletions/ insertions and methods for analysis of short tandem repeats (STR) will not be discussed.

Automated DNA sequencing for mutation analysis

The development of automated DNA sequencers with up to 384 capillaries in parallel has dramatically improved the throughput of DNA sequencing, thus it may seem logical simply to use automated DNA sequencing for large-scale mutation detection. However, automated DNA sequencing requires expensive equipment and the technology is still relatively expensive with respect to reagent cost. The major disadvantage is perhaps that detection of heterozygote mutations using automated DNA sequencing is not a trivial process, which often requires extensive manual data analysis. Automated scoring of heterozygotes using computer software is difficult due to the uneven incorporation of the fluorescently labelled dideoxynucleotides which are used in the sequencing reaction (Figure 5.1,

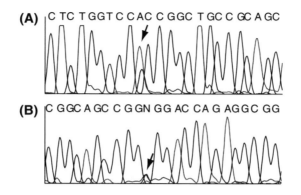

Figure 5.1   Detection of a heterozygote mutation using automated DNA sequencing. Uneven peak height is caused by uneven incorporation of fluorescent dideoxynucleotides. (A). Sense strand. The automated base calling has missed the mutation. However, the mutation was detected by visual inspection of the peak pattern, and was confirmed by sequencing of the anti-sense strand (B). Sequencing of the anti-sense strand alone results in automated calling of the mutation, but the peaks are almost indistinguishable from the background. The mutation, c532 G > A, was found in the *KCNQ1* gene, in a patient diagnosed with the inherited cardiac arrythmogenic disorder Long QT syndrome

upper panel) and even manual detection of heterozygotes becomes difficult in samples with high background (Figure 5.1, lower panel). These problems may in part be overcome by sequencing both strands or by using labelled primers instead of labelled nucleotides. However, both of these solutions require that more sequencing reactions are performed (i.e. two reactions for sequencing both strands using labelled dideoxynucleotides and four reactions for sequencing using labelled primers). Furthermore, the use of labelled primers adds dramatically to the reagent cost and is only practically suitable for projects involving analysis of a relatively small number of mutations in a large series of samples.

Methods for detection of specific mutations

The large interest in studying the association between SNPs and complex traits has spurred the development of novel methods for detection of specific mutations as well as improvements of known methods with respect to speed, automation and throughput.

   In the following we will explain the principle behind a number of the commonly used techniques for high-throughput detection of known mutations. Not all methods are mentioned here. For a recent update in the field, please read the excellent and thorough reviews by Syvänen (2001) and Gut (2001; also see Chapter 4).

DNA microarrays

DNA microarrays consist of micrometre-scale arrays of DNA fragments attached to a solid support, usually a glass plate. Each microarray can be smaller than a microscope slide but may include thousands of different DNA fragments, which are arranged in an ordered pattern. The DNA fragments are usually applied to the solid support by robotic spotting (Figure 5.2) or by *in situ* synthesis on the solid support using light-directed chemical synthesis (Figure 5.3). See Chapter 7.

In principle DNA microarrays are ideal for all aspects of high-throughput mutation screening due to the possibility of performing thousands of analyses in parallel. High-density oligonucleotide arrays containing hundreds of thousands of probes are commercially available from Affymetrix, Santa Clara, CA (see below) and have been used for large-scale detection of specific mutations. However, the use of high-density oligonucleotide arrays is quite expensive and less flexible, thus a number of alternative DNA microarray-based techniques have been developed among which the most widely used is arrayed primer extension (reviewed in Larsen *et al.* 2001a and Syvänen 2001).

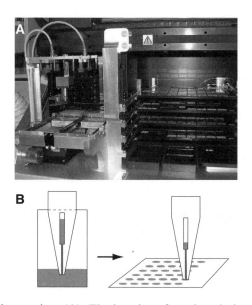

Figure 5.2   Robotic spotting. (A). The interior of a robot designed for manufacturing of microarrays (Microgrid II, Biorobotics, UK). The samples are transferred from microtitre plates (left) to the surface of microscope slides (right) using a pin-tool attached to a robotic arm. (B). Cartoon showing the principle in contact-spotting. The sample is applied to a cavity in the pin-head by capillary forces (left) and a small amount of sample is delivered to the slide by surface tension forces when the pin touches the slide (right)

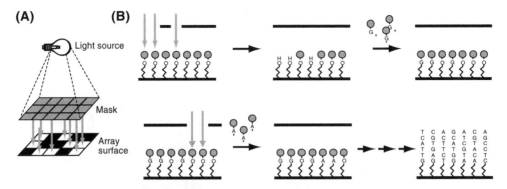

Figure 5.3 Light-directed chemical synthesis. Linkers with light-sensitive protection groups are attached to the chip surface. Photolithography (A) is used to direct light on specific areas of the chip, in which the linkers are de-protected. The first of four hydroxyl-protected deoxynucleosides is incubated with the surface (B), and covalent attachment occurs to the deprotected linkers. After washing, light is directed to other areas of the surface and a second deoxynucleoside is chemically attached to the unprotected groups. By repeated cycling simultaneous synthesis of thousands of oligonucleotides, each with a length of 20–25 bp, is possible

High-density oligonucleotide arrays (HDOA)

High-density oligonucleotide arrays (HDOA) (Lipshutz *et al.*, 1999) consist of 300 000–400 000 oligonucleotide probes arrayed on a chip surface of approximately 1.6 cm^2. HDOAs are produced by light-directed chemical synthesis, which combines photolithography and solid-phase oligonucleotide synthesis (Figure 5.3).

The advantage of using light-directed chemical synthesis is that a large number of different probes can be incorporated in each array at a low relative production cost per oligonucleotide. However, due to the high number of oligonucleotides per chip, HDOAs are still relatively expensive to produce. Thus, they are mass-produced and manufactured for specific applications. For mutation detection the target is usually fluorescently labelled PCR amplicons, which are hybridized to the probes on the chip. After stringent washing, the amount of fluorescent dye hybridized to each chip position is measured in a high-resolution fluorescence scanner. The data is collected, stored and analysed in a computer.

One of the major challenges related to mutation detection using HDOAs is to obtain a high sensitivity without loss of specificity. Due to the large number of probes, signals generated from unspecific cross-hybridization may result in a large number of false positives. For example, 1% cross-hybridization will result in 3000 false-positive results using a typical array of 300 000 probes. However, this problem has been solved using several probes to interrogate each mutation (Figure 5.4).

(A)

Reference sequence allele A:
GGTGATTATG**A**AACCTACTAT

Reference sequence allele B:
GGTGATTATG**G**ACCTACTAT

Probes with SNP in position 0

GGTGATTATG**A**ACCTACTAT   Mismatch

GGTGATTATG**T**ACCTACTAT   Perfect match
                            allele A

GGTGATTATG**C**ACCTACTAT   Perfect match
                            allele B

GGTGATTATG**G**ACCTACTAT   Mismatch

Probes with SNP in position -1

GTGATTATG**A**ACCTACTATC

GTGATTATG**T**ACCTACTATC

GTGATTATG**C**ACCTACTATC

GTGATTATG**G**ACCTACTATC

(B)

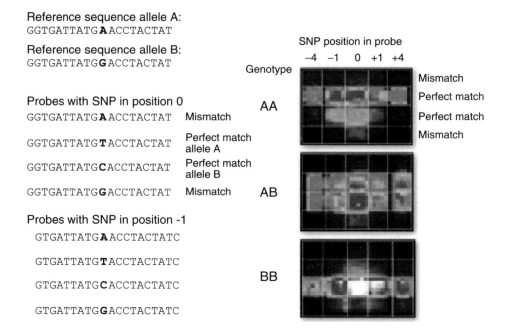

Figure 5.4 Detection of specific mutations using Affymetrix GeneChip® arrays. (A) Five perfect match probes and five probes with a single base mismatch are designed for each allele. The 20-mer probes are designed with the SNP in different positions. Examples of probes with SNP in position 0 and position −1 are shown. (B) Examples of hybridization patterns of three genotypes. Top panel: AA homozygote, middle panel: AB heterozygote, low panel: BB homozygote. Fluorescence intensity is shown in shades of grey. White is highest intensity and black is lowest intensity. Reproduced in part from Lipshutz *et al.* (1999) with permission

Affymetrix GeneChip® arrays are commercially available for simultaneous detection of 18 specific mutations in the human *CYP2D6* and *CYP2C19* encoding two cytochrome P450 enzymes involved in drug metabolism and may be used for pharmacogenetic studies.

Recently a chip has been developed for simultaneous detection of 1494 bi-allelic markers in the human genome. This chip, the HuSNP™ assay, can be used to perform genome-wide linkage analysis with a theoretical resolution of approximately 2 Mbp, given that all markers are informative. Thus, in some cases the chip may be used to perform a rapid initial genome scan, which must be followed by higher-resolution linkage analysis using conventional methods. It is likely that additional arrays with markers for high-resolution genotyping will be developed in the near future.

The HuSNP™ GeneChip® array has been used for genome-wide detection of loss-of-heterozygosity (LOH) involved in cancers (Lindblad-Toh *et al.* 2000; Schubert *et al.* 2002). For these applications, approximately 350 SNPs were found to be informative, giving an average resolution of 8.5 Mb. The use of the GeneChip® array dramatically decreased the workload and the amount of DNA required for genotyping 1500 loci, compared with conventional methods for detection of LOH. Furthermore, genotyping was easily performed using formalin-fixed pathological samples (Schubert *et al.* 2002).

Arrayed primer extension (APEX)

A goal of most mutation detection techniques is a high sensitivity combined with a minimum of false-negative and false-positive results. It is difficult to obtain this goal if discrimination between normal samples and mutants is based on differential hybridization as in the GeneChip system. This is because the optimal hybridization temperature of each probe is dependent on the DNA sequence surrounding the mutation. Thus, it is impossible to obtain optimal hybridization conditions for each probe in an array of thousands of probes. This problem has partly been solved using a highly redundant set of probes in the GeneChip system.

An alternative approach has been to combine DNA hybridization with an enzymatic reaction performed by a DNA polymerase, DNA ligase or reverse transcriptase in order to increase the method's power of discrimination. Enzyme-assisted discrimination has been shown to increase the power of discrimination at least 10-fold compared with methods based on hybridization alone.

Arrayed primer extension (APEX) or arrayed 'minisequencing' (Syvanen 2001) combines DNA microarray hybridization with a primer extension reaction performed by a DNA polymerase (Figure 5.5). In APEX, 5' modified DNA oligos are arrayed on the surface of a microscope slide using a robotic spotter. The probes are attached to the surface of the slide using different chemistries, e.g. attachment of a 5' aminolinker to slides coated with aminosilane or epoxysilane. The target DNA is amplified by PCR and after purification the amplicon is hybridized to the probe array. A DNA polymerase performs extension of the probe. The probe is designed to hybridize to the DNA sequence immediately upstream from the mutation. The discrimination of the mutation may rely on two principles: primer extension using labelled dideoxynucleotides or allele-specific primer extension using labelled deoxynucleotides, in which discrimination relies on the inability of DNA polymerase to extend a primer with a mismatch in the 3' end. The advantage of the former method is the possibility of performing multiplex analyses using different labels on each dideoxynucleotide. This is not possible with allele-specific extension, but this method allows for incorporation of more than one labelled nucleotide, which results in a higher signal intensity. However, a

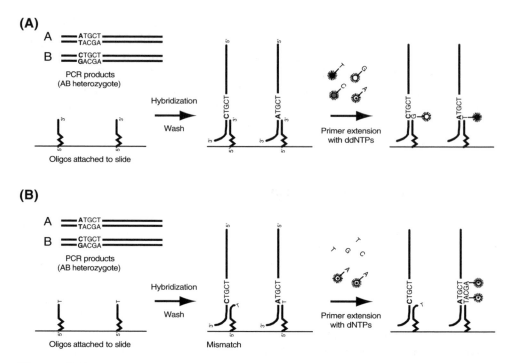

Figure 5.5 Arrayed primer extension (APEX). PCR products are produced using primers flanking the mutation. The PCR amplicons are hybridized to 5′ modified oligonucleotides probes which are attached to the surface of a microscope slide. The probes hybridize to the sequence immediately upstream from the mutation, and primer extension is performed using a DNA polymerase. Discrimination of the mutation may rely on (A) incorporation of fluorescently labelled dideoxynucleotides (ddNTPs) or (B) allele-specific primer extension using a mixture of deoxynucleotides (dNTPs) of which one nucleotide is labelled with a fluorephor

disadvantage of allele-specific extension is that in some cases the DNA polymerase can extend primers with 3′ mismatches (Ayyadevara *et al.* 2000).

An example of genotyping of 84 SNPs by APEX using fluorescently labelled ddGTP is shown in Figure 5.6. APEX have been used both for detection of disease causing mutations and for SNP analysis (Larsen *et al.* 2001a; Syvanen 2001).

The main advantage of APEX compared with high-density oligonucleotide array hybridization is that the method is more flexible and, in principle, assays can be designed for analysis of any combination of SNPs in any gene by all laboratories equipped with a microarray spotter and scanner. The main disadvantage is that the density of the microarrays used in APEX is small compared with HDOAs, thus the throughput is lower.

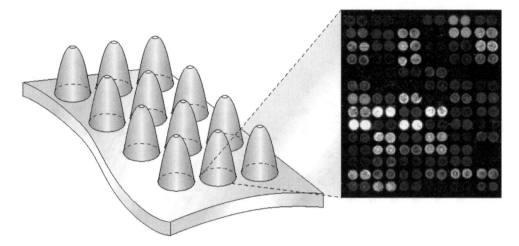

Figure 5.6 Genotyping of 84 SNPs using arrayed primer extension. Several small arrays of primers designed for detection of 84 SNPs in duplicate was spotted on the slide surface. Each array is separated from each other by a silicone rubber cone which forms an individual reaction chamber for the array. Reproduced from Syvanen (2001) with permission

Methods based on mass spectrometry

During the last decade, a number of methods for detection of specific mutations using matrix-assisted laser desorption/ionization time-of-flight mass spectometry (MALDI-TOF MS) have been developed (reviewed in Pusch *et al.*, 2002).

In MALDI-TOF MS the samples are mixed with a crystallized matrix consisting of a laser light absorbing compound (usually an organic acid) which protects the sample from photo-dissociation (Figure 5.7). Excitation of the matrix by a high-intensity laser pulse leads to evaporation of the matrix and desorption and ionization of the sample. The sample ions, which are predominately single charged, are accelerated by high voltage and separated in a field-free flight tube. The time-of-flight from when the ions enter the tube to when they hit the detector at the end of the tube is related to the mass-to-charge (m/z) ratio of each ion. Thus, measuring the time-of-flight is used to generate a mass spectrum.

Detection of specific mutations using MALDI-TOF MS has several advantages. First, ionization, separation and detection is performed in milliseconds, thus an average of several measurements are performed in seconds. Secondly, the detection is based on measurements of the mass of the analyte, thus the result is absolute and does not require radioactive or fluorescence labelling. Thirdly, for many applications, automation of most analysis steps is feasible.

(A)

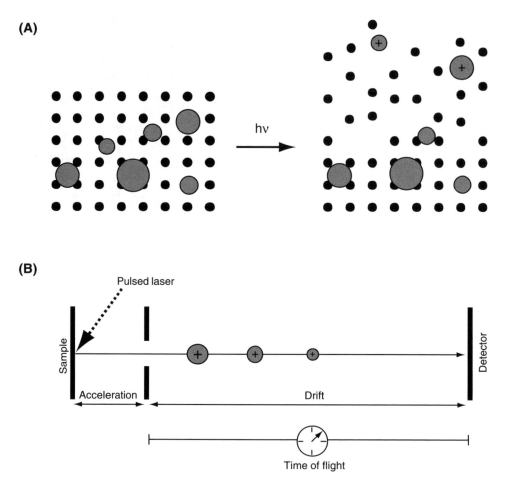

(B)

Figure 5.7 Matrix-assisted laser desorption/ionization time-of-flight mass spectometry (MALDI-TOF MS). (A) The sample (grey) is mixed with a crystallized matrix (black). Energy from a high-intensity laser pulse is absorbed by the matrix. This leads to evaporation of the matrix and desorption and ionization of the sample. (B) The sample ions are accelerated by high voltage and separated in the flight tube. Measurement of the time-of-flight, which is related to the mass-to-charge ratio, is used to generate the mass spectrum

Most MALDI-TOF MS genotyping methods are based on a primer extension reaction (see above) followed by detection of the incorporated nucleotide by measuring the related change in mass. In primer extension assays involving extension by a single dideoxynucleotide, each possible point mutation will result in a specific mass difference between the extended oligonucleotides (see example in Figure 5.8). For some mutations, the mass difference is as little as 9 Da.

Positively charged ions tend to build salts with the negatively charged sugar–phosphate backbone of the DNA, which may interfere with the determination of allele-specific changes in mass. Thus, in order to avoid interference from ions

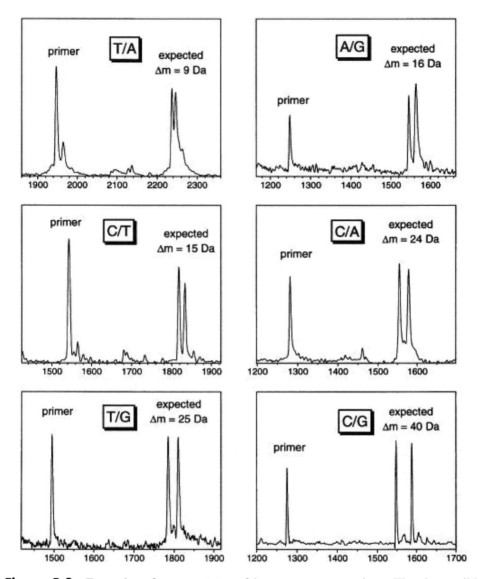

Figure 5.8 Examples of mass spectra of heterozygous mutations. The six possible combinations of dideoxynucleotides are shown (T/A, C/T, A/G, C/A, T/G, and C/G) with the expected mass differences between them. The SNP genotype is determined for each sample by measuring the mass difference between the primer and the extension products. Reproduced from (Li *et al.* 1999) with permission

present in the reaction buffer it is important that the reaction products are purified before the MS analysis.

Several attempts have been made to circumvent the problems related to measurement of very small mass differences and interference from contaminating ions (Pusch *et al.* 2002). The former have been solved by cleavage of the oligonucleotide before MS analysis (Figure 5.8) or the use of specialized oligonuleotide primers and modified nucleotides in the primer extension reaction (Sauer *et al.* 2000). These modifications result in smaller reaction products, for which small mass changes are better resolved. Translation of the DNA sequence to protein before MS analysis or use of the neutrally charged nucleic acid analogue peptide nucleic acid (PNA) for allele-specific hybridization has been used to solve the problems associated with interference from contaminating ions. MALDI-TOF MS analysis is an attractive technology for high-throughput detection of specific mutations because the MS analysis is performed in seconds. However, it should be taken into consideration that a substantial amount of sample preparation and purification has to be performed prior to the analysis step.

Oligonucleotide ligation assay (OLA)

Oligonucleotide ligation assay (OLA) takes advantage of a thermostable Taq DNA ligase that will ligate two oligonucleotides (primers) when the 3′ end of one fragment is adjacent to the 5′ end of the other and both are annealed to a complementary strand (Grossman *et al.* 1994). For detection of a SNP, each SNP requires three primers – one homologous to the normal sequence, one to the variant (where the SNP is the 3′-nucleotide) and one that can ligate to either of the first two. Ligation is only accomplished when there is a perfect annealing. The length of the oligonucleotides may be varied and therefore can be separated by simple electrophoresis. This may be done in a multiplex using different dyes and different oligonucleotide lengths (or by adding different lengths of oligomeric pentaethylene oxide (PEO) tails at the 5′-end of the hybridizing oligonucleotide) and subsequently analysed by capillary electrophoresis (see below).

OLA can be performed in high throughput by multiplexing and the assay time is very short. It is a simple assay, but requires a PCR template and may require quite a bit of optimization in order to obtain multiplex analysis.

Methods based on microtitre plate formats

Today, a number of mutation detection methods are available that use a microtitre format either in a fluorescence reader or in instruments dedicated to the specific method such as the pyrosequencing techniques described below. In the following we will discuss some of the most important methods based on microtitre formats. Each method has its own advantages and drawbacks, but a common feature is easy integration with common liquid handling robots and the possibility of performing parallel analyses in a scalable format (i.e. 96 wells, 384 wells or 1536 wells). In theory, probes and other reaction components can be pre-dried in the microtitre plate and the reaction can be initiated by addition of target DNA and enzyme.

Molecular beacons

Molecular beacons (Tyagi and Kramer 1996) are oligonucleotide probes that are designed to posses a stem-and-loop conformation under conditions where hybridization to the target sequence is unfavourable (Figure 5.9). The loop portion of the probe is complementary to the target sequence, while two 'arms' complementary to each other form the stem. A fluorophore is attached to the end of one arm and a 'quencher' molecule is attached to the end of the other arm. When the stem-and-loop structure is formed, the fluorophore and the quencher are held in close proximity to each other, thus the fluorophore is quenched by fluorescence resonance energy transfer, FRET (i.e. energy is transferred from the fluorophore to the quencher and dissipated as heat instead of being emitted as light).

Because the loop sequence is longer (e.g. 12–16 nucleotides) than each of the arm sequences (5–6 nucleotides), hybridization to the target sequence is more stable than formation of the stem structure. The stem will not form during hybridization because the structure of the double-stranded DNA is relatively rigid and is only formed when the probe is unable to hybridize to the target, e.g. due to a mismatch caused by a mutation. Thus, hybridization to the target sequence will cause the fluorophore and the quencher to move away from each other, which will result in the emission of light from the fluorophore when it is excited.

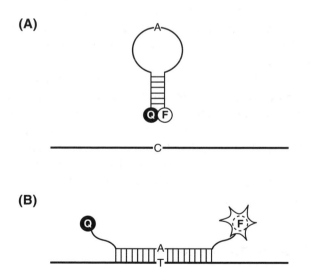

Figure 5.9 Mutation detection using molecular beacons. (A) The probe obtains a stem and loop structure under conditions unfavourable for hybridization (i.e. the probe contains a single base mismatch). The quencher is in close proximity to the fluorophor and no light is produced. (B) The probe is hybridized to a perfect match target. The fluorophore is no longer quenched and light is produced when the fluorophore is excited

The advantage of techniques based on molecular probes is that it is unnecessary to remove unhybridized probes because the fluorescence is quenched. This means that the probes can be applied directly to a PCR reaction and used to detect mutations during the amplification process (real-time PCR; see Chapter 2) obviating the need for post PCR manipulations. Colour multiplexing is possible using fluorophores with different emission wavelengths and the same quencher molecule (Marras *et al.* 1999). The disadvantage of the molecular beacon technique is that the probes are more difficult to design compared with conventional hybridization probes and they are relatively expensive to manufacture.

Flap endonuclease assay: the Invader® technique

Structure-specific endonucleases have been applied for very high throughput screening for SNPs in either a combined PCR-Invader® technique or a FRET-based combined Invader® technique that is based on linear amplification, which obviates the need for PCR. Basically, the technique is an isothermal enzymatic method where a structure-specific endonuclease cleaves a structure formed when two separate but overlapping oligonucleotides anneal to a target strand (Lyamichev *et al.* 1999). The method is a two-step assay (Figure 5.10). Three oligonucleotides are used, normal probe, variant probe and invader probe, together with target DNA. The cleaved flap (5′ end of the original probe normal or variant) subsequently serves as invader probe in a second reaction where an allele specific probe containing a quencher and a signal dye is used to generate a fluorescent signal when the probe is cleaved. The resulting signal can be read in any fluorescence detection system.

The advantages of the Invader® assay are that it allows for direct assay on templates without the need for PCR amplification. It is a fast assay with a high throughput that requires no expensive instrumentation. However, the system is still in its infancy and validation of the system remains to be performed in order to determine sensitivity. The major disadvantage is the need to assay the two alleles of each SNP in separate reactions. Also, assay for each SNP needs to be optimized in respect to assay temperature, which may be a laborious job when a large number of SNPs are to be assayed.

TaqMan® allelic discrimination assay

Real-time PCR (see Chapter 2) is a widely used for detection of SNPs. One of the most popular methods is the TaqMan® allelic discrimination assay (Livak 1999).

The TaqMan® allelic discrimination assay exploits the fact that the Taq DNA polymerase has an intrinsic 5′-nuclease activity. The allelic discrimination assay uses three or four oligonucleotides, two for PCR amplification and one or two

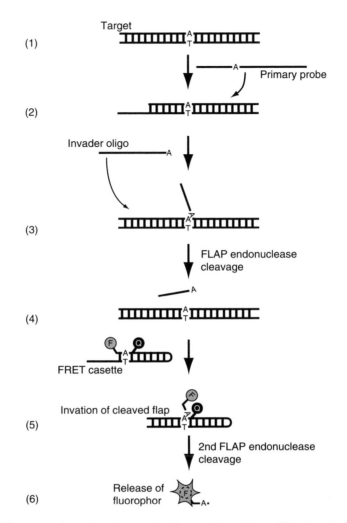

Figure 5.10 Mutation detection using flap endonuclease. The Invader® technique. The target DNA (1) is denatured and the primary probe (2) is hybridized to the target. The invader oligo invades the double strand at the site of the mutation and a flap is created by the un-hybridized part of the primary oligo (3). The flap is recognized by the flap endonuclease and cleaved exactly at the 3′ side of the mutation, creating a short oligonucleotide (4). The cleaved flap invades the FRET casette and a new flap is created (5). The quenched fluorophore is released by the 2nd FLAP endonuclease cleavage, and fluorescence is produced (6)

probe(s) targeted at the polymorphic site, the TaqMan® probe (Figure 5.11). The TaqMan® probe is a site-specific 20–30 bp oligonucleotide, which contains a reporter fluorescent dye at its 5′ end and a quencher dye attached to the 3′ end. The probe is disabled from extension at its 3′ end. When the probe is intact, the fluorescence is quenched by FRET because the quencher is very close to the dye.

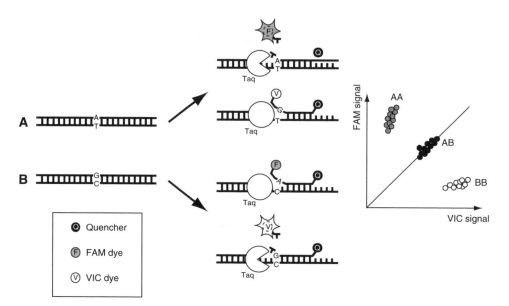

Figure 5.11 The Taqman® allelic discrimination assay. The TaqMan® probes contain a reporter fluorescent dye at their 5′end and a quencher dye attached to the 3′ end. A perfect match probe is designed for each allele, each probe contains a different fluorophor (i.e. FAM for the 'A' allele and VIC for the 'B' allele). Only perfect match probes are cleaved by the 5′ nuclease activity of the Taq DNA polymerase, resulting in light emission from the fluorophore now released from the probe. The theoretical possible readout from a marker with the genotypes A and B is shown on the right. The genotype AA will result in emission of light from the FAM dye only, the genotype BB will result in emission of light from the VIC dye only, while an AB heterozygote will result in emission of light with both wavelengths

If two allele-specific TaqMan® probes are used, one probe contains fluorophore A and the other fluorophore B. The probe containing the perfect match will be cleaved during strand displacement, whereas a mismatch probe will greatly reduce probe hybridization due to lower melting temperature (T_m) and subsequently reduce efficiency of cleavage. This assay may be performed during PCR amplification or by end-point signal analysis. Instruments have been developed to assay in a 96 or 384 micro-well format and the throughput is relatively high. Methods have been developed to improve TaqMan® probes in order to shorten the length of the probes so the difference in T_m between the perfect match and the mismatch probes and thus improve mismatch discrimination. So any system that increases T_m of a probe could be useful. Recently, addition of a minor groove binder (MGB) has resulted in marked T_m increase and enabling use of shorter probes (13–20 bp) (de Kok et al., 2002).

The allelic discrimination assay is a very well documented method and it does not require post-PCR processing, a very important advantage which reduces labour and material costs. Also, the two alleles of each SNP may be assayed in the same reaction. The disadvantages include the design of the probe, the fact that the assay cannot be multiplexed and it is sensitive to multiple SNPs in a narrow region as each SNP requires a specific probe.

Pyrosequencing

Pyrosequencing (Fakhrai-Rad *et al*. 2002; also see Chapter 6) is a novel technique that is very promising and in many ways an alternative to DNA sequencing by the dideoxy chain termination method. Like dideoxy chain termination sequencing, pyrosequencing is an enzyme-based technique. However, the present state of the art does not allow for high-throughput DNA sequencing due to the limited read length of the technique. Today, only less than 100 bp can be read. Nevertheless, instruments have been developed to perform high-throughput SNP detection. It is a real-time technique, where a light signal is generated after each addition of a nucleotide. The light signal is generated through coupled enzymatic reactions where polymerization of DNA using nucleotides results in formation of pyrophosphate. A coupled enzyme reaction with sulfurylase and luciferase converts the pyrophosphate to ATP and subsequently to light as shown in Figure 5.12. Each light signal is proportional to the number of nucleotides incorporated. The light produced in the luciferase-catalysed reaction is detected by a charge coupled device (CCD) camera and seen as a peak in a pyrogram™. Each light signal is proportional to the number of nucleotides incorporated. At the same time the enzyme apyrase competes with the luciferase reaction for the generated ATP and any excess dNTP in the solution. When the reaction is terminated, the next addition may follow. Pyrosequencing is a method that has a high throughput. It can be multiplexed up to a three-plex and an 96-sample assay can be performed in less than 20 min. By connecting a feeder device this may result in a throughput of up to 30 000 SNPs in 24 h.

It is a simple assay to perform and the results are in real-time. It can identify multiple mutations in a narrow region (e.g. HLA), which is something other methods cannot. The assay also has a very low cost.

The disadvantages of the system are the very short reads that result from the analysis of purified PCR products, which makes it an inadequate substitution for traditional DNA sequencing. Furthermore, the assay requires a dedicated instrument.

Methods for mutation scanning

Detection of unknown mutations is fundamentally different and often much more complicated than testing for the presence or absence of a known mutation.

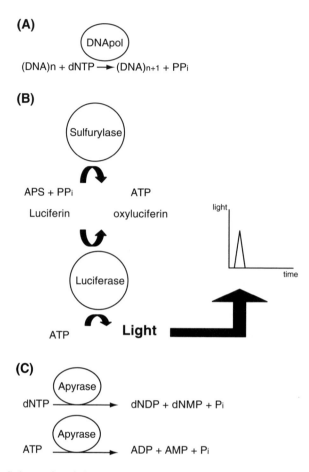

Figure 5.12 Schematic of the pyrosequencing method. A single stranded DNA template of a PCR fragment, a sequencing primer, a mix of four enzymes (DNA-polymerase, ATP sulfurylase, Luciferase and Apyrase) together with the substrates (deoxynucleotide, adenosine phosphosulfate and luciferin) are all added to a reaction tube in the pyrosequencing instrument. (A) DNA polymerase (DNApol) catalyses the polymerization of DNA, where the DNA strand is extended by addition of dNTPs; the products of the reaction are an extended DNA strand and an equimolar amount of pyrophosphate (PPi) ion. B. The generated pyrophosphate (PPi) ion reacts with adenosine 5'-phosphosulfate catalysed by ATP sulfurylase generating an equimolar amount of ATP. This ATP is used by the luciferase enzyme, which catalyses the reaction of luciferin and ATP to generate oxyluciferin and light. Each light signal is proportional to the number of nucleotides incorporated. (C) Apyrase degrades nucleoside tri- and diphosphates and deoxynucleoside tri- and diphosphates to the corresponding monophosphate. When degradation is complete another dNTP is added

Consequently, the number of different methods for scanning of unknown mutations is small compared with the methods for detection of specific mutations.

Nevertheless, detection of a single point mutation among several candidate genes in a large number of samples is often the goal in areas as, for example, molecular diagnostics, disease gene discovery and pharmacogenomics. Thus, high-throughput mutation scanning is becoming increasingly important. In the following we will explain the principles behind three fundamentally different technologies for mutation screening: capillary electrophoresis, denaturing high-performance liquid chromatography and high-density oligonucleotide array hybridization.

Methods based on capillary electrophoresis

High-throughput screening for unknown mutations is usually a laborious search for a single mutation in one of many DNA fragments. This task has been made more easily attainable through the development of automated capillary electrophoresis (CE) instruments supplied with capillary arrays of up to 384 capillaries. Thus, the throughput may be boosted from around 50–100 assays per day to at least several hundreds per day, depending on the number of capillaries in the array and the method used. Most gel-based mutation detection systems can be converted into capillary electrophoresis based systems with only minor changes (Larsen *et al.* 2000). Photolithography has been used for production of micro-fabricated capillary electrophoresis chips. Preliminary results with these chips are very promising and this is an area where throughput can be increased radically as the electrophoresis time is reduced approximately 10-fold. The basic principles of CE are briefly outlined below. Michelson and Cheng have thoroughly reviewed the theory behind (Michelson and Cheng 2001b) and the applications for CE (Michelson and Cheng 2001a).

The outline of a capillary electrophoresis instruments is shown in Figure 5.13. Fused silica capillaries (30–50 cm in length and a diameter of 50–100 μm) containing a separation matrix, consisting of a polymer solution, are used to separate the DNA fragments. The instrument consists of a sample tray from which samples are injected into the capillaries (either electrokinetically or by pressure) at the anode end of the instrument; subsequently the capillaries are transferred to an anode buffer chamber to which high power current (5–30 kV) is applied. The cathode end of the instrument contains a buffer chamber and has a pump to supply fresh polymer between each separate run. The separated molecules are detected at the cathode end of the capillary by UV absorption or laser-induced fluorescence detection. All signals are collected and stored in a dedicated computer, from which they can be analysed.

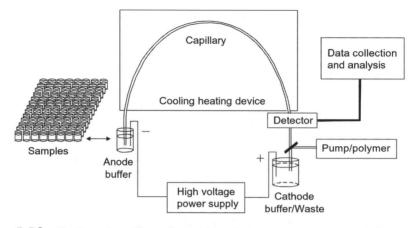

Figure 5.13 Outline of capillary electrophoresis instruments. The DNA fragments are separated according to size inside a fused silica capillary with an inner diameter of 50–100 μm under high voltage. As a fluorescently labelled DNA fragment pass the detector, the fluorophore is exitated by a laser and emitted light is measured. Sample injection and refilling of separation matrix before each run, data acquisition and storage is controlled by a computer

Mutation detection by single strand conformation polymorphism

One of the most widely used mutation detection methods is single strand conformation polymorphism (SSCP), which is a method first described by Orita and co-workers (Orita *et al.* 1989). The method has been used in many guises ever since due to its simplicity and high sensitivity.

The theory behind SSCP is that any given single strand DNA fragment may form a sequence-specific tertiary structure (Figure 5.14). Double stranded DNA may be separated by denaturation to form two complementary single strand DNA strands (ssDNA). If the ssDNA is allowed to re-nature without re-annealing to the complementary strand, the ssDNA folds into a tertiary structure, which is dependent on the nature (sequence) of the DNA. If two ssDNA fragments only differ by one base, the tertiary structure will differ. This postulate does not always hold true, but numerous studies have shown that in most cases it is true. Therefore, this method may be applicable to screening for heterozygotes or even homozygous mutations in diploid organisms (e.g. human). If ssDNA samples are analysed in a native gel or polymer, it is possible to separate the two variants from each other (Figure 5.15). The conformation of ssDNA is highly temperature dependent and differences may sometimes be more apparent at one temperature than another. By labelling the PCR fragments with fluorescent dyes (either using dye-labelled primers or post-PCR end-labelling) it is possible to detect the ssDNA structures using capillary electrophoresis (CE-SSCP).

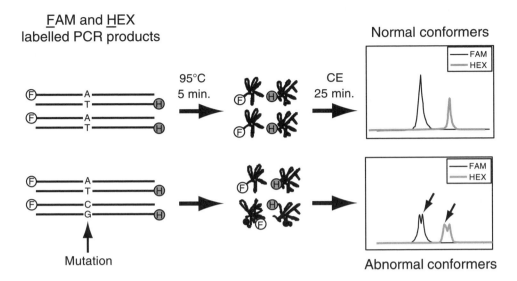

Figure 5.14 Principle of capillary electrophoresis based single strand conformation polymorphism analysis (CE-SSCP). DNA fragments are amplified by PCR using primers flanking the mutation. The PCR fragments are diluted in formamide and NaOH and denatured by heat. The single stranded DNA fragments fold in sequence-dependent secondary or tertiary structures by internal base-pairing as they enter the non-denaturing matrix in the capillary. The folded DNA fragments are separated by capillary electrophoresis. The migration in the capillary is dependent on folding of the DNA. Thus, abnormal folding due to a mutation in the DNA fragment is detected as a different peak pattern in the chromatogram (arrows)

CE-SSCP is a qualitative assay and is dependent on a range of compositions and conditions. First, assay temperature is an important parameter. It is well known that the conformation of DNA is highly temperature dependent. The question is then if two different conformations can be distinguished equally well at any given temperature. The answer to that has obviously been no for a long time, and most labs have therefore routinely performed SSCP assays at two or more temperatures in order to obtain high sensitivity. Many studies show that sensitivity of CE-SSCP is higher at ambient or sub-ambient temperatures; this may often cause a problem as not all CE-instruments have cooling capability. However, as seen below this should not be a hindrance for performing SSCP at higher temperatures.

The length of the PCR fragments is also an important parameter in SSCP analysis. In gel-based systems it was shown that the sensitivity of SSCP was drastically reduced when fragments longer than 2–300 bp were used (Sheffield *et al.* 1993). A similar study has not been done for CE-SSCP, where the sensitivity generally is better than for gel-based SSCP, so the upper fragment size limit is still unknown for CE-SSCP. However, generally mutation detection up to 400 bp can be performed, which is acceptable as most exons that are analysed rarely extend

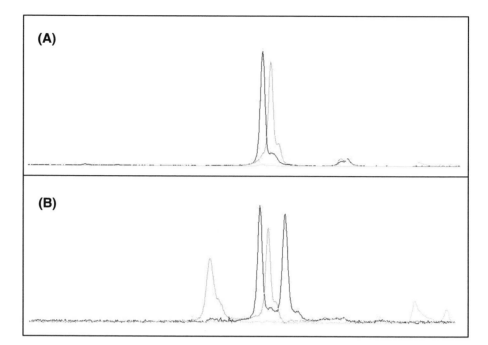

Figure 5.15   CE-SSCP example. Dark grey peaks forward strand and light grey peaks reverse strand. (A) Wild-type homozygous sample single peaks of both forward and reverse strands. (B) Heterozygous sample; both forward and reverse strand show two distinct peaks – one resulting from the wild-type fragment the other from the variant fragment

beyond that length. Also, fragments of that length rarely contain more than one common polymorphism. Polymorphisms can be a problem during mutation scanning as this increases the number of different SSCP patterns, which results in more difficult data analysis.

Polymer and buffer composition are very important determinants of good CE-SSCP results. A number of different polymers have been suggested for the assay. Commercial polymers supplied by the instrument manufacturers are available, but are not always the best choice and much can be gained by studying the literature. Recently it was shown that SSCP run at a single temperature was attained with high sensitivity by using dimethylpolyacrylamide polymer (Kukita *et al.* 2002).

The sensitivity of CE-SSCP has been found to be between 96% and 100% (Andersen *et al.* 2003). Apart from the high sensitivity, the advantages of the method are also a high specificity and the method is very easy to perform reproducibly. The main disadvantages are the need for a PCR fragment labelled with a fluorescent dye. In addition, the PCR fragments should not exceed 500 bp in length and it is not applicable in regions with many mutations. The high sensitivity is in part dependent on experienced personnel.

Methods based on variations of SSCP

Formation of heteroduplexes (HD) is a situation that has been exploited in a number of mutation detection techniques like simple non-denaturing gel-based HD analysis and more complex but highly sensitive DGGE or enzymatic/chemical cleavages of heteroduplex structures. Today, these methods have been converted to automated methods where the method either uses a capillary electrophoresis based system (CE-HD and Temperature Gradient Capillary Electrophoresis) or a HPLC system as described below. These automated methods are currently used in many diagnostics laboratories.

Heteroduplex formation is often seen in gel-based systems where both SSCP and HD bands are visible. This is not normally seen in CE-SSCP. However, if both the dilution mixture is changed to only Tris-HCl (pH 8.5) and the denaturing step at high temperature is followed by direct transfer of samples to the CE-instrument rather than snap cooling on ice, the heteroduplex (HD) pattern may also be obtained (Kozlowski and Krzyzosiak 2001). Theoretically, this should increase sensitivity as the SSCP and HD methods are based on different principles. In practice a more complex peak pattern could in some cases have an adverse effect in that the peak pattern becomes more difficult to read. Capillary electrophoresis dideoxyfingerprinting (CE-ddF) is a combination of chain termination DNA sequencing and SSCP analysis (Larsen *et al.* 2001b). First, a sequencing reaction is performed on a PCR product using only one dideoxy nucleotide. Then, the chain termination products are analysed by CE-SSCP as described above. The result is a more complex peak pattern than is seen for conventional SSCP, commonly called a 'fingerprint'. The method is very sensitive even using a single electrophoresis temperature. However, the optimization of the assay is very time consuming. Furthermore, the addition of the enzymatic step before SSCP analysis and interpretation of the more complex peak pattern is more time consuming compared with CE-SSCP.

Constant denaturing capillary electrophoresis (CDCE) and temperature gradient capillary electrophoresis (TGCE)

Constant denaturing CE (CDCE) (Khrapko *et al.* 1994) and temperature gradient CE (TGCE) (Gelfi *et al.* 1994) are variants of the gel-based denaturing gradient gel electrophoresis (DGGE) (Figure 5.16). Both methods are used for detecting mutations based upon the melting properties of DNA molecules as a separation

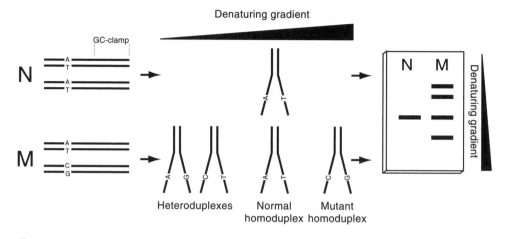

Figure 5.16   Denaturing gradient gel electrophoresis (DGGE). Normal (N) and mutated (M) DNA fragments are amplified by PCR using primers flanking the mutation. The primers are designed so they will lead to the production of an amplicon with a GC-rich region at one end (GC-clamp). The PCR products are denatured and allowed to re-anneal before electrophoresis. The fragments are separated in a gel with increasing denaturant. At a certain denaturing concentration the strands will separate in the low melting part of the DNA fragment, but not in the CG-rich region. The partly melted DNA fragments migrate slower than double-stranded DNA during gel-electrophoresis, thus the position of the fragments in the gel are dependent on the melting properties of the low-melting region. Heteroduplex DNA has the lowest melting temperature due to mispairing between the mutation and the normal nucleotide

mechanism as a partially denatured DNA molecule travels much more slowly than a native DNA molecule. By using DNA with a low-melting and a high-melting domain (either a naturally occurring DNA fragment or *in vitro* constructed by adding a so-called GC-clamp to one end of the PCR product), a situation may be obtained with partial denaturation as double stranded DNA melts in domains rather than in a zipper-like fashion. If a person is heterozygous in a given DNA region, DGGE of this particular DNA should result in four peaks, two resulting from the two homoduplexes and two from the two heteroduplexes. In CE-based systems the CDCE is used where DNA fragments are fluorescently labelled, and after pre-calculation of strand separation temperatures of the low-melting domain (lower T_m) and of the high-melting domain (upper T_m), test runs are performed using various denaturing conditions. A similar method where the denaturant is at a standard concentration a temperature gradient in the CE-run is applied. The pre-programmed temperature gradient is applied to the run and the samples are separated and the various hetero- and homoduplexes are detected.

Advantages of the two similar systems are – at least based on results from DGGE – a high sensitivity, and many CE-instruments allow for applying a temperature gradient. However, the methods have not been validated in a CE-system. The disadvantage of this system lies in the difficulty in designing appropriate primers and in case of TGCE only instruments that have the temperature gradient application available may be used. As for CGCE the need optimization of each assay is also a drawback of the method.

Denaturing high performance liquid chromatography (DHPLC)

With the invention of denaturing high performance liquid chromatography (DHPLC) in 1995 (Oefner and Underhill 1995) many criteria for an ideal mutation screening technology were met and consequently the technology is now one of the most popular methods for automated mutation screening.

The principle behind mutation screening by DHPLC is simple (Figure 5.17): PCR fragments are denatured and allowed to re-anneal slowly. Under such conditions heterozygous mutations will lead to the formation of heteroduplexes. The PCR fragments are loaded on a heated (50–70°C) column packed with alkylkated poly(styrene-divinylbenzene) particles or alkylkated silica particles. Chromatographic separation of the DNA fragments is performed using triethylamonium acetate (TEAA) as the mobile phase against a gradient of acetonitrile. The elevated column temperature ensures that the PCR fragments are partly denatured during the chromatographic separation. Under the right temperature conditions, heteroduplex DNA will be retained in shorter time on the column than the homoduplexes. Thus samples with heterozygous mutations will lead to a different peak pattern on the chromatogram compared with normal controls (Figure 5.18).

The advantages of DHPLC are that no labelling or purification of the samples is needed, the chromatographic separation is fast and the sensitivity is high. DHPLC is performed using a dedicated instrument in which the samples are

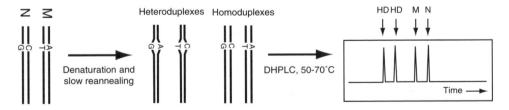

Figure 5.17 Denaturing high performance liquid chromatography (DHPLC). Normal (N) and mutated (M) DNA fragments are amplified by PCR using primers flanking the mutation. The PCR products are denatured and allowed to re-anneal in order to allow formation of heteroduplexes and subsequently loaded on a column packed with alkylated particles. The PCR products are separated by HPLC under partly denaturing conditions. Heteroduplex DNA will be retained less on the column than the homoduplexes and will display a different peak pattern on the chromatogram

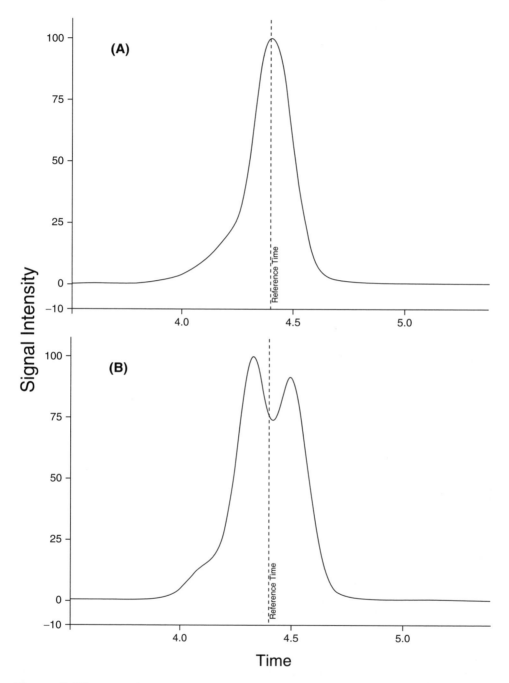

Figure 5.18 Mutation screening using DHPLC. Detection of a mutation (249T > A) in *HOXD13*. (A) Normal control. (B) Heterozygous mutant. Data was kindly provided by Claus Hansen, Wilhelm Johannsen Centre for Functional Genome Research, Department of Medical Genetics, University of Copenhagen

loaded in microtiter plates. All additional handling of the samples and data collection is performed automatically. The separation time is proportional to the length of the PCR fragment, which is normally between 200–1000 bp. However, a typical analytical cycle, including column pre-equilibration takes about 5 min, which gives a throughput of 288 samples/24 h. At least two different DHPLC instruments are commercially available, with detection based on UV-absorption or laser-induced fluorescence (LIF) detection. The latter requires labelling of each DNA fragment with a different fluorescent dye, but allows for simultaneous detection of four samples.

The sensitivity of the method (i.e. the percentage of mutations in a given material, which are detected by the method) is between 96–100% (Oefner and Underhill 1995). The main disadvantage associated with DHPLC is that the correct column temperature needs to be optimized for each different DNA fragment, e.g. 10 optimization experiments needs to be performed initially, if a gene containing 10 exons are screened for mutations. This somewhat elaborate procedure is necessary because melting temperatures of DNA double strands are dependent on the DNA sequence, thus in principle PCR amplicons of 10 exons may have 10 different melting temperatures. However, computer software for prediction of the correct column temperature based on the DNA sequence is publicly available on the Internet (http://insertion.stanford.edu/melt.html) and is useful for targeting the optimal column temperature.

High-density oligonucleotide arrays

High-density oligonucleotide arrays can also be used for detection of unknown mutations. Such arrays are designed to contain a probe for each nucleotide in the reference sequence and a probe for each possible base substitution. Detection of a base substitution is based on a loss of fluorescence in the position of the 'reference' probe and a gain of signal in the position of the 'mutation' probe. Thus for detection of an unknown point mutation in a DNA fragment of N bp in length, 8N probes are typically used. In principle, this strategy may also be used for detection of small insertions and deletions. However, for detection of deletions and insertions of up to Z bp in size, an additional 2ZN and 2(4Z)N probes, are needed, respectively (e.g. for analysis of a 1000 bp DNA fragment a total of (8*N) + (2* Z*N) + (2*4*Z*N) = (1000*8) + (2*1*1000) + (2*[4*1]*1000) = 18000 probes are necessary for detection of all single base-pair substitutions, deletions and insertions). If deletions and insertions of up to, for example, 5 bp are included in the analysis, the number of necessary probes increases dramatically to a total of 2 066 000 probes. Thus, mutation scanning using HDOAs is normally restricted to single base substitutions.

The sensitivity of HDOA analysis has been reported to be between 81% and 84% with a specificity between 86% and 100% (Ahrendt et al. 1999; Wikman et al.

2000). The low sensitivity and specificity is caused by sub-optimal hybridization kinetics of some probes under the given hybridization conditions (see above).

Nevertheless, the technology has proved to be very efficient for large-scale discovery of SNPs where a sensitivity well below 100% is satisfactory (Lindblad-Toh *et al.* 2000; Wang *et al.* 1998). HDOAs have also been used to mutation scanning in the *CFTR* gene, *BRCA1* gene, HIV-1 protease gene, *ATM* gene and *TP53* gene (reviewed in Larsen *et al.* 2001a). An *Affymetrix* GeneChip® array is commercially available for mutation screening of the *TP53* gene.

References

Ahrendt, S.A., Halachmi, S., Chow, J.T., Wu, L., *et al.* (1999) Rapid p53 sequence analysis in primary lung cancer using an oligonucleotide probe array. *Proc Natl Acad Sci USA* **96**: 7382–7387.

Andersen, P.S., Jespersgaard, C., Vuust, J., Christiansen, M., and Larsen, L.A. (2003) High-throughput single strand conformation polymorphism mutation detection by automated capillary array electrophoresis: validation of the method. *Hum Mutat* **21**: 116–122.

Ayyadevara, S., Thaden, J.J. and Shmookler Reis, R.J. (2000) Discrimination of primer 3′-nucleotide mismatch by taq DNA polymerase during polymerase chain reaction. *Anal Biochem* **284**: 11–18.

Bader, J.S. (2001) The relative power of SNPs and haplotype as genetic markers for association tests. *Pharmacogenomics* **2**: 11–24.

de Kok, J.B., Wiegerinck, E.T., Giesendorf, B.A. and Swinkels, D.W. (2002) Rapid genotyping of single nucleotide polymorphisms using novel minor groove binding DNA oligonucleotides (MGB probes). *Hum Mutat* **19**: 554–559.

Fakhrai-Rad, H., Pourmand, N. and Ronaghi, M. (2002) Pyrosequencing: an accurate detection platform for single nucleotide polymorphisms. *Hum Mutat* **19**: 479–485.

Gelfi, C., Righetti, P.G., Cremonesi, L. and Ferrari, M. (1994). Detection of point mutations by capillary electrophoresis in liquid polymers in temporal thermal gradients. *Electrophoresis* **15**: 1506–1511.

Grossman, P.D., Bloch, W., Brinson, E., Chang, C.C., *et al.* (1994) High-density multiplex detection of nucleic acid sequences: oligonucleotide ligation assay and sequence-coded separation. *Nucleic Acids Res* **22**: 4527–4534.

Gut, I.G. (2001) Automation in genotyping of single nucleotide polymorphisms. *Hum Mutat* **17**: 475–492.

Khrapko, K., Hanekamp, J.S., Thilly, W.G., Belenkii, A., *et al.* (1994) Constant denaturant capillary electrophoresis (CDCE): a high resolution approach to mutational analysis. *Nucleic Acids Res* **22**: 364–369.

Kozlowski, P., and Krzyzosiak, W.J. (2001) Combined SSCP/duplex analysis by capillary electrophoresis for more efficient mutation detection. *Nucleic Acids Res* **29**: E71.

Kukita, Y., Higasa, K., Baba, S., Nakamura, M., *et al.* (2002) A single-strand conformation polymorphism method for the large-scale analysis of mutations/polymorphisms using capillary array electrophoresis. *Electrophoresis* **23**: 2259–2266.

Larsen, L.A., Christiansen, M., Vuust, J., and Andersen, P.S. (2000) High throughput mutation screening by automated capillary electrophoresis. *Comb Chem High Throughput Screen* **3**: 393–409.

Larsen, L.A., Christiansen, M., Vuust, J. and Andersen, P.S. (2001a) Recent developments in high-throughput mutation screening. *Pharmacogenomics* **2**: 387–399.

Larsen, L.A., Johnson, M., Brown, C., Christiansen, M., *et al.* (2001b) Automated mutation screening using dideoxy fingerprinting and capillary array electrophoresis. *Hum Mutat* **18**: 451–457.

Li, J., Butler, J.M., Tan, Y., Lin, H., *et al.* (1999) Single nucleotide polymorphism determination using primer extension and time-of-flight mass spectrometry. *Electrophoresis* **20**: 1258–1265.

Lindblad-Toh, K., Tanenbaum, D.M., Daly, M.J., Winchester, E., *et al.* (2000) Loss-of-heterozygosity analysis of small-cell lung carcinomas using single-nucleotide polymorphism arrays. *Nat Biotechnol* **18**: 1001–1005.

Lipshutz, R.J., Fodor, S.P., Gingeras, T.R., and Lockhart, D.J. (1999) High density synthetic oligonucleotide arrays. *Nat Genet* **21**: 20–24.

Livak, K.J. (1999). Allelic discrimination using fluorogenic probes and the 5' nuclease assay. *Genet Anal* **14**: 143–149.

Lyamichev, V., Mast, A.L., Hall, J.G., Prudent, J.R., *et al.* (1999) Polymorphism identification and quantitative detection of genomic DNA by invasive cleavage of oligonucleotide probes. *Nat Biotechnol* **17**: 292–296.

Marras, S.A., Kramer, F.R., and Tyagi, S. (1999) Multiplex detection of single-nucleotide variations using molecular beacons. *Genet Anal* **14**: 151–156.

Michelson, K.R. and Cheng, J. (2001a) *Capillary Electrophoresis of Nucleic Acids Volume II: Practical Applications of Capillary Electrophoresis.* Totowa, Humana Press.

Michelson, K.R. and Cheng, J. (2001b) *Capillary Electrophoresis of Nucleic Acids. Volume I: Introduction to the Capillary Electrophoresis of Nucleic Acids.* Totowa, Humana Press.

Oefner, P.J. and Underhill, P.A. (1995) Comparative DNA sequencing by denaturing high-performance liquid chromatography (DHPLC). *Am J Hum Genet* **57**: A266.

Online Mendelian Inheritance in Man OT. McKusick-Nathans Institute for Genetic Medicine, Johns Hopkins University (Baltimore, MD) and National Center for Biotechnology Information, National Library of Medicine (Bethesda, MD), 2002. Ref Type: Electronic Citation.

Orita, M., Iwahana, H., Kanazawa, H., Hayashi, K., and Sekiya, T. (1989) Detection of polymorphisms of human DNA by gel electrophoresis as single-strand conformation polymorphisms. *Proc Natl Acad Sci USA* **86**: 2766–2770.

Pusch, W., Wurmbach, J.H., Thiele, H. and Kostrzewa, M. (2002) MALDI-TOF mass spectrometry-based SNP genotyping. *Pharmacogenomics* **3**: 537–548.

Sauer, S., Lechner, D., Berlin, K., Lehrach, H. *et al.* (2000) A novel procedure for efficient genotyping of single nucleotide polymorphisms. *Nucleic Acids Res* **28**: E13.

Schubert, E.L., Hsu, L., Cousens, L.A., Glogovac, J., *et al.* (2002) Single nucleotide polymorphism array analysis of flow-sorted epithelial cells from frozen versus fixed tissues for whole genome analysis of allelic loss in breast cancer. *Am J Pathol* **160**: 73–79.

Sheffield, V.C., Beck, J.S., Kwitek, A.E., Sandstrom, D.W. and Stone, E.M. (1993) The sensitivity of single-strand conformation polymorphism analysis for the detection of single base substitutions. *Genomics* **16**: 325–332.

Syvanen, A.C. (2001) Accessing genetic variation: genotyping single nucleotide polymorphisms. *Nat Rev Genet* **2**: 930–942.

Tyagi, S. and Kramer, F.R. (1996) Molecular beacons: probes that fluoresce upon hybridization. *Nat Biotechnol* **14**: 303–308.

Wang, D.G., Fan, J.B., Siao, C.J., Berno, A., *et al.* (1998) Large-scale identification, mapping, and genotyping of single-nucleotide polymorphisms in the human genome. *Science* **280**: 1077–1082.

Wikman, F.P., Lu, M.L., Thykjaer, T., Olesen, S.H., *et al.* (2000) Evaluation of the performance of a p53 sequencing microarray chip using 140 previously sequenced bladder tumor samples. *Clin Chem* **46**: 1555–1561.

6

Determination of Nucleic Acid Sequences by Pyrosequencing

Elahe Elahi and **Mostafa Ronaghi**

Introduction

DNA sequencing became one of the most important tools for analysis of bio-logical systems very soon after the introduction of reliable protocols by Maxam and Gilbert (Maxam and Gilbert 1977) and by Sanger in the 1970s (Sanger *et al.* 1977). Its continued importance is reflected in the achievements of the Human Genome Project. In the post-genomic era, the ability to sequence only short stretches of DNA, in the range of 50 nucleotides to one nucleotide, will become the major need of scientists and clinicians. The vast majority of sequences, which will need to be determined, are those which show variation among related individuals and organisms. They will serve in linkage and association studies for genetic disorders and drug efficacy and toxicity, ultimately leading to the identification of relevant genes. They will also serve in the classification of organisms for the purposes of taxonomy, evolutionary studies, microbial typing and population and ecological studies. The sequencing protocols most appropriate for these objectives will be those which are accurate, sensitive, inexpensive, robust, easy to perform and have high throughput and potential for automation. Pyrosequencing is a sequencing protocol which fulfils these criteria and is among the few which hold the greatest potential for the future (Franca *et al.* 2002). The general principles, methodology, some specific applications and future scopes of pyrosequencing will be presented here.

Molecular Analysis and Genome Discovery edited by Ralph Rapley and Stuart Harbron
© 2004 John Wiley & Sons, Ltd   ISBN 0 471 49847 5 (cased)   ISBN 0 471 49919 6 (pbk)

General principles

Pyrosequencing is a DNA sequencing by synthesis technique employing biolu-minometric detection (Ronaghi *et al.* 1996, 1998). It relies on real-time detection of PPi released upon addition and incorporation of a known nucleotide during DNA synthesis; detection is luminometric because PPi affects the generation of photons. As generally performed today, a pyrosequencing cycle takes about one minute and involves a cascade of four enzymatic reactions performed in a homogeneous liquid medium (Figure 6.1). As all four enzymes are present together in the reaction mixture from the start and throughout the entire sequencing protocol, considerable effort was put into defining reaction condi-tions conducive to all four. Given the opposing nature of some of the reactions (nucleotide incorporation vs. nucleotide degradation), kinetic properties (K_M, K_{cat}) and empirical data were used to define appropriate absolute and relative concentrations of the enzymes and substrates. The cascade of enzymatic reactions starts with a nucleotide polymerization step upon addition of only the complementary nucleotide as dictated by template sequence and release of PPi. The addition of any of the non-complementary nucleotides would clearly not result in the release of PPi. The DNA polymerase (exonuclease-deficient Klenow or Sequenase) included in the reaction mixture to catalyse this reaction lacks 3′ to 5′ exonuclease activity so that synchronous synthesis on all template molecules can be achieved. Even without the proofreading

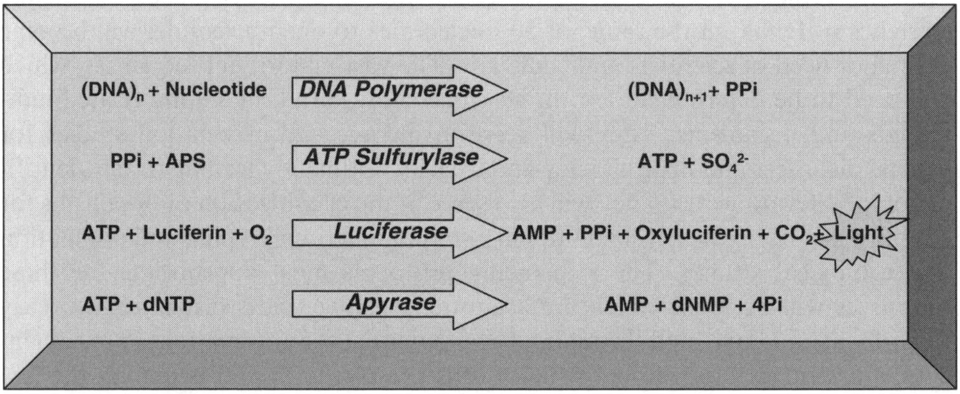

(DNA)$_n$ + Nucleotide	*DNA Polymerase*	(DNA)$_{n+1}$ + PPi
PPi + APS	*ATP Sulfurylase*	ATP + SO$_4^{2-}$
ATP + Luciferin + O$_2$	*Luciferase*	AMP + PPi + Oxyluciferin + CO$_2$ + Light
ATP + dNTP	*Apyrase*	AMP + dNMP + 4Pi

Figure 6.1 The general principle behind pyrosequencing reaction. A DNA polymerase catalyzes incorporation of nucleotide(s) into a nucleic acid chain. As a result of the incorporation, PPi molecule(s) is released and subsequently converted to ATP by ATP sulfurylase. Light is produced in the luciferase reaction during which a luciferin molecule is oxidized. Apyrase degrades unused ATP and dNTP prior to the addition of the next nucleotide

activity, a misincorporation rate of less than $10^{-5} - 10^{-6}$ is achieved (Wong *et al.* 1991).

The PPi released during incorporation of a nucleotide quickly reacts to produce ATP in a reaction catalysed by ATP sulfurylase. This reaction is very efficient. The ATP sulfurylase used in pyrosequencing is a recombinant version of the enzyme from *Saccharomyces cerevisiae* (Karamohamed *et al.* 1999). The ATP generated is used as energy source in the third enzymatic reaction in which luciferase oxidizes luciferin and in the process generates light. The source of the luciferase is the firefly insect. The final enzymatic reaction in the pyrosequencing cascade is catalysed by apyrase, a nucleotide degrading enzyme from potato. Low amounts of this enzyme in the reaction mixture efficiently degrade unincorporated deoxynucleoside triphosphates and unused ATP. The time course of the first three reactions and the elimination of unused nucleotides by apyrase allows for iterative addition of different nucleotides in a programmed order to attain the desired length of sequence read.

The entire process from polymerization to light detection takes 3–4 s at room temperature (Ronaghi 2001a; Ronaghi *et al.* 2001b). In the presence of one pmol template, one pmol nucleotide incorporation yields 6×10^{11} ATP molecules, which in turn, generate about 6×10^9 photons at a wavelength of 560 nanometres (Ronaghi *et al.* 2002a). This amount of light is easily detected by a photodiode, photomultiplier tube, or a CDD-camera (Charge Coupled Device camera).

Methodology

Amplicons generated in PCR are currently used as templates in pyrosequencing reactions. The amplicons need to be purified as unincorporated nucleotides, unused primers and salts in the PCR reaction mix will interfere with the pyrosequencing enzymatic reactions. For the purification protocol most commonly used (Fakhrai-Rad *et al.* 2002), one of the PCR primers needs to have been biotinylated at its 5′ terminus. After the PCR, streptavidin-coated magnetic beads or Sepharose beads are used to capture the biotinylated amplicons. Unwanted components of the PCR are removed by sedimentation of the beads and washings. The captured double stranded DNA amplicons are then denatured by alkaline treatment and the immobilized biotinylated strands are separated from the non-immobilized ones. Both strands can be used as pyrosequencing templates. This preparation protocol gives high quality sequence data. The Sepharose beads result in a somewhat better signal-to-noise ratio because of their relative transparency and higher capacity for immobilization as compared with the magnetic beads. These same properties allow the use of less PCR product as template with the Sepharose beads.

Most users perform the pyrosequencing reactions using automated commercial versions of the pyrosequencing machine (www.pyrosequencing.com). In these formats, purified templates are placed into the wells of a 96-well or 384-well microtitre plates and the plate is placed into a dark chamber in the machine. Concurrently, the pyrosequencing enzymes, substrates and each of the four nucleotides are placed into separate compartments of an inkjet cartridge and the cartridge is placed into a dark chamber in the machine. The cartridge has the capacity for iterative and precise delivery of a small volume (200 nl) of each of the reagents. After an initial dispensation of substrates and enzymes into each well, nucleotides are dispensed with an appropriate time lapse according to an order previously specified for each well. The reactions take place at room temperature and the microtitre plate is continuously agitated to enhance the reaction rate. A lens array efficiently focuses the luminescence generated in each well onto the chip of a CDD camera. The camera images the plate every second. A data acquisition module and an interface for PC-connection are used in the instrument. The light signals are recorded in a graph of time versus relative light intensity known as a pyrogram (Figure 6.2). As the dispensation time between nucleotides is constant, the time axis is annotated with the identity of the nucleotide dispensed at each time point. The signals in a program show high quality sequence data, which are easily and immediately interpretable. The height of the peaks is proportional to the number of incorporated nucleotides. The slopes of the ascending and descending legs of the curves provide information on the kinetics of all four enzymes in the reaction.

Applications

Pyrosequencing has facilitated many sequence-based analyses and is generally considered one of the most useful and promising of sequencing techniques (Franca *et al.* 2002). Perhaps its most significant contribution has been in the arena of single nucleotide polymorphism (SNP) genotyping (Fakhrai-Rad *et al.*

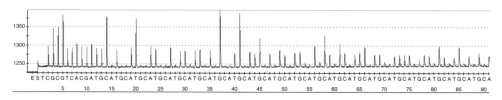

Figure 6.2 Pyrogram of the raw data obtained from pyrosequencing. The order of nucleotide addition is indicated below the pyrogram. Y-axis indicates the luminescence intensity

2002; Ronaghi 2003). Availability of information on nearly 2 million SNPs of the human genome in The SNP Consortium (TSC) database has rendered these the most important markers for analyses of variation in the human genome. For analyses of SNPs by pyrosequencing, the 3′ end of a sequencing primer is designed to hybridize one or a few bases before the polymorphic site. The status of each sample is rapidly established by sequencing of only a few nucleotides. Pyrograms of the three possible genotypes, two homozygous and one heterozygous, are easily distinguished by visual inspection (Figure 6.3). This feature lends itself to the use of pattern recognition software. Pyrosequencing of SNPs results in high call rates and very good accuracy. Accuracy of more than 1200 SNP genotypes determined

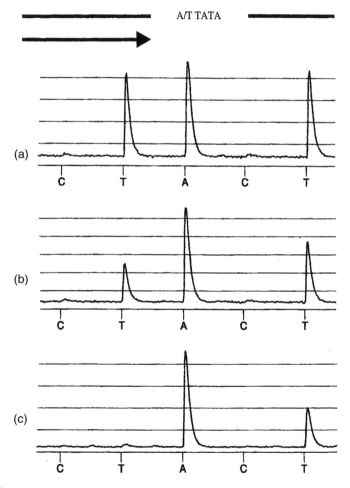

Figure 6.3 SNP data obtained by pyrosequencing. Three different SNPs are distinguishable by pyrosequencing. (a) T homozygous, (b) A/T heterozygous, (c) A homozygous

by pyrosequencing was confirmed by comparison to genotype analysis performed by an alternative protocol (Nordfors *et al.* 2002). Recently, SNP analysis by pyrosequencing was carried out on DNA of ovarian cancer (Hefler *et al.* 2002), prostate cancer (Kittles *et al.* 2001) and idiopathic generalized epilepsy patients (Sander *et al.* 2002). The technique has also been used in plant systems for gene localization (Ching and Rafalski 2002) and for genotype determination in a polyploid species (Rickert *et al.* 2002).

Most importantly, SNP allele frequencies have been established by pyrosequencing on the pooled DNA of 140–370 individuals (Gruber *et al.* 2002; Wasson *et al.* 2002; Neve *et al.* 2002). When compared with frequencies assessed on the basis of analysis of individual DNAs, the maximum divergence was $1.5 + / - 0.9$ per cent (Neve *et al.* 2002). Allele frequency differences of less than 2 per cent to 4 per cent between pools were detectable (Gruber *et al.* 2002; Neve *et al.* 2002). In a study comparable to using pooled DNA of different individuals, genotyping of SNPs was performed to determine the extent of haematopoietic chimerism between donor and recipient DNA after stem cell transplantation (Hochberg *et al.* 2003). The presence of 5 per cent donor cells was reported to be detectable in replicate assays. The potential to reliably use pooled and heterogeneous DNA samples affects cost, time and labour and, therefore, has great implications for the many investigations in which data on numerous samples is required for meaningful analysis.

Microbial typing is another area in which pyrosequencing has proven to be effective (Ronaghi *et al.* 2002b; Elahi *et al.* 2003; Gharizadeh *et al.* 2002). The general approach is sequence determination of one or more short variable region(s) embedded within conserved or semi-conserved sequences. An oligonucleotide primer complementary to an adjacent and more conserved sequence is designed so as to allow reading into the variable region. The sequences determined can be used to place each organism into one of previously known groupings using the concept of signature matching or to define hitherto unknown groups. In bacteria, classification based on variable regions of the 16S rRNA gene is quite reliable (Jonasson *et al.* 2002; Monstein *et al.* 2001). Non-ribosomal RNA genes have been used for typing of *Bordetella pertussis, Bordetella parapertussis and Bacillus anthracis* (Nygren *et al.* 2000). Pyrosequencing of SNPs has recently been used for grouping of *Listeria monocytogenes* strains (Unnerstad *et al.* 2001). Sequence determination of a variable region in the 18S rRNA gene has been used to type yeast strains. Pyrosequencing of stretches as short as 14 nucleotides has allowed correct classification of human papillomaviruses (Gharizadeh *et al.* 2002). HCV typing and subtyping required pyrosequencing of several short variable regions of the viral genome (Elahi *et al.* 2003; Pourmand *et al.* 2002).

Pyrosequencing is also an appropriate protocol for resequencing of known sequences, for example, for mutation detection. Mutations will be readily detected by comparison of pyrograms to those of known wild-type sequences. Achievable read lengths are considerably longer than *de novo* sequencing because the nucleotide dispensation order can be programmed according to the known sequence, resulting in incorporation with every nucleotide dispensation. This strategy has been successfully used for mutation detection in the p53 gene (Garcia *et al.* 2000; Ahmadian *et al.* 2000). Pyrosequencing was also used for assessment of frequency of known mutations in Connexin 26 and in mitochondrial 12S RNA associated with hereditary hearing loss in different populations (Ferraris *et al.* 2002).

Pyrosequencing is ideal for rapid and efficient Tag or signature sequencing of cDNA libraries in order to study genes represented in those libraries (Nordstrom *et al.* 2001). Short reads of 15 to 30 nucleotides of mRNA sequences preceding the poly(A) tail suffice to identify genes by homology searches of DNA sequences in databases. Unidentified tag sequences can be further investigated in the cDNA libraries and this may lead to gene discovery. This sort of analysis has been carried out on cDNA libraries representing transcription in macrophage and foam cells in a study on arteriosclerosis (Agaton *et al.* 2002). The genes identified matched very well with genes found using expressed sequence tag (EST) sequencing.

Sequences which lend themselves to hairpin structures have proven to be difficult to sequence by conventional methods due to problems which arise during gel electrophoresis. Such sequences have been successfully sequenced by pyrosequencing. G/C rich sequences are also difficult to sequence by Sanger sequencing (Ronaghi *et al.* 1999). Regions having greater than 70 per cent G/C composition have been accurately sequenced by pyrosequencing (Ronaghi and Elahi, unpublished results).

Finally, pyrosequencing holds great promise for clinical and forensic applications (Berg *et al.* 2002; Andreasson *et al.* 2002). With the great wealth of sequence information becoming available, demand for sequencing of individual samples in these settings is increasingly felt. As powerful as Sanger sequencing is, the cost and expertise associated with its implementation render it prohibitive for routine use. The speed and user-friendly qualities of pyrosequencing, without sacrifice of accuracy, qualify it as a very viable substitute.

Future perspectives

Several modifications to the original pyrosequencing protocol which are now routinely used have ameliorated its performance. The first were substitutions of dATP by dATPαS as precursor nucleotide for polymerization and the use of apyrase rather than washing between nucleotide additions (Ronaghi *et al.* 1996). dATP created non-specific signals because it could be used by luciferase. While dATPaS is efficiently incorporated by DNA polymerases, it is inert for luciferase. Use of apyrase allowed iterative addition of nucleotides into a reaction mixture without the need to repeatedly add the other components (Ronaghi *et al.* 1998). More recently, the inclusion of single stranded DNA-binding protein (SSB) has resulted in yet longer reads and has facilitated sequencing of difficult templates (Ronaghi 2000a). The introduction of the purified Sp isomer of dATPaS was reported to increase read length to 50–100 bases on different types of templates (Gharizadeh *et al.* 2002). In a publication on HLA typing using pyrosequencing, it was reported that more than 750 samples could be processed in a working day, providing sequence data for more than 50 000 bases (Ringquist *et al.* 2002). With the availability of two new machines from Pyrosequencing, up to 50 000 samples per day can be analyzed.

Multiplexing using a three primer system and an enzymatic protocol for template preparation are more recently introduced modifications. For multiplexing, multiple sequence primers are hybridized adjacent to multiple variable positions on a single template. A fingerprint reflecting the sequence of the variable positions is obtained. This protocol has been used for HCV typing and for multiple clustered SNP analysis (Figure 6.4). In the case of SNP analysis, multiplexing is more appropriate for homozygous DNA templates, allowing analysis of inbred organisms. Clearly, multiplexing results in more rapid analysis and reduces cost.

A three primer system of pyrosequencing permits yet further reduction of cost. It is a system which allows the use of a single biotinylated primer for template preparation rather than target specific biotinylated primers (Fakhrai-Rad *et al.* 2002). Two unlabelled target specific primers are used for the PCR amplification reaction, one of which has a handle sequence complementary to the universal biotinylated primer. This strategy produced accurate data on 555 SNPs of the human genome.

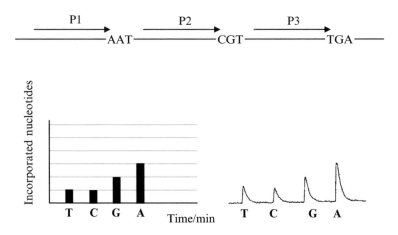

Figure 6.4 Theoretical pattern (left panel) and pyrogram presenting the raw data obtained from one of the subtypes using multiplex pyrosequencing. Height of Y-axis is determined by the number of incorporated nucleotides and X-axis is determined by time. Nucleotides are added according to the specified order at one-minute intervals. The added nucleotide is indicated below the diagrams

A need for biotinylated primers and solid supports is completely eliminated in the enzymatic protocol for template preparation (Nordstrom *et al.* 2000a, 2001). Automation of template preparation is thus facilitated. A nucleotide degrading enzyme and exonuclease I are added to the PCR mix after amplification to remove unused nucleotides and primers. The degrading enzymes are heat inactivated prior to addition of sequencing primers. Dilution of template is required to compensate for incompatible amplification and sequencing buffer conditions. As such, a very sensitive light detecting system is required in this protocol.

A solid phase pyrosequencing format using immobilized DNA templates in a three enzyme system is an extremely promising procedure being developed. A washing step dependent on microfluidic technology is employed to eliminate unincorporated nucleotides, AMP and dAMPαS. Accumulated AMP and dAMPαS in the four enzyme system inhibit luciferase activity, resulting in shorter reads. The microfluidic approach allows faster reads, use of less enzymes and integration of PCR amplification, template preparation and pyrosequencing analysis. Use of microarrays, immobilized enzymes and a piezoelectric ultrasonic sprayer for nucleotide delivery is envisioned. The throughput with this system will be very high.

References

Agaton, C., Unneberg, P., Sievertzon, M., Holmberg, A., *et al.* (2002) Gene expression analysis by signature pyrosequencing. *Gene* **289**: 31–39.

Ahmadian, A., Lundeberg, J., Nyrén, P., Uhlén, M. and Ronaghi, M. (2000) Analysis of the p53 Tumor Suppressor Gene by Pyrosequencing. *BioTechniques* **28**: 140–147.

Andreasson, H., Asp, A., Alderborn, A., Gyllensten, U. and Allen, M. (2002) Mitochondrial sequence analysis for forensic identification using pyrosequencing technology. *BioTechniques* **32**: 124–133.

Berg, L.M., Sanders, R. and Alderborn, A. (2002) Pyrosequencing technology and the need for versatile solutions in molecular clinical research. *Expert Rev Mol Diagn* **2**: 361–369.

Ching, A. and Rafalski, A. (2002) Rapid genetic mapping of ests using SNP pyrosequencing and indel analysis, *Cell Mol Biol Lett* **7**: 803–810.

Elahi, E., Pourmand, N., Cheung, R. Rofoogaran, A., *et al.* (2003) Determination of hepatitis C virus genotype by pyrosequencing. *J Virol Methods* **109**(2): 171–176.

Fakhrai-Rad, H., Pourmand, N. and Ronaghi, M. (2002) Pyrosequencing: an accurate platform for analysis of single nucleotide polymorphism. *Human Mutation* **19**: 479–485.

Ferraris, A., Rappaport, E., Santacroce, R., Pollak, E., *et al.* (2002) Pyrosequencing for detection of mutations in the connexin 26 (GJB2) and mitochondrial 12S RNA (MTRNR1) genes associated with hereditary hearing loss. *Hum Mutat* **20**: 312–320.

Franca, L.T., Carrilho, E. and Kist, T.B. (2002) A review of DNA sequencing techniques, *Q Rev Biophys* **35**: 169–200.

Garcia, C. A., Ahmadian, A., Gharizadeh, B., Lundeberg, J., *et al.* (2000) Mutation detection by pyrosequencing: sequencing of exons 5 to 8 of the p53 tumour suppressor gene. *Gene* **253**: 249–257.

Gharizadeh, B., Nordstrom, T., Ahmadian, A., Ronaghi, M. and Nyren, P. (2002) Longread pyrosequencing using pure 2′-deoxyadenosine-5′-O′-(1-thiotriphosphate) Sp-isomer, *Anal Biochem* **301**: 82–90.

Gruber, J.D., Colligan, P.B. and Wolford, J.K. (2002) Estimation of single nucleotide polymorphism allele frequency in DNA pools by using Pyrosequencing. *Hum Genet* **110**: 395–401.

Hefler, L.A., Ludwig, E., Lampe, D., Zeillinger, R., *et al.* (2002) Polymorphisms of the endothelial nitric oxide synthetase gene in ovarian cancer, *Gynecol Oncol* **86**: 134–137.

Hochberg, E.P., Miklos, D.B., Neuberg, D., Eichner, D.A., *et al.* (2003) A novel rapid single nucleotide polymorphism (SNP) based method for assessment of hematopoietic chimerism after allogeneic stem cell transplantation. *Blood* **101**: 363–369.

Jonasson, J., Olofsson, M. and Monstein, H.J. (2002) Classification, identification and subtyping of bacteria based on pyrosequencing and signature matching of 16S rDNA fragments, *APMIS* **110**: 263–272.

Karamohamed, S., Nilsson, J., Nourizad, K., Ronaghi, M., *et al.* (1999) Production, purification, and luminometric analysis of recombinant *Saccharomyces cerevisiae* MET3 adenosine triphosphate sulfurylase expressed in *Escherichia coli*. *Prot Exp & Purif* **15**: 381–388.

Kittles, R.A., Panguluri, R.K., Chen, W., Massac, A., *et al.* (2001) Cyp17 promoter variant associated with prostate cancer aggressiveness in African Americans. *Cancer Epidemiol Biomarkers Prev* **10**: 943–947.

Maxam, A.M. and Gilbert, W. (1977) A new method for sequencing DNA. *Proc Natl Acad Sci* **74**: 560–564.

Monstein, H., Nikpour-Badr, S. and Jonasson, J. (2001) Rapid molecular identification and subtyping of *Helicobacter pylori* by pyrosequencing of the 16S rDNA variable V1 and V3 regions. *FEMS Microbiol Lett* **199**: 103–107.

Neve, B., Froguel, P., Corset, L., Vaillant, E., *et al.* (2002) Rapid SNP allele frequency determination in genomicx DNA pools by pyrosequencing, *BioTechniques* **32**: 1138–1142.

Nordfors, L., Jansson, M., Sandberg, G., Lavebratt, C., *et al* (2002) Large-scale genotyping of single nucleotide polymorphisms by Pyrosequencing and validation against the 5′nuclease Taqman assay, *Hum Mutat* **19**: 395–401.

Nordström, T., Ronaghi, M., Forsberg, de Faire, L., *et al.* (2000a) Direct analysis of single nucleotide polymorphism on double-stranded DNA by pyrosequencing. *Biotechnol Appl Biochem* **31**: 107–112.

Nordström, T., Nourizad, K., Ronaghi, M. and Nyrén P. (2000b) Method enabling pyrosequencing on double-stranded DNA. *Anal Biochem* **282**: 186–193.

Nordstrom, T., Gharizadeh, B., Pourmand, N., Nyren, P. and Ronaghi, M. (2001) Method enabling fast partial sequencing of cDNA clones. *Anal Biochem* **292**: 266–271.

Nygren, M., Reizenstan, E., Ronaghi, M. and Lundeberg, J. (2000) Polymorphism in the pertussis toxin promoter region affecting the DNA-based diagnosis of *Bordetella* infection. *J Clin Microbiol* **38**: 55–60.

Pourmand, N., Elahi, E., Davis, R. W. and Ronaghi, M. (2002) Multiplex Pyrosequencing. *Nucl Acids Res* **1**: E31–1.

Rickert, A.M., Premstaller, A., Gebhardt, C. and Oefner, P.J. (2002) Genotyping of SNPs in a polyploid genome by pyrosequencing. *BioTechniques* **32**: 592–600.

Ringquist, S., Alexander, A.M., Rudert, W.A., Styche, A. and Trucco, M. (2002) Pyrosequencing-based typing of alleles of the HLA-DQB1 gene, *BioTechniques* **33**: 166–175.

Ronaghi, M. (2000a) Improved performance of Pyrosequencing using single-stranded DNA-binding protein. *Anal Biochem* **286**: 282–288.

Ronaghi, M. (2000b) Improved performance of Pyrosequencing using single-stranded DNA-binding protein. *Anal Biochem* **286**: 282–288.

Ronaghi, M. (2001) Pyrosequencing sheds light on DNA sequencing. *Genome Res* **11**: 3–11.

Ronaghi, M. (2003) Pyrosequencing for SNP genotyping. *Methods Mol Biol* **212**: 189–195.

Ronaghi, M. and Elahi, E. (2002a) Discovery of single nucleotide polymorphisms and mutations by Pyrosequencing. *Comparative and Functional Genomics* **3**: 51–56.

Ronaghi, M. and Elahi, E. (2002b) Pyrosequencing for microbial typing. *J. Chromatography B Analyt Technol Biomed Life Sci* **782**: 67–72.

Ronaghi, M., Karamohamed, S., Pettersson, B., Uhlen, M. and Nyren, P. (1996) Real-time DNA sequencing using detection of pyrophosphate release. *Anal Biochem* **242**: 84–89.

Ronaghi, M., Uhlen, M. and Nyren, P. (1998) A sequencing method based on real-time pyrophosphate. *Science* **281**: 363–365.

Ronaghi, M., Nygren, M., Lundeberg, J. and Nyren, P. (1999) Analyses of secondary structures in DNA by Pyrosequencing *Anal Biochem* **267**: 65–71.

Ronaghi, M., Pourmand, N. and Stolc. V. (2001) Pyrosequencing a bioluminometric method of DNA sequencing. *Adv Luminescence Biotechnol* CRS press, pp. 247–253.

Sander, T., Toliat, M.R., Heils, A., Becker, C. and Nurnberg, P. (2002) Failure to replicate an allelic association between an exon 8 polymorphism of the human alpha(1A) calcium channel gene and common syndromes of idiopathic generalized epilepsy, *Epilepsy Res* **49**: 173–177.

Sanger, F., Nicklen, S. and Coulson, A.R. (1977) DNA sequencing with chain-terminating inhibitors. *Proc Natl Acad Sci* **74**: 5463–5467.

Unnerstad, H., Ericsson, H., Alderborn, A., Tham, W., *et al.* (2001) Pyrosequencing as a method for grouping of *Listeria monocytogenes* strains on the basis of single-nucleotide polymorphisms in the inlB gene. *Appl Environ Microbiol* **67**: 5339–5342.

Wasson, J., Skolnick, G., Love-Gregory, L. and Permutt, M.A. (2002) Assessing allele frequencies of single nucleotide polymorphisms in DNA pools by pyrosequencing technology, *BioTechniques* **32**: 1144–1150.

Wong, I., Patel, S. S. and Johnson, K. A. (1991) An induced-fit kinetic mechanism for DNA replication fidelity: direct measurement by single-turnover kinetics. *Biochemistry* **30**: 526–537.

7
An Introduction to DNA Chips

Magdalena Gabig-Ciminska and **Andrzej Ciminski**

The GeneChip, which is gaining popularity throughout the field of molecular biology and biotechnology, is a recently developed technique which provides efficient access to genetic information using miniaturized, high-density arrays of DNA or oligonucleotide probes. Both fundamental and applied research benefit greatly from the specific advantages of this method. Initially developed to enhance genome sequencing projects, the technology is rapidly evolving and has been adapted to a large set of applications. DNA chips are powerful tools to study the molecular basis of interactions on a scale that would be impossible using conventional analysis. The recent development of the microarray technology has greatly accelerated the investigation of gene regulation. Arrays are mostly used to identify which genes are turned on or off in a cell or tissue, and also to evaluate the extent of a gene's expression under various conditions. Indeed, this technology has been successfully applied to investigate simultaneous expression of many thousands of genes and to the detection of mutations or polymorphisms, as well as for their mapping and sequencing. In this chapter, we discuss the format of DNA chips, the technology used to fabricate and read them, and their applications.

Introduction

The occurrence of DNA chips in the second half of 1990s became a new biotechnical revolution equivalent in importance to the decoding of DNA structures in the 1950s, research of the fundamental laws of molecular genetics, such as genetic code, and the major dogma of molecular biology in the 1960s, discovery of

Molecular Analysis and Genome Discovery edited by Ralph Rapley and Stuart Harbron
© 2004 John Wiley & Sons, Ltd   ISBN 0 471 49847 5 (cased)   ISBN 0 471 49919 6 (pbk)

reverse transcription, the creation of first recombinant constructions in the 1970s and research of enzyme methods manipulating genetic material *in vitro*, like amplification technologies in the 1980s. This technology has come a long way since the first official meeting of scientists working in this field, which took place in Moscow in 1991, comprising about 50 scientists from Europe and the USA. At that time, the initial aim was the development of a technology for high-throughput sequencing of the human genome, named sequencing by hybridization. The field quickly branched out, however, in its methodology and application. Today, DNA chips seem to be a common commodity in biological sciences.

DNA chips have been greatly successful in a variety of genomic analyses (Debouck and Goodfellow 1999), ranging from detecting SNPs (Wang *et al.* 1998; see also Chapter 4) to functional genomics (Lockhart *et al.* 1996). There are different names for the chips, such as DNA/RNA Chips, BioChips, Gene-Chips or DNA arrays. The array, described as a macroarray or microarray, can be defined as an ordered collection of macrospots or microspots, respectively, each spot containing single defined species of a nucleic acid. Large sets of these nucleic acid probe sequences are immobilized in defined, addressable locations on the surface of a substrate capable of accessing large amounts of genetic information from biological samples in a single hybridization assay. Each spot represents the equivalent of a conventional analysis in a test tube. The matrix of spots applied to the biochip reproduces each of these analyses several times – up to a total of three or four – and thus increases the reliability of the test results.

The 'key and lock principle' of complementary hybridization is utilized in biochip analysis. In a typical application, DNA or RNA target sequences of interest are isolated from a biological sample using standard molecular biology protocols. The sequences are fragmented and labelled with reporter molecules for detection, and the mixture of labelled sequences ('key') is applied to the array, under controlled conditions, for hybridization with the surface probes ('lock'). Sequence complementarity leads to the hybridization between two single-stranded nucleic acid molecules, one of which is immobilized on a matrix (Southern *et al.* 1999). The array is then imaged with a reader to locate and measure the binding of target sequences from the sample to complementary probe on the array, and software reconstructs the sequence data and presents it in a format determined by the application. In this way highly parallel DNA hybridization assays are possible on miniaturized flat substrates or 'chips'.

There exist two variants of the chips, in terms of the property of arrayed nucleic acid sequence with known identity: Format I – cDNA microarrays and Format II – oligonucleotide arrays (Lockhart *et al.* 1996). Although both the cDNA and oligonucleotide chips can be used to analyse patterns of gene expression, fundamental differences exist between these methods. The microarrays are subsequently exposed to a labelled sample, hybridized and the identity/abundance of complementary sequences are determined.

Moreover, there are two basic approaches to GeneChip creation. Historically the first was to deposit on the chip previously synthesized oligonucleotides or cDNA fragments. The second approach is *in situ* synthesis using photolithography or piezoelectric printing that uses inkjet printer-like technology.

The chapter is organized as follows. First, the basic structure and design of DNA chips, including their production process and functioning is presented. Requirements for DNA chip handling, pointing out advantages and disadvantages of the technique, are then described. Finally, showing prospectives and directions of the future research, practical values and possible fields of its application are suggested.

DNA chip structure and operating principles

Integrated circuits technology

An integrated circuit (IC) called a chip or microchip in electronic technology is an electronic circuit built on a semiconductor substrate, usually one of a single crystal of silicon. The circuit is packaged in a hermetically sealed case or a non-hermetic plastic capsule, with leads extending from it for input, output and power-supply connections, and for other connections that may be necessary when the device is put to use. There is also a different scale of integration of such circuits. In the case of very large-scale integration, an IC can contain more than 1000 transistors. Intel founder Gordon Moore observed that the number of transistors per semiconductor chip about doubles every 18–24 months. The present microprocessor chips (used in personal computers) contain more than 100 million transistors per several square centimetre. Such huge integration scale, which has almost reached manufacturing technology limits, helps to make modern computers very fast, compact and relatively inexpensive.

A comparable phenomenon is observed in molecular biology. The miniaturization of certain tools is suitable for the construction of a smart and portable device – the spotted array system, which offers the pharmaceutical, biotechnology and agriculture industries more efficient and economical solutions. DNA microarrays built using photolithography have been on a similar pathway as integrated circuits. In 1998, an Affymetrix array contained less than 1000 genes; by 2000, it boasted 12 000.

In the process of manufacturing BioChips, semiconducting materials are widely used. A semiconductor is a solid material whose electrical conductivity at room temperature is between that of a conductor and an insulator. At high temperatures its conductivity approaches that of a metal, and at low temperatures it acts as an insulator. The substances first used for semiconductors were the elements germanium, silicon and grey tin. Although other materials can be used, chips are usually fabricated on wafers of single-crystal silicon, that is, silicon in which the orientation of all of the crystal is the same. The major fabricating steps include film formation, impurity doping, photolithography and packaging. Silicon oxide, also called silica, is grown on the surface of the silicon during the film-formation stage. Photolithographic methods are used selectively to remove the oxide from areas of the silicon. A layer of photoresist is added to the oxide layer and then exposed to ultraviolet light through a mask. After exposure, the silicon wafer is developed, and an etching process removes the unwanted areas of the oxide. Impurity doping adds charge carriers to the silicon; this process produces the unique electronic properties of semiconductors. Additional layers of silicon are deposited to create for instance bipolar transistors.

DNA chip design

There are several steps in the design and implementation of a DNA chip technology: probe creation, chip fabrication, sample preparation, assay, readout, and software (Table 7.1). Many strategies have been investigated for each of these steps.

Table 7.1   Steps in the design and implementation of a DNA chip experiment

1. Probe creation	2. Chip fabrication	3. Target (labelled sample preparation)	4. Assay	5. Readout	6. Informatics
capture probe choice: • oligonucleotides • PCR from genomic DNA and from cDNA	putting probes on the chip: • spotting (contact printing with needles; non-contact printing with piezo elements) • on-chip *in situ* synthesis (mask controlled synthesis; maskless synthesis)	• RNA extraction & purification => total RNA and mRNA • RNA amplification => cDNA double strand synthesis and *in vitro* transcription	• pre-hybridization • sample hybridization	• optical detection transmission (phosphoimager, fluorescence reader) • electrical detection (electrical readout by redox-recycling)	• data output, analysis, interpretation and visualization

Probe creation

Production of microarrays begins with the selection of probes to be printed on the array. In most cases, these are chosen directly from databases including Gene-Bank (Benson *et al.* 1997) and UniGene (Schuler *et al.* 1996).

Two commonly used types of chips, Format I and Format II, differ in the size of the arrayed nucleic acids. In Format I, cDNA probes, which are relatively long DNA molecules (500–5000 bases long), are immobilized to a solid surface such as membranes, glass or silicon chips. The probes are often single-stranded DNA fragments processed with the help of polymerase chain reaction (PCR). These PCR products amplified with gene-specific primers are generated using chromosomal DNA as a template, and subsequently purified by precipitation or gel-filtration, or both. Spotting cDNAs or PCR products representing specific genes onto a matrix produces DNA arrays. The deposition of a few nano-litres of purified material usually at $100 - 500\,\mu g/ml$ generates each array dot. The printing is carried out either by a high-precision robotic system, i.e. by high-speed robot spotting (arrayer) or utilizing inkjet technology that spots a sample of each gene product onto a number of matrices in a serial operation or printouts, respectively. These types of arrays are used mostly for large-scale screening and expression studies.

In Format II, an array of oligonucleotide probes is fabricated either by *in situ* light-directed chemical synthesis or by conventional synthesis followed by on-chip immobilization (McGall *et al.* 1996). Those with short nucleic acids (oligonucleotides up to 25 nt) are useful for the detection of mutations and expression monitoring, gene discovery and mapping.

The membranes commonly used for nucleic acid arrays are commercially available nitrocellulose and charged nylon that are employed in standard blotting assays (Southern blot, colony and plaque blot, dot and slot blot). The disadvantages of this method are that the genetic material is non-covalently attached which may result in its loss from the support, and that only a small amount of the DNA is available for hybridization. Glass-based arrays are most often made on microscope slides. They are coated with poly-lysine, amino silanes or amino-reactive silanes, which enhance both the hydrophobicity of the slide and the adherence of the deposited DNA (Schena *et al.* 1996). In most cases, DNA is cross-linked to the matrix by ultraviolet irradiation. After fixation, residual amines on the slide surface are reacted with succinic anhydride to reduce the positive charge at the surface. As the final step, the deposited DNA is split into single strands by heat or alkali. Adapting semiconductor photolithography to synthesize oligonucleotide probes *in situ* on glass or membrane substrate produces oligonucleotide chips. These chips are designed and produced on the basis of sequence information alone, without the need for any clones, PCR products, DNA and so on. Probe arrays are manufactured by light-directed chemical synthesis, which combines solid-phase chemical synthesis with photolithographic fabrication techniques employed in the semiconductor industry.

Silanol groups of silicate glass cannot couple oligonucleotides directly to the surface or to most plastics. It is necessary to create the surface with a group from which the growth of the oligonucleotide chain can be initiated. Such spacers also help to overcome steric interference, the ends of the probes closest to the surface being less accessible than the ends further away. Oligonucleotides on long spacers extend away from their neighbours and from the surface, and thus they allow more efficient interaction with the target. The spacer's length has a marked effect on hybridization yield. It has been shown that the optimal spacer length gives up to 150-fold increase in the yield of hybridization (Southern *et al.* 1999).

In addition to the flat-surface glass or silicon chips, supporting materials such as microscopic beads, nanochannel glass, 96-well microtitre plates, microelectrode array and phototransistor arrays are also used for depositing nucleic acid material. One of the most promising approaches is the microscopic bead-based chip technology, as it offers high sensitivity, flexibility and many replicates in one assay. These non-array methods are used efficiently to score large numbers of probes. They do not use spatial location as the key for probe identity, in contrast to flat-surface chips. Another interesting development is the use of nanochannel glass slides for array printing. Nanochannel glass materials are unique glass structures containing a regular geometric array of parallel holes or channels as small as 33 nm in diameter or as large as several micrometres in diameter (Watson *et al.* 2000). As a result, the surface area of nanochannel glass is much greater than that of regular glass, enabling larger amounts of nucleic acid material to be deposited in each spot. The hybridization kinetics are also greatly improved in this case.

DNA chip fabrication

DNA chip technology has evolved along two major paths (Table 7.2). In one method, nucleic acids (previously chemically synthesized oligonucleotides or single-stranded DNA fragments, i.e. cDNA) are immobilized on the chip surface sequentially to form capturing probes. Using either printing pins or inkjets, many copies of nucleic acid sequence can be attached to a substrate (Figure 7.1). Contact printing involves wetting a printing pin with the nucleic acid solution and tapping it on to the microarray surface. Inkjetting ejects uniform droplets of solution onto the substrate.

An alternative, known as *in situ* fabrication, builds the oligonucleotide sequence at each site one nucleotide at a time. This is done using either inkjets or photolithography (Figure 7.1). Construction of such chips begins with a substrate slide that has been chemically primed with sites ready to bind nucleotides. In inkjetting, solutions of nucleotides are ejected from the nozzle onto the substrate,

Table 7.2   BioChip basics

Method	Features and/or applications
I) accommodation on chip previously arranged nucleic acids:	
—spotting long DNA fragments (Stanford University)	• array of spotted PCR products • gene expression analysis
—array of prefabricated oligonucleotides, i.e. gel pads (Motorola) and microelectrodes (Nanogen)	• oligonnucleotides are attached to patches of activated polyacrylamide • controlled electric fields for immobilization
II) *in situ* synthesis of oligonucleotides:	
—photolithography (Affymetrix)	• light-directed oligonucleotide-synthesis on chip • adapted from semiconductor industry
—inkjet technology (Agilent)	• oligonucleotides are synthesized drop-by-drop • adapted from the technique used in ink-jet printers

and then chemically fixed to the surface. The next set of nucleotides are jetted onto the first and chemically fixed to those. The process is repeated until the desired set of nucleic acid is complete. In photolithography, a photosensitive chemical that detaches under illumination caps the sites. Light at 365 nm is shone through a patterned mask onto the chip, causing the capping chemical to break away from the areas it strikes, thus exposing the primed spots. A solution containing one of the four types of nucleotides (each molecule of which is itself attached to a capping molecule) is then washed over the chip. The nucleotides bond only to the areas that have been exposed, and add a capping layer themselves. As the procedure can be repeated with another mask and different nucleotide, a variety of DNA sequences can be built on the chip. Multiple probe arrays are synthesized simultaneously on a large wafer. This parallel process enhances reproducibility and helps achieve economies of scale (Southern *et al.* 1999). Production-scale chips can pack 400 000 probes in 20 μm patches. One weakness of the current photolithography method is that a new set of masks must be produced for every new type of array. There is a maskless technique that uses an array of micromirrors that reflect onto the appropriate spots on the chip.

The technology for spotting arrays is undoubtedly simpler than that for *in situ* fabrication. Simultaneous production of many arrays with the same set of probes makes the deposition more economical than *in situ* synthesis. Moreover, deposition is also a method of choice for long sequences, which are available as PCR products.

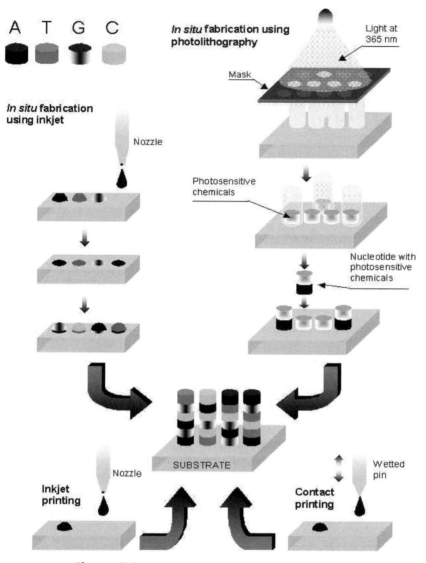

Figure 7.1 Methods for creating a DNA array

However, comparing the two types of nucleic acid arrays, arrays of prefabricated oligonucleotides or cDNAs and *in situ* (on-chip) synthesis of oligonucleotides, the latter has some advantages over deposition of pre-synthesized oligonucleotides. It is not profitable to make large arrays using pre-synthesized oligonucleotides attached to the surface. On the other hand, it is difficult to assess the quality of the oligonucleotides made on a surface. Therefore, this technique could be used for a quality control, but it is not available for most biological laboratories. In contrast, the pre-synthesized oligonucleotides can be assessed before they are attached to the surface.

Sample preparation and hybridization assay

There is still some confusion in the nomenclature of the target and probe. It is commonly accepted that the target determines the labelled material (RNA or DNA), and immobilized oligonucleotide, cDNA or PCR product is the probe. But sometimes probe is also defined as a piece of labelled RNA or DNA which is used in a hybridization assay.

The targets for arrays are usually labelled representations of cellular RNA or DNA pools. It has been found that the short targets can interact more efficiently with oligonucleotide or cDNA arrays than large ones. Ideally, target and probe should have the same length.

Lots of analyses are directed to complex targets, for example, human genomic DNA. In general, it is preferable to reduce sequence complexity to produce a good hybridization signal within a reasonable hybridization time. RNA as a target has a stable secondary structure, which can interfere with hybridization. To reduce this effect, RNA can be fragmented, preferably to a size close to that of the probes on the arrays. Secondary structure is less of a problem with DNA targets and PCR products.

The effects of different charged groups in the spacer were also examined in the hybridization process. It was shown that both positively and negatively charged groups in the spacer diminish the yield of hybridization. Additionally, the base composition and sequence of the oligonucleotides have also a large effect on duplex formation. The effect is of course due to the lower stability of A : T than G : C pairs. According to this, the oligonucleotides of the same length have correspondingly different T_m values. Adding an A : T base pair increases T_m by roughly 2°C, compared with 4°C for a G : C pair.

Readout and data analysis

The detection of the nucleic acid hybridization relies on the signal generated by the binding event. Scanning or imaging the chip surface is essential for obtaining the complete hybridization pattern. Fluorescence imaging is commonly used for such 'reading' of the chips (Figure 7.2). Although the predominant method for array signal detection is still based on fluorescence, many other new methods also show promise. Radioactive materials have the advantage of high incorporation efficiency, high sensitivity and low cost. They were not used in high-density arrays due to the lack of high-resolution imaging methods. Detection methods based on oxidation–reduction reaction, resonance light scattering, capacitance change after hybridization, resonance ionization mass spectrum methods and the nanoparticle-promoted silver staining detection have been also reported (Watson *et al.* 2000).

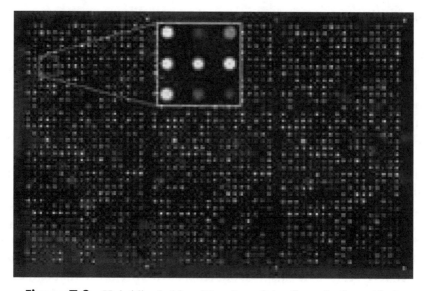

Figure 7.2 Hybridized chip with enlarged detail, ready for analysis

The use of biochip technology enables researchers to obtain experimental data using a highly parallel approach. Such experiments provide unprecedented quantities of data. Bioinformatics tools are used to relate the complexity of the data into useful information. Their interpretation should therefore be based on three main processes: statistical quality analysis, data interpretation and a presentation that makes it accessible to human thinking. Because of the shear wealth of information, only statistical procedures allow an assessment and filtering of microarray data. Data integration is essential, making connections that make subsequent interpretation feasible. This requires a modular data warehouse concept combining the storage of data such as raw signal intensities, gene annotations and their functional categories as well as experimental annotations. For later queries, the annotations should be determined by a pre-defined and catalogued vocabulary. Sophisticated computational tools for data visualization and reduction of data complexity are important to make the information accessible to a human mind, at least as long as there are no automatic expert systems.

Generally speaking, data collection and image analysis are the relatively mature part of the microarray data analysis problem. There are already many commercially available software packages or free software packages that can handle these issues reasonably well.

Achievements and future research directions

There was also a patent battle over the GeneChip (Table 7.3). This is understandable because this technology has already been applied to a diverse set of applica-

Table 7.3   The patent battle over the BioChip

Year	Event
1987	Patent filed on Sequencing By Hybridization (SBH) R. Drmanac, Belgrade Univ. → Argonne National Laboratory → HySeq
1988–1991	Several groups published reports on SBH E. Southern, Oxford University (Oxford Gene Technology) A. Mirzabekov, Engelhard Institute, Moscow → Agronne National Laboratory S. Fodor, Affymetrix W. Bains, Bath University
1989	European patent granted to Southern on 'Oligonucleotide arrays as a testing platform'
1993	US patent on SBH granted to HySeq
1997–1998	HySeq accuses Affymetrix of patent infringement 'We are not doing sequencing, but mutation detection'
1998	Courtcase between Southern and seven chip companies (Affymetrix, HySeq, Hoffman La Roche, Abbot etc.)
1998	US patent to Incyte (Synteni) on technology to print microarrays with density higher than 100 polynucleotides per cm^2
1998–1999	Affymetrix and Incyte (and others) accuse each other of patent infringements

tions. DNA microarrays continue to gain popularity as a number of biotechnology companies aggressively pursue DNA chip technology enhancement and cost reduction.

In recent years a number of different methods for the microarrays have been developed. Electric fields have been used greatly to accelerate the hybridization of labelled target to immobilized sample oligonucleotides. The microarray was fabricated on a $1\,cm^2$ silicon chip. The silicon substrate was thermally oxidized and then platinized to form a $1\,mm \times 1\,mm$ array of 25 microelectrodes. The electrodes were covered with a permeation layer of streptavidin-agarose, to which biotinylated DNA sample was coupled under a positive potential. Such a use of electric field to increase the transport rate of negatively charged probe leads to a ten-fold increase in the hybridization rate.

Another approach to perform and improve gene expression analysis is dynamic DNA hybridization (DDH) on a chip using paramagnetic beads (Fan *et al.* 1999). The advantages of this method are the dynamic supplies of both nucleic acid sample (target) and probe, and also use of the paramagnetic beads as a transportable solid support. It reduces hybridization time and makes the reaction more efficient. The magnetic coated beads are loaded with labelled capture sample or the capture samples are coated onto activated magnetic beads. In such a microfabricated device, simultaneous analysis of many samples is possible. The microfluidic platform of the device is developed for automated analysis of nano-litre volumes. A pneumatic pumping apparatus transports probes and other reagents into the microfluidic device while hydrostatic pumping is used for the introduction of beads with samples. The sample/bead complex is introduced into the device in which hybridization takes place with a complementary probe. At the present time, the paramagnetic beads are extensively used for the preparation, separation and detection of biological molecules such as DNA, RNA and proteins. Their efficiency, simplicity and low cost are very favourable.

The DNA chip is simple in concept, but generating probes on a solid array surface requires considerable expertise and technical expertise. Key matters in creating DNA arrays include: fidelity, reproducibility, ease of synthesis, flexibility, shelf-life, cost, hybridization conditions and steric considerations. As mentioned before, the major advantage of this technique is its high flexibility, which allows creation of a chip with any necessary sequences. But this approach has disadvantages that are difficult to resolve and which limit its usage. First is the time and resources needed to synthesize the required number of different oligonucleotides. Oligonucleotide arrays have longer shelf-lives, compared with cDNA chips, which may only be useable for a few weeks. Another difficulty is to extend the capability of the chip so that thousands of genes can be detected simultaneously. Further, fluorescence technology, which is the most commonly used detection method for array readouts, is reproducible, but is limited in sensitivity. That is why chemiluminescence, diode array detectors, direct electrical charge detection and piezoelectric readout are all being developed as alternative detection methods. Accelerated hybridization techniques are being developed by using electric field control.

Pointing out advantages and disadvantages of this technique reveals perspectives and directions of future research. The innovation of DNA chips opens up a window into the complex world of biology and provides a good tool for the forthcoming post-genomic project research.

Applications and potential use of the DNA chips

The GeneChip technology may be employed in diagnostics (mutation detection), gene discovery, gene expression and mapping (Wang *et al.* 1998). It is used to measure expression levels of genes in bacteria, plant, yeast, animal and human samples (Iyer *et al.* 1997; Oh and Liao 2000).

At the present time, the main large-scale application of microarrays is comparative expression analysis (see Chapters 8 and 9). The microarray technology enables the analysis of expression profiles for thousands of genes in parallel. Another application is the analysis of DNA variation on a genome-wide scale. Both of these applications have many common requirements.

By hybridization with labelled mRNA, cDNA, arrayed PCR products or oligonucleotides on a substrate have been successfully used for monitoring transcript levels, single nucleotide polymorphism (SNP) (Wang *et al.* 1998) or genomic variations between different strains (Behr *et al.* 1999). One of the most significant applications of this technique is, as mentioned above, gene expression profiling on the whole genomic scale. For example, the expression levels of the genes in the *Saccharomyces cerevisiae* genome have been successfully determined with both the DNA and oligonucleotide microarray technology. This technique has also been used to investigate physiological changes in human cells (Iyer *et al.* 1997). DNA microarray technology was applied to detect differential transcription profiles of a subset of the *Escherichia coli* genome.

The microarray technology is still a powerful economical tool for characterizing gene expression regulation and will prove to be useful for strain improvement and bioprocess development. It may prove to be useful for process diagnosis and process monitoring in bioreactors. Information obtained from DNA chip analysis may enable researchers to determine the impact of a drug on a cell or group of cells and consequently to determine the drug's efficacy or toxicity. Knowledge of gene expression profiles can also help researchers to identify new drug targets (Debouck and Goodfellow 1999).

The BioChip opens a new world of diagnostics based on genetics. This technology may be adequate to answer many medical questions. For example, gene expression profiles can be used for classification of tumours and for prognosis.

The technology finds increasing application in fundamental and applied research. The major feature of this technique is that it allows the simultaneous analysis of a great number of DNA sequences. As the Human Genome Project nears completion, a new era of genomic science is beginning. The GeneChip technology is a new technique that undoubtedly will substantially increase the speed of molecular biology research.

Furthermore, when sample preparation time and complexity decreases, it will be possible to evolve current microarrays into more complete on-chip laboratories capable of performing all the necessary procedures to extract genetic materials from tissue or blood samples and analyse it as well. Perhaps more other applications of this technology will be achievable in the near future.

Concluding remarks

Over a decade of rapid advances in biology has swept an avalanche of genetic information into scientists' laps. But analysis of so vast an input, whether to deduce the inner workings of cells or to diagnose diseases, would be impractical without high-throughput technologies. Of these, DNA microarrays are in the lead. They allow scientists to look for the presence, productivity or sequence of thousands of genes at a time. Currently the power and breadth of DNA microarrays technology has made it a predominant factor in genomics, transcriptomics, pharmacogenomics and system biology, simultaneously becoming ever more important in pre-clinical research and even clinical study. Innovative efforts, coupling fundamental biological and chemical science with technological advances in the fields of micromachining and microfabrication should lead to even more powerful devices that will accelerate the realization of large-scale genetic testing. While offering remarkable tools for genetic analysis, proper applications of these new devices still require a solid intellectual input.

References

Behr, M.A., Wilson, M.A., Gill, W.P., Salamon, H., *et al.* (1999) Comparative genomics of BCG vaccines by whole genome DNA microarray. *Science* **284**: 1520–1523.

Benson, D.A., Boguski, M.S., Lipman, D.J. and Ostell, J. (1997) GenBank. *Nucleic Acids Res* **25**: 1–6.

Debouck, C. and Goodfellow, P. (1999) DNA microarrays in drug discovery and development. *Nat Genet* **21**: 48–50.

Fan, Z.H., Mangru, S., Granzow, R., Heaney, R., *et al.* (1999) Dynamic DNA hybridisation on a chip using paramagnetic beads. *Anal Chem* **71**: 4851–4859.

Iyer, V., DeRisi, J., Eisen, M., Ross, D., *et al.* (1997) Use of DNA microarrays to monitor differential gene expression in yeast and humans. *FASEB J* **11**: 1126–1131.

Lockhart, D.J., Dong, H., Byrne, M.C., Follettie, M.T., *et al.* (1996) Expression monitoring by hybridisation to high-density oligonucleotide arrays. *Nat Bitechnol* **14**: 1675–1680.

McGall, G., Labadie, J., Brook, P., Wallraff, G., *et al.* (1996) Ligh-directed synthesis of high-density oligonucleotide arrays using semiconductor photoresists. *Proc Natl Acad Sci USA* **91**: 5022–5026.

Oh, M.K. and Liao, J.C. (2000) Gene expression profiling by DNA microarrays and metabolic fluxes in *Escherichia coli. Biotechnol Prog* **16**: 278–286.

Schena, M., Shalon, D., Heller, R., Chai, A., *et al.* (1996) Parallel human genome analysis: microarray-based expression monitoring of 1000 genes. *Proc Natl Acad Sci USA* **93**: 10614–10619.

Schuler, G.D., Boguski, M.S., Stewart, E.A., Stein, L.D., *et al.* (1996) A gene map of the human genome. *Science* **274**: 540–546.

Southern, E., Mir, K. and Shchepinov, M. (1999) Molecular interactions on microarrays. *Nat Genet* **21**: 5–9.

Wang, D.G., Fan, J., Siao, C., Berno, A., *et al.* (1998) Large-scale identification, mapping and genotyping of single-nucleotide polymorphisms in the human genome. *Science* **280**: 1077–1082.

Watson, S.J., Meng, F., Thampson, R.C. and Akil, H. (2000) The 'chip' as a specific genetic tool. *Biol Psychiatry* **48**: 1147–1156.

8

Overview of Microarrays in Genomic Analysis

Janette K. Burgess

List of Abbreviations

aRNA	amplified RNA
bp	base pairs
CCD	Charge-Coupled Device
cDNA	complementary DNA
dCTP	deoxycytosine triphosphate
DNA	deoxyribonucleic acid
dNTP	deoxyribonucleotide triphosphate
dUTP	deoxyuracil triphosphate
ESTs	Expressed Sequence Tags
HuGE Index	Human Gene Expression Index
IMAGE	Integrated Molecular Analysis of Genomes and their Expression
kb	kilobase
Mb	Megabase
MGED	Microarray Gene Expression Data

Molecular Analysis and Genome Discovery edited by Ralph Rapley and Stuart Harbron
© 2004 John Wiley & Sons, Ltd   ISBN 0 471 49847 5 (cased)   ISBN 0 471 49919 6 (pbk)

MIAME	Minimum Information About a Microarray Experiment
mRNA	messenger RNA
ng	nanogram
nl	nanolitre
nts	nucleotides
PCR	Polymerase Chain Reaction
pg	picogram
RNA	ribonucleic acid
RT-PCR	Reverse Transcriptase – Polymerase Chain Reaction
SNP	single nucleotide polymorphism
μg	microgram
μl	microliter

Introduction

In any living organism hundreds or thousands of genes interact at any one time to create the cellular outputs that can be measured. Traditional molecular biological techniques that are used for the measurement of gene expression, are limited by the amount of data that can be obtained from one experiment and by the amount of time and materials required to perform these techniques. These approaches have either involved a gene targeted approach (e.g. RT-PCR or Northern blot (Alwine *et al.* 1977)), which allows the study of one or a few genes at a time, or they result in the isolation of groups of genes with differential regulation that have unknown sequence and identity (e.g. differential display (Liang and Pardee 1992), subtractive hybridization (Kuang *et al.* 1998) or representational difference analysis (Lisitsyn and Wigler 1993)). Recently, sequencing of the complete genomes of many organisms has been accomplished including the human genome (Venter *et al.* 2001; Lander *et al.* 2001). This rapid expansion of the quantity of sequence data available is currently revolutionizing the science of genomics, which aims to understand how genes function in the context of a whole organism. The next major step after identifying the sequence of the genes is to develop assays that allow the examination of the interaction of all of the genes as they occur in the normal function of the organism. Advances in the development of high sensitivity assays on solid surfaces have made possible the analysis of the expression levels of large numbers of genes (>10 000) in a single experiment using a technology known as DNA microarrays.

What is a microarray?

A DNA microarray, also called a biochip, DNA chip, gene array, gene chip® (a registered trademark belonging to Affymetrix, Inc) and genome chip, is an ordered arrangement of samples on a solid support that can be used for identifying RNA or DNA sequences in a test population. The principle behind a microarray involves matching unknown RNA or DNA and known DNA based on base pairing (A–T and G–C for DNA, A–U and G–C for RNA) or hybridization to identify the unknowns.

There are two types of arrays in general, microarrays or macroarrays. The difference between these arrays is the size and density of the sample spots. Microarrays have spots of 100 microns or less in diameter on a glass support. Macroarrays usually have spots of 250–300 micrometres in diameter and are usually prepared on a nylon support. The methodology for performing an experiment with either a microarray or a macroarray is the same.

To perform a microarray experiment, two populations of cells, which are identical except for the treatment or disease state of interest, are prepared. RNA is isolated from the cells and each sample labelled with a different fluorochrome or radioisotopic tag. The fluorescently labelled samples are pooled and hybridized to the microarray, the radioisotopic tagged samples are bound to separate macroarrays, non-specific probe is washed away and the fluorescent/radioactive image is detected using a scanner (Figure 8.1). An automated analysis process is then used to identify differential expression of any or all of the genes present on the array.

The convention adopted by the laboratories that have led the development in the field refers to the immobilized cDNA or oligonucleotide on the microarray as the 'probe' and the labelled RNA or DNA in the sample as the 'target' (Phimister 1999); this chapter also uses the adopted standard terminology.

History

The initial report that labelled nucleic acids could be bound to nucleic acid molecules attached to a solid support and used to monitor gene expression occurred in the mid 1970s (Southern 1975). Since that observation, several groups have explored the possibility of hybridizing mRNA to cDNA libraries arrayed on nylon filters (Lennon and Lehrach 1991; Kafatos *et al.* 1979). The major drawback of this approach was the cumbersome implementation of these techniques.

Immunoassays have employed antibodies attached to solid supports since the 1960s (Ekins and Chu 1999). In the early 1980s scientists working in this field began to explore the possibility of using microarray methods for immunodiagnostic purposes (Ekins *et al.* 1989). One very important concept for the

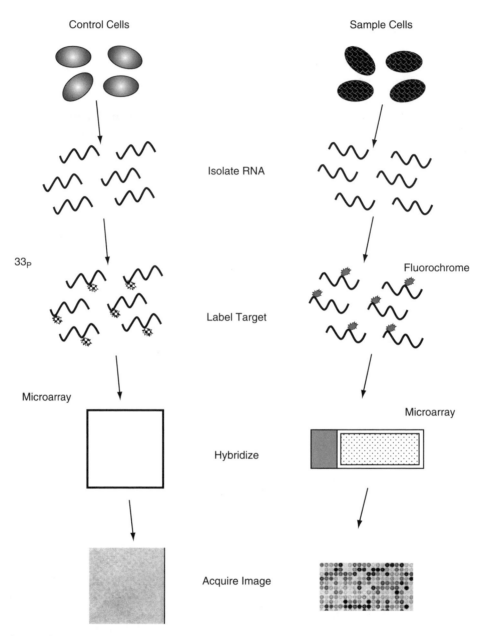

Figure 8.1   A microarray experiment. Two populations of cells that are identical except for the treatment or disease state of interest are prepared. RNA is isolated from the cells and each sample labelled with a radioisotopic tag or one of two different fluorochromes. The radioactive sample is hybridized with a nylon macroarray and the fluorogenic labelled samples are pooled and hybridized to a microarray. The unbound non-specific sample is washed away and the image is captured using a phosphorimager for the radioactive signal and a scanner for the fluorescent image

development of microarray technology was that high sensitivities could be achieved by using small amounts of a binding agent anchored on a solid support. It had been thought that it was necessary to 'bind the majority of the analyte present in a sample' to achieve a high sensitivity (Hay *et al.* 1991). Ekins and Chu, in collaboration with Boehringer Mannheim, described and demonstrated the construction and use of microarray-based assays for protein and small molecule analytes such as glycoprotein and steroid hormones, viral and allergen antibodies as well as oligonucleotide-based microarrays for RNA and DNA detection (Ekins 1998; 1996).

Advances in array printing technology (Schena *et al.* 1998), including the development of automated spotting devices with very precise movement controls on the x-, y- and z-axes and the improvement of the printing head technology to provide very accurate, reproducible spots on the arrays, have sparked a renewed interest in microarrays within the wider research community (Schena *et al.* 1995). The use of microarrays to study gene expression profiles is a developing technology that is rapidly being implemented in many fields of research.

Production of a cDNA microarray

cDNA microarrays are usually created using an automated precise x y z robot that is programmed to spot probes on to a solid substrate, usually a glass microscope slide, in a high-density pattern (Cheung *et al.* 1999; Duggan *et al.* 1999; Southern *et al.* 1999). The spots created by the special pins that are fitted to the head of the array robot currently range between 80 and 150 μm in diameter (depending on the pin used) and are placed at 100 μm intervals on the slide. The number of different probes spotted per movement is dependent on the size of the head fitted to the printing arm and the number of pins mounted on the printing head. There are several designs of spotting pins (Figure 8.2). Spots can be administered using either active delivery using high-frequency solenoid valves with a stepper motor-driven syringe pump, or passive dispensing (Franssen-van Hal *et al.* 2002). The different styles of pins are able to print different numbers of slides (depending on the size of the reservoir in the pin) before they need to be returned to the stock plate to be refilled. Alternative printing technologies are offered on some microarrayers which include different ink jet printing techniques (Hughes and Shoemaker 2001; Hughes *et al.* 2001).

On a standard glass microscope, slide arrays that contain at least 80 000 spots can be created, although arrays containing 10 000–20 000 spots are more usual. Macroarrays have a larger format, with spots of slightly larger size in a lower density pattern. They are usually produced on a nylon membrane and can have as few as 500 or as many as 18 000 spots. These arrays can be produced using the same robotic technology from the cDNA source plates. Very early generation macroarrays were spotted with bacterial colonies that were lysed *in situ*.

Solid Pin

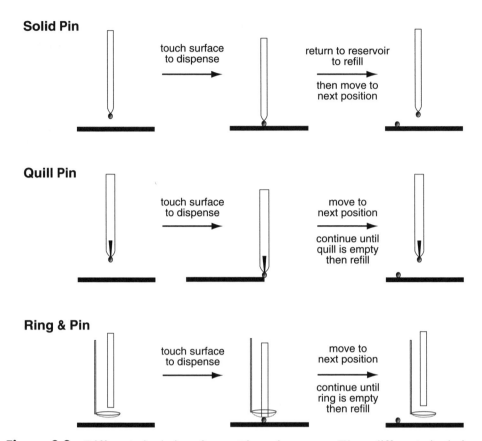

Quill Pin

Ring & Pin

Figure 8.2 Different pin designs for spotting microarrays. Three different pin designs used for spotting microarrays: Solid pin, Quill pins and the Ring and pin system. In each case the pin is positioned above the site where the cDNA or oligonucleotide is to be spotted. Touching the pin to the surface of the array substrate applies the spot. The solid pin then returns to the liquid reservoir for replenishment of the DNA before moving to the next position; the quill pin can apply multiple spots from the quill reservoir before it needs to return to the liquid reservoir for replenishment; the ring and pin system can apply several spots from the liquid meniscus held in the ring before it needs to return to the liquid reservoir

The sample-spotting pattern on the arrays is controlled from a computer attached to the robotic arrayer, which ensures that the precise location of each cDNA probe is recorded to allow identification of genes with expression changes during the analysis phase of the experiment (see Image analysis). Patrick Brown's laboratory at Stanford University created the first x y z arrayer and instructions on how to build an arrayer can be found on their website (http://cmgm.stanford.edu/pbrown/) (Table 8.1). Many companies now produce microarrayers commercially, using a number of variations on the way the spot is placed on the substrate (Table 8.2).

Table 8.1 Microarray related websites

General Information
http://edtech.clas.pdx.edu/gene_expression_tutorial/

http://highveld.com/f/fgenetics.html
http://www.bioplanet.com/index.php
Genamics.com   http://genamics.com/

Microarray Information Sites
http://www.gene-chips.com/
http://www.bsi.vt.edu/ralscher/gridit/index.htm
http://www.cs.washington.edu/homes/jbuhler/research/array/
http://tagc.univ-mrs.fr/microarrays/
http://www.arrayit.com/e-library/
http://www.mpiz-koeln.mpg.de/~weisshaa/Adis/DNA-array-links.html
http://www.microarrays.org/
http://stat-www.berkeley.edu/users/terry/zarray/Html/
http://ihome.cuhk.edu.hk/~b400559/index.html

Institutes, Organizations and Databases
National Human Genome Research Institute
   Microarray project   http://research.nhgri.nih.gov/microarray/
The Institute for Genomic Research   http://www.tigr.org/
Vanderbilt University Microarray Core Facility   http://www.microarrays.com
National Centre for Biotechnology Organisation   http://www.ncbi.nlm.nih.gov/
European Bioinformatics Institute   http://www.ebi.ac.uk/
Microarray Gene Expression Data Society   http://www.mged.org/
University of Manchester   http://www.bioinf.man.ac.uk/microarray/
Patrick Brown's Lab at Stanford University   http://cmgm.stanford.edu/pbrown/
Human Gene Expression Index (HuGE Index)   http://server3.mgh.harvard.edu/hio/welcome/
   index.html
Genbank   http://www.ncbi.nlm.nih.gov/Genbank/GenbankOverview.html
Unigene   http://www.ncbi.nlm.nih.gov/entrez/query.fcgi?db = unigene
LifeSeq   www.incyte.com/index.shtml
IMAGE human cDNA clone set   http://image.llnl.gov/
RIKEN Gene Bank   http://www.rtc.riken.go.jp
Flybase   http://flybase.bio.indiana.edu/
Microarray Gene Expression Data group   http://www.mged.org

Microarray Forums
http://cmgm.stanford.edu/pbrown/mguide/logbook.html
GENE-ARRAYS@ITSSRV1.UCSF.EDU (instructions for joining are explained on the
   gene-chips web page)
http://www.egroups.com/group/microarray/
http://forum.arrayit.master.com/texis/master/search/msgbd.html

This list contains a selection of websites related to the topic, there are other websites that may also contain information related to microarrays. All website addresses were correct at the time of writing this chapter.

Table 8.2   Microarray related companies' websites

Affymetrix	http://www.affymetrix.com/
Amersham	http://www5.amershambiosciences.com
Arrayit	http://arrayit.com/Products/
Axon Instruments Inc	http://www.axon.com/GN_Genomics.html
Biodiscovery	http://www.biodiscovery.com
Bio-Rad	http://www.bio-rad.com
Biorobotics	http://www.biorobotics.com/
Cartesian Technologies	http://www.cartesiantech.com/
Clontech	http://www.clontech.com
Corning	http://www.corning.com/lifesciences/Greater_Europe/en/
DNA microarray.com	http://DNAmicroarray.com
Engineering Services Inc	http://www.esit.com/
Fuji	http://home.fujifilm.com/products/science/index.html
GeneMachines	http://www.genemachines.com/index.html
GeneScan Europe	http://www.genescan-europe.com/intro/
	intro_xx_start_fs.php3?nav_lang = us
Genetix	http://www.genetix.co.uk/
Genomic Solutions Inc	http://www.genomicsolutions.com/
Genomic Solutions Inc	http://www.genomicsolutions.com/
Grace Bio Labs	http://www.gracebio.com/
Imaging Research Inc	http://www.imagingresearch.com/
IncyteGenomics	http://reagents.incyte.com/
Intelligent Bio-Instruments	http://www.intelligentbio.com/
Mergen Ltd	http://www.mergen-ltd.com/
Molecular Probes	http://www.probes.com/
Perkin Elmer life Sciences	http://www.packardbiochip.com/
Phase 1 Molecular	http://www.phase1tox.com/
Toxicology Inc	
Qiagen	http://oligos.qiagen.com/index.php
Radius Biosciences	http://www.ultranet.com/~radius/
Research Genetics	http://www.resgen.com/
Schleicher & Schuell, Inc	http://www.s-and-s.com/index.htm
Silcon Genetics	http://www.sigenetics.com/
Silicon Graphics	http://www.sgi.com/
Spotfire	http://www.spotfire.com/
Super Array Inc	http://www.superarray.com/
SurModics	http://www.surmodics.com/
Xenomatrix	http://www.xeno.com/

This list contains a selection of companies' websites related to the topic, there are other company websites that may also contain information related to microarrays. All website addresses were correct at the time of writing this chapter.

Sources of cDNA

cDNA microarrays are usually spotted with cDNAs amplified from an IMAGE (Integrated Molecular Analysis of Genomes and their Expression) cDNA clone set or a custom cDNA library (Duggan *et al.* 1999). The clones are generally amplified by the polymerase chain reaction (PCR) to produce 0.5 to 2 kb products for arraying. Each spot on the array is created by depositing between 0.1 nl and 6 nl (depending if active delivery or passive dispensing is used) of a 200 ng/μl solution of DNA from the tip of the pin onto the substrate. A spot of 250 pg to 1 ng of DNA is enough to provide the signal recorded by the array scanner (see Image acquisition). Currently, the IMAGE human cDNA clone set (http://image.llnl.gov/) contains about 40 000 sequence-verified known genes (genes that have a function and/or identity assigned) and Expressed Sequence Tags (Adams *et al.* 1991) (ESTs – segments of genes that have been sequenced but have no known function /identity to date). Information about the clones and current knowledge about gene sequences is provided in public genomic databases such as Genbank (http://www.ncbi.nlm.nih.gov/Genbank/GenbankOverview.html), Unigene (http://www.ncbi.nlm.nih.gov/entrez/query.fcgi?db = unigene) and Life-Seq (Incyte, Inc) (www.incyte.com/index.shtml) (see Table 8.1). With the pace of the human genome project, which has completed the sequence for chromosomes 21 and 22 and recently released a working draft of the remainder of the human genome sequence (Venter *et al.* 2001; Lander *et al.* 2001), this set of clones is expanding rapidly. The human cDNA set is available commercially from Research Genetics or Incyte (see Table 8.2). Although the full genome sequence has not been completed for the mouse, a clone set containing 18 000 full-length cDNA clones is available from RIKEN (http://www.rtc.riken.go.jp). Clone sets containing the majority of the genes expressed by the yeast *Saccharomyces cerevisiae* (6307 open reading frames), and clones from a variety of other organisms including rat, *Caenorhabditis elegans* and *Arabidopsis thaliana* are available for arraying from a variety of sources (Bowtell 1999). Many international collaborations have been established to increase the rate of sequencing the genome for a wide variety of organisms and these efforts are providing valuable material that is being applied to microarrays. The complete genome sequence for *Drosophila melanogaster* was completed by such an international collaboration (http://flybase.bio.indiana.edu/). There are now 40 or more organisms whose complete sequence is known. The complete sequences for many other organisms are rapidly being completed including a number of important human pathogens. The availability of complete sequences is enabling researchers to monitor global gene expression patterns for a whole organism simultaneously in a single experiment.

The same sources of cDNAs, as described for the microarrays are used to produce macroarrays.

Production of an oligonucleotide microarray

An alternative technology for the production of microarrays uses a technique that has been adapted from the computer chip production industry referred to as miniaturized photolithography. This technique is used to assemble the oligonucleotides (usually 20–30 nts) directly on the glass surface (see Figure 8.3) (Lipshutz *et al.* 1995; 1999; Singh-Gasson *et al.* 1999). The process of photolithography requires a series of film-like templates (masks) and a light source. The surface of the platform that the array will be assembled on is coated with a covalent linker molecule (which has the capacity to bind nucleotides) coated with a light-sensitive agent. When the array surface is overlaid with the first mask and exposed to the light source, deactivation of the light-sensitive compound at the positions exposed by the mask allows binding of allocated nucleotides to those positions (see Figure 8.3). The nucleotides that are bound to the array substrate are pre-linked to the light sensitive compound, which prevents further nucleotides binding to them until they are subsequently exposed to light. A second mask is then used to expose a different pattern of binding sites to light for deactivation of the light sensitive compound and subsequent binding of a second nucleotide in these positions. The process of selective light exposure using masks to deactivate the light sensitive compound, allowing nucleotide coupling to the array substrate is repeated to add nucleotides one at a time to growing chains at each position on the array (*in situ* oligonucleotide synthesis). This process is continued until the *in situ* synthesis of the 20–30 nts oligonucleotides designed to encode the genes represented on that array is completed.

Affymetrix pioneered the use of this form of array production with the development of the 'Genechip' (Lipshutz *et al.* 1995; Lockhart *et al.* 1996). The Genechip series are designed with 16–20 different oligonucleotides, from different regions of the gene, to represent each gene on the array. To monitor the degree of non-specific binding occurring on the array, each of the oligonucleotides present on the array also has a partner oligonucleotide on the array that has a single base change (mismatch) which should not bind the specific gene (Lipshutz *et al.* 1999) (see Figure 8.4) (http://www.affymetrix.com/products/exp_ov.html). This technology can also be used for the detection of single nucleotide polymorphisms in the genes on the array (http://www.affymetrix.com/products/gc_HUSNP1.html).

Many other companies now market their own versions of synthetic oligonucleotide arrays. A maskless array synthesizer method using digital micromirrors has been established at the University of Wisconsin (Singh-Gasson *et al.* 1999) and is being developed by NimbleGene Systems (Madison WI) (Nees and Woodworth 2001). Rosetta Inpharmatics Inc have developed Flexjet technology (adapted from Hewlett Packard) which produces arrays of tens of thousands of oligonucleotides synthesized in situ by an ink jet printing using phosphoramidite chemistry (Hughes *et al.* 2001).

Recently arrays have also been produced by spotting 25 to 80 base single-stranded oligonucleotides onto the array substrates in the same way that cDNAs have been spotted (Nees and Woodworth 2001). These spotted oligonucleotide arrays will become more important as the costs for bulk orders of small-scale synthesis of custom oligonucleotides decrease. Results comparing spotted oligonucleotide arrays with cDNA arrays show close correlation for most genes.

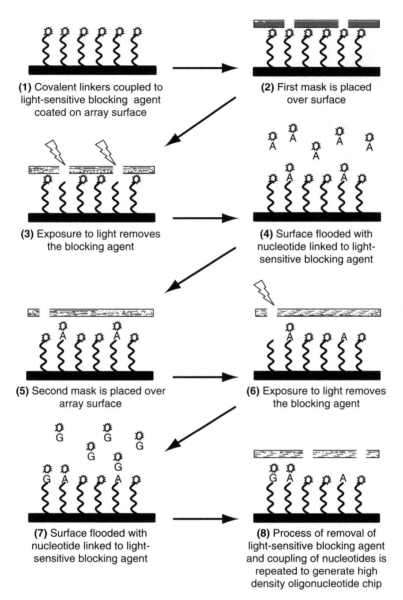

Figure 8.3 Photolithographic synthesis of an oligonucleotide chip. (1) The array surface is coated with a light sensitive agent at the sites where the oligonucleotides are to be synthesized. (2) A mask that has been designed to expose the positions at which the first nucleotide is to be attached, is placed over the array surface and (3) the appropriate sites are activated by exposure to light. (4) The array surface is flooded with the nucleotide linked to the light sensitive agent, resulting in its coupling to the array surface. (5) The process is repeated with the positioning of the next designed mask and (6) the activation of the second series of sites. (7) The next nucleotide linked to the light-sensitive agent is coupled to the substrate. (8) The whole process is repeated using a compete set of masks designed for the in situ synthesis of 20–30 nts oligonucleotides

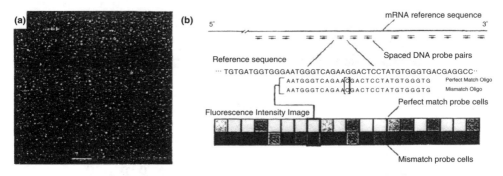

Figure 8.4 Oligonucleotide microarray probe an array design. Oligonucleotide probes are chosen based on uniqueness criteria and composition design rules. For eukaryotic organisms, probes are chosen typically from the 3′ end of the gene or transcript (nearer to the poly A tail) to reduce problems that may arise from the use of partially degraded mRNA. The use of the perfect match minus mismatch differences averaged across a set of probes greatly reduced the contribution of background and cross-hybridization and hence increases the quantitative accuracy and reproducibility of the measurements. Reproduced from Lipshutz *et al.* (1999) with permission

Variations usually occur in those gene families that have alternative mRNA processing. The differential expression of these splice variants can be detected only with the oligonucleotide arrays (Shoemaker *et al.* 2001; Clark *et al.* 2002).

Array substrates

Glass has a naturally low autofluorescence, is inert to high ionic strength buffers and can be treated to enhance the immobilization of the DNA on the surface (e.g. poly-L-Lysine or amino-silane coating), which makes it a suitable substrate for the production of microarrays (Schena *et al.* 1996). Silicon wafers have also been used as supports for oligonucleotide arrays (Lipshutz *et al.* 1999; Lockhart *et al.* 1996). The low background fluorescence of the substrate is an important feature as it enables highly sensitive detection of the fluorescent signals that are used to label the targets applied to the microarray (see Target labelling).

Nylon filter membranes are used as the substrates that support macroarrays.

Commercial sources of arrays

Establishing a microarray facility is an expensive undertaking that is beyond the reach of many smaller research institutions. There are several alternatives available that still enable these researchers to utilize array technology (see Table 8.2). Pre-made arrays can be purchased from many companies. These microarrays are being provided as general screening arrays, which are usually spotted with known and unknown genes, or as directed arrays, which are spotted with a defined group of genes usually associated with a particular disease (e.g. cancer) or function (e.g. cytokines or transcription factors). The microarrays are usually supplied as a pair of arrays in a kit with two practice arrays and the solutions required for the experiment. Each microarray can only be hybridized once. Most molecular based companies are now launching their own version of the microarray but some of the first companies to provide glass arrays include NEN Life Sciences, Radius Biosciences, Mergen Ltd, Incyte, Operon Technologies Inc, Phase I Molecular Toxicology Inc, Display Systems Biotech, Biochip Technologies and Scimatrix.

Spotted oligonucleotide arrays are also available. The first full genome array for yeast using spotted oligonucleotides is produced by Operon Technologies. Rosetta Inpharmatics Inc recently released the first complete genome chip for humans using spotted oligonucleotides (Shoemaker *et al.* 2001).

Nylon macroarrays are also available for researchers to purchase from Research Genetics, Incyte, Super Array Inc or Clontech. The range of genes represented on these arrays is very similar to those available on the microarrays. These arrays are usually supplied in pairs with a test membrane provided for establishing hybridization conditions. With care, each macroarray can be stripped and reprobed up to five times allowing multiple sample comparisons to be performed with the same set of macroarrays.

Another approach that researchers can undertake is to submit their samples to companies like DNA Microarray, Clontech or Incyte who will perform the labelling, hybridization and image analysis and provide the data in a spreadsheet form to the researcher. The premade arrays and services are currently available only for the more common genetic models (human, mouse and rat mainly).

Researchers can access the Affymetrix Genechip technology only by purchasing the complete system including the scanner, software and other components from Affymetrix.

Isolation of RNA

The quality of the RNA, the initial material required for the microarray experiment, plays a large part in determining the final outcome (Duggan *et al.* 1999). Many of the standard RNA isolation techniques can be used to produce the high quality RNA needed for a microarray experiment provided that care is taken during the preparation.

It is very important that when a microarray experiment is designed, care is taken to ensure that the only potential difference between the control and the treated samples reflects the question being asked (Wildsmith *et al.* 2001; Wildsmith and Elcock 2001; Hegde *et al.* 2000). The sensitivity of this technique has been demonstrated as increases and decreases in gene expression have been observed when the only differences between batches of cultured cells were the media or serum batches in which they were grown (Lander 1999). Therefore these experiments need to be stringently controlled to ensure that the control and test samples are treated in exactly the same way, thereby reducing false results.

Microarray experiments currently require a large amount of good quality RNA to provide an adequate signal at the time of image acquisition (generally more than 50 μg). Acquiring this quantity of high quality RNA can be a challenge but as the labelling protocols develop, this quantity may decrease. To enable the use of this technology in instances where the quantity of RNA available is low, techniques such as linear amplification (Eberwine et al. 1992; Luo et al. 1999) and alternative direct labelling protocols (Lockhart et al. 1996) are used to enhance the quantity of mRNA in vitro to allow the gene expression profiles to be monitored.

Amplification

The need for large amounts of RNA as starting material for a microarray experiment limits the type of samples that can be used as sources of material for application to arrays. To overcome this problem and allow samples which yield only small amounts of RNA (for example fine needle aspirates (Ohnmacht et al. 2001) or samples isolated using laser capture microdissection (Luo et al. 1999; Nagle 2001; Ohyama et al. 2000)) mRNA amplification has been developed. This technique involves the conversion of mRNA to cDNA, production of double-stranded cDNA which is transcribed, in vitro, into amplified RNA (aRNA). This process is usually repeated for one or two more rounds to increase the amount of RNA available for labelling and application to the microarray. It has been reported that between 200 (Nagle 2001) and 1000 (Luo et al. 1999) cells are enough for RNA linear amplification and subsequent microarray analysis. Following the linear amplification of the mRNA the aRNA can be labelled using a standard microarray labelling reaction before hybridization with the microarray using a standard protocol for a microarray experiment. In a comparison of array results from amplified RNA and total RNA, Feldman and colleagues found that the fidelity of the array results using one or two rounds of RNA amplification was preserved (Feldman et al. 2002). They also observed that the quality of the arrays results was improved by the RNA amplification.

Target labelling

cDNAs from the sample in which gene expression is to be measured are labelled with a fluorescent or a radioactive marker to create the targets for the micro/macroarray and spotted oligonucleotide arrays. The RNA is reverse transcribed to cDNA, either directly or indirectly, in the presence of nucleotides modified with a fluorescent or a radiolabelled tag (Schena et al. 1995; Richter et al. 2002) using one of several methods (see Figure 8.5). These methods are similar to those used

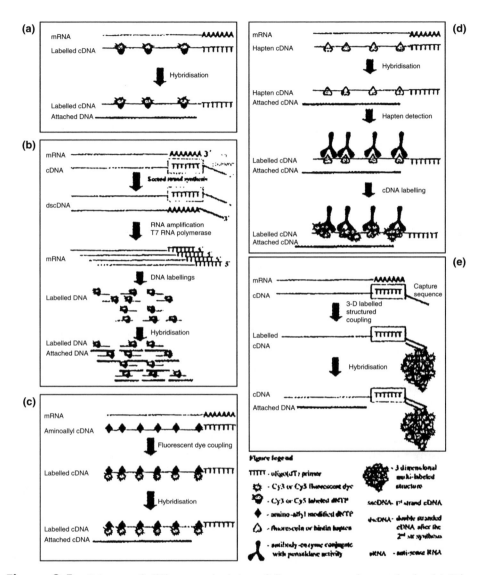

Figure 8.5 Scheme of different principles of fluorescent probe synthesis. (a) Direct labelling method. (b) T7 RNA polymerase amplification method. (c) Aminoallyl labelling method. (d) Hapten-antibody enzymatic labelling method. (e) Labelling method using 3-D multi-labelled structures. Reproduced from Richter *et al.* (2002) with permission.

for the creation of a Northern or Southern blot probe. The most commonly used fluorophores to date are Cy3-dNTP and Cy5-dNTP dyes (Amersham) or Cyanine3-dNTP or Cyanine5-dNTP (NEN Life Sciences) but other dyes such as the Alexa series (Molecular Probes) are being developed for this application. The most common radionucleotide currently used for macroarrays is P-33 (^{33}P – αdCTP) but Rajeevan and colleagues (Rajeevan *et al.* 1999) have reported the use of digoxigenin-labelled probes. After synthesis the labelled targets are purified, denatured and applied to the micro/macroarray.

For Affymetrix oligonucleotide arrays the sample mRNA is reverse transcribed to cDNA in the absence of labelled nucleotides. It is then converted into double-stranded cDNA which is, in turn, transcribed into complementary RNA in the presence of nucleotides modified with a biotin tag. This additional step in the labelling reaction is similar to the amplification process described above. It increases the amount of target by approximately 50-fold (Schulze and Downward 2001) which enables much smaller quantities of starting RNA to be used for these arrays.

Hybridization

The hybridization of a microarray generally occurs in the minimum volume possible, usually measured in microlitres, under a coverslip. Many laboratories have adopted special chambers to enclose the microarrays to enable their immersion in a water bath to keep them at a constant temperature during the hybridization step. In a manner similar to protocols employed during other hybridization-based techniques, the incubation of the target with the microarray is typically carried out at 65°C for aqueous hybridization buffers and at 42°C if 50 per cent (v/v) formamide is present in the hybridization solution. It is common for two different samples (a control and a test sample) labelled with two different fluorophores to be hybridized to a microarray at one time, which enables the measurement of the relative levels of gene expression in the two target populations.

To ensure that the concentration of the target in the sample largely determines the rate of binding to the array during the hybridization reaction, the probe (the spot for binding) on the microarray should be at least ten times more concentrated than in the sample. Usually the spot on the array contains enough DNA copies for both samples' targets to bind without competition. Factors that can influence the concentration of the DNA in the spot on the array include the robustness of the coupling chemistry on the array and the efficiency of the probe delivery to the glass surface during the creation of the array (Adessi *et al.* 2000; Cheung *et al.* 1999). The size, concentration, composition and number of replicates of the PCR products spotted onto the array can alter the final amount of target that hybridizes with the probe (Stillman and Tonkinson 2001; Franssen-van Hal *et al.* 2002). Other factors that can influence the rates of hybridization include the presence or absence of monovalent cations and/or formamide, the hybridization time, temperature and the length and specific activity of the probes (Chan *et al.* 1995; Shchepinov *et al.* 1997; Peterson *et al.* 2001; Southern *et al.* 1999; Mir and Southern 1999; Beier and Hoheisel 1999). Following hybridization, the microarray is washed repeatedly to remove the unbound and non-specific signal and it is then ready for the acquisition of the image.

Hybridization of oligonucleotide arrays is carried out under similar conditions to those used with cDNA microarrays. It is usual to hybridize only one sample to a single oligonucleotide array because of the difference in the labelling protocol. In this system the control sample and the test sample are hybridized to different arrays. After the hybridization step an avidin-conjugated fluorophore is added for the detection of the bound target. The oligonucleotide microarray is also washed repeatedly to remove the unbound target and non-specific signal from the unbound avidin-conjugated fluorophore before it is then ready for the acquisition of the image.

Hybridization of the macroarrays is usually carried out in roller bottles in a standard hybridization oven using a few millilitres of hybridization solution. Radioactive targets (Cheung *et al.* 1999) are used with macroarrays which means only one sample can be applied to each array at a time. Arrays probed with radioactive targets can be stripped and re-hybridized following acquisition of the image (Coombes *et al.* 2002).

Image acquisition

The output from a microarray experiment is acquired using a scanner for fluorescent signal detection. The microarray scanners are designed to excite the fluorochromes adhered to the probes on the array and measure the intensities of the light emissions from the field of the microarray. The data are interpreted by a specialized software package that computes the intensities of the fluorochrome emissions at each location and provides a readout of the data (generally as a spreadsheet) that is usually linked to a colour-coded image map on which the colour of the spot indicates the gene expression profile (Figure 8.6) (Iyer *et al.* 1999). Traditionally one sample is assigned a green colouring and the other a red colouring on the image map. If there is higher expression of a gene in sample one, that spot will be green in colour. If the expression is higher in sample two, the spot will be red. If there is equal expression in both samples the spot is usually coloured yellow on the image map. The level of gene expression is indicated by the degree of colour assigned to the spot. Intense colour indicated a high level of gene expression whilst a weaker coloured spot would indicate only a low level of gene expression. The scanners currently available use either scanning confocal laser technologies or a CCD camera. Most current scanners are designed to detect two wavelengths of light emission, but scanners with the capacity to detect multiple wavelength light emissions on any one slide have been released. The image analysis software package for the quantification of the light emission intensities for each spot on the microarray is now either supplied with, or provided as an option during the purchase of the scanner.

The data from nylon filters hybridized with targets with a radiolabel are acquired as images captured using a phosphorimager from Molecular Dynamics,

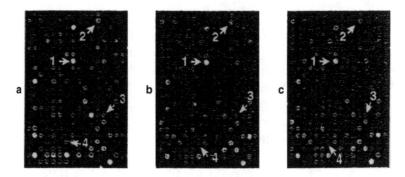

Figure 8.6 A microarray image. The same section of a microarray is shown for three independent hybridizations comparing RNA isolated at the 8-h time point after serum treatment to RNA from serum deprived cells. Each microarray contained 9996 elements, including 9804 human cDNAs, representing 8613 different genes. mRNA from serum deprived cells was used to prepare cDNA labelled with Cy3-deoxyuridine triphosphate (dUTP), and mRNA harvested from cells at different times after serum stimulation was used to prepare cDNA labelled with Cy5-dUTP. The two cDNA probes were mixed and simultaneously hybridized to the microarray. The image of the subsequent scan shows genes whose mRNAs are more abundant in the serum-deprived fibroblasts (that is, suppressed by serum treatment) as green spots and genes whose mRNAs are more abundant in the serum treated fibroblasts as red spots. Yellow spots represent genes whose expression does not vary substantially between the two samples. The arrows indicate the spots representing the following genes: 1, protein disulfide isomerase related protein P5; 2, IL-8 precursor; 3, EST AA057170; and 4, vascular endothelial growth factor. Reproduced from Iyer *et al.* (1999) with permission

Canberra Packard or Fuji (Pietu *et al.* 1996; Zhao *et al.* 1995). These images can also be analysed using specialized software to compute the intensity values for each spot (Figure 8.7). In these experiments the data from two separate analyses must be compared to try to calculate the differences in gene expression between two samples, which introduces a number of other issues about the normalization of the two separate hybridizations. The ideal method of acquiring information using the macroarrays is to perform every step of the comparison at the same time. The control and test samples should be labelled at the same time, hybridized to two separate membranes that were spotted in the same batch for the same length of time before exposing them at the same time to the same phosphorscreen.

Differential labelling and hybridization efficiencies of the targets affect both the micro-and macroarray experiments. Traditional 'housekeeping genes' are usually included on all arrays to assist in normalizing individual hybridization differences. The list of genes that can be classified as 'housekeeping genes' is dependent on the genetic material being used and the treatment tested. Affymetrix have identified and catalogued a large number of potential housekeeping genes (Warrington *et al.* 2000). The Human Gene Expression Index (HuGE Index) has a list

Figure 8.7 A macroarray image. A macroarray is shown for RNA isolated at the 24-h time point after serum treatment. Each macroarray contained 36 864 elements, representing 18 372 nonredundant human cDNA clones from I.M.A.G.E. library. mRNA from serum treated cells was used to prepare cDNA labelled with ^{33}P-dUTP. The cDNA probe was hybridized to the macroarray. The image of the subsequent scan shows genes whose mRNAs were abundant in the serum-treated cells as larger, darker spots

of 451 potential housekeeping genes for the human genome (http://server3.mgh. harvard.edu/hio/databases/index.html). Some manufacturers spot total genomic DNA on the arrays to be used as orientation and normalization guides. Another approach that can be used is to spike each sample with a cDNA that has been spotted on the array but is not present in either the control or the test sample, to act as an internal control for the labelling and hybridization processes (Schuchhardt *et al.* 2000). This is not always possible as the range of RNAs in the samples is usually unknown.

In experiments where multiple samples are to be compared (for example a time course experiment) each mRNA sample can be compared with a reference mRNA pool which is composed of an equal mix of all the experimental RNA populations (Scherf *et al.* 2000; Eisen and Brown 1999; Puskas *et al.* 2002; Alizadeh *et al.* 2000). The reference pool is labelled with one fluorophore and the target of interest with another fluorophore. The two samples are then hybridized to the array and the intensity readings for each probe read. Using this sort of reference many samples that are hybridized to separate arrays can be compared as they will all have the same reference.

Another possible reference for cDNA hybridizations is to pool a small amount of all of the cDNA PCR products (the probes) before they are printed onto the microarray and use this pool as a reference (Sterrenburg *et al.* 2002).

Image analysis

After the acquisition of the image data a further software package – the image analysis program – is usually employed to calculate a quantitative ratio of the level of gene expression between the two samples being compared. The first step in the image analysis is the overlaying of a grid on the image of the microarray to map the location of the pixels representing each spot. In some cases the software has an automated function for the aligning of this grid but in other versions this is a completely manual process. Once the grid in is place the user annotates the image. Some of the programs have functions that enable the user to indicate spots that they would like left out of the final analysis. This function of the software is useful if the reproducibility of the arrayer is poor and the spotting pattern is uneven. It is also of use if there are dust specks on the array that may interfere with the outline of a spot and distort the intensity readout. Other functions include the ability to indicate that a spot is of interest (possibly because of a high expression change) and this indication of interest will be carried through to the resulting data spreadsheet.

Most of the different versions of the software packages used for image analysis incorporate some degree of normalization process. This allows the user to account for differences in the labelling efficiencies with the two fluorophores and for any background signal before the intensity ratios are calculated (Lipshutz *et al.* 1999). There are many different methods for normalization which are beyond the scope of this chapter (refer to Chen *et al.* 1997; Duggan *et al.* 1999; Hess *et al.* 2001; and Quackenbush 2002 for further information). The software package produces an output of data that usually includes the calculated intensity ratios which provide an indication of the relative amounts of any given RNA species in the two samples applied to the microarray.

Most scanners are sold with an option of an image analysis software package such as Imagene (Biodiscovery) or GenePix (Axon Instruments) and there are also many stand-alone image analysis packages emerging on the market (see Table 8.2). A software package for the initial image analysis that was developed at Stanford University (http://rana.stanford.edu/software/) is freely available to academic users (Eisen *et al.* 1998). Many of these packages also include functions for further downstream analysis of gene expression profiles from multiple experiments. Some of these packages use cluster analysis algorithms to group the genes on an array according to their expression profiles over a number of experiments (Figure 8.8) (Iyer *et al.* 1999). The recently published Primer on Medical Genomics Part III Microarray Experiments and Data Analysis (Tefferi *et al.* 2002) provides a good overview of many forms of cluster analysis. A version of this type of program, called Cluster, is also available from Stanford University (see Table 8.1) (Eisen *et al.* 1998). There are many resources available for the analysis of microarray data, some being developed by commercial entities but many are provided on the internet with free access for academic users.

Most microarray experiments will produce hundreds of genes that are up or down regulated, unless very specific arrays with only a few hundred genes are employed. Determining which gene expression changes are physiologically relevant in the system that is being studied, and therefore worth further investigation, is part of the analysis process that accompanies the use of microarrays. The efficiency of data retrieval after the initial experiment will play a large part in the power of microarray analysis. To be able to gain the maximal amount of information from each experiment it is important to have bioinformatics tools that can link into other databases (e.g. Genbank or Unigene) to obtain the information that is already known about the genes of interest in an automated fashion. The development of more powerful and sophisticated mathematical modelling is occurring to enable researchers to perform multi-dimensional analyses of multiple experiments (Sherlock 2000; Brazma and Vilo 2000; Brazma and Vilo 2001; Hegde *et al.* 2000; Quackenbush, 2001).

Data storage

Thousands of data points are generated very rapidly when a researcher commences microarray experiments. This amount of data provides a series of new challenges in terms of data storage, manipulation and presentation. The management and analysis of these volumes of data constitute one of the major bottlenecks currently associated with this type of technology. Bioinformatics teams around the world are developing databases that can be used to record all of the information relevant to a microarray experiment which will allow results to be combined and queried across these large data sets (Brazma *et al.* 2002).

One example of a microarray gene expression database has been developed at the European Bioinformatics Institute (EBI) in collaboration with the German

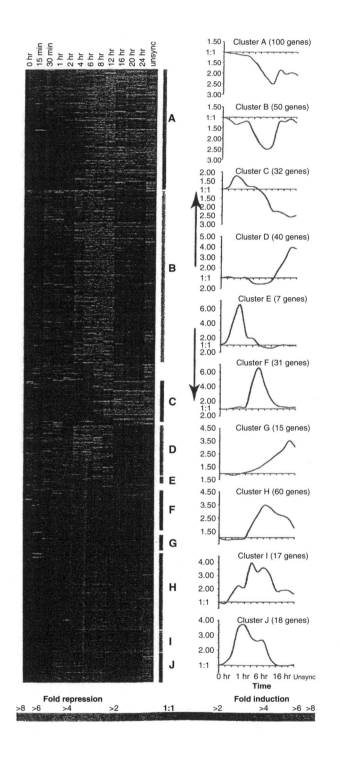

Cancer Research Centre. ArrayExpress was originally posted on the internet in November 1999 (http://www.ebi.ac.uk/arrayexpress/). Another resource for handling all aspects of microarray data has been developed at the University of Manchester, UK. This suite of software (http://bioinf.man.ac.uk/microarray/resources.html) is provided free to academic users and is widely used for microarray laboratory informatics support. Many other institutions have developed their own bioinformatics support to handle the databases that are an essential part of any large-scale microarray facility. There are also several commercial packages that can handle the data storage needed in this environment.

Another example is ArrayDB which has been developed at the National Human Genome Research Institute (http://genome.nhgri.nih.gov/arraydb/) (Ermolaeva *et al.* 1998). This relational database stores information including image data, experiment data (printing arrays, probes), and clones (IMAGE Clone ID, Title, UniGene cluster). This software/database is freely available to the public.

Reporting standards

As microarrays are implemented in many different fields a need has arisen to standardize the way the results are reported. In an effort to provide some guidelines to authors, editors and reviewers of papers describing microarray gene expression, members of the international Microarray Gene Expression Data (MGED) group published an open letter to scientific journals proposing standards for publication of microarray experiments (Brazma *et al.* 2001). This letter attempts to define the Minimum Information About a Microarray Experiment (MIAME) that should be included in a publication in an effort to provide some

Figure 8.8 Cluster image showing the different classes of gene expression profiles. Five hundred and seventeen genes whose mRNA levels changed in response to serum stimulation were selected for cluster analysis. These genes were clustered hierarchically into groups on the basis of the similarity of their expression profiles by the procedure of Eisen *et al.* (1998). The expression pattern of each gene in this set is displayed here as a horizontal strip. For each gene, the ratio of mRNA levels in fibroblasts at the indicated time after serum stimulation ('unsync' denotes exponentially growing cells) to its level in the serum-deprived (time zero) fibroblasts is represented by a colour, according to the colour scale at the bottom, The graphs show the average expression profiles for the genes in the corresponding 'cluster' (indicated by the letters A to J and the colour coding)' In every case examined, when a gene was represented by more than one array element, the multiple representations in this set were seen to have identical or very similar expression profiles, and the profiles corresponding to these independent measurements clustered either adjacent or very close to each other, pointing to the robustness of the clustering algorithm in grouping genes with very similar patterns of expression. Reproduced from Iyer *et al.* (1999) with permission

standards for microarray data. The MGED have established a website (http://www.mged.org) that includes a checklist of the points considered essential MIAME (see Table 8.3). An updated version of this checklist can be found at http://www.mged.org/Workgroups/MIAME/miame_checklist.html.

The international scientific community has welcomed these attempts to provide standards for the reporting of microarray experiments and most of the top international journals have adopted these standards (Anon 2002a, 2002b; Ball *et al.* 2002).

Applications to genomic analysis

Microarrays are being implemented in a wide range of fields for the analysis of gene expression, genomic alterations, mutations in DNA or single nucleotide polymorphisms to answer many varied questions. The initial publication that reported the use of microarrays analysed glass arrays containing 45 spots of PCR product from *Arabidopsis thaliana* cDNA clones hybridized with fluorescently-labelled cDNA (Schena *et al.* 1995). Since this original publication the size of the arrays has progressively increased. Differential gene expression, measured by microarrays, has been reported in many cell lines including the response of human fibroblasts to serum (Iyer *et al.* 1999) and the heat shock response of human T (Jurkat) cells (Schena *et al.* 1996). An array of 6400 elements representing the whole genome of *Saccharomyces cerevisiae* (Spellman *et al.* 1998) has been used to provide a measurement of the levels of the expression of each gene as the cells go through the cell cycle (see also Chapter 9).

In fields where the genome sequence is only partially complete, array technology is also being implemented. A shotgun DNA microarray was constructed from random inserts from a *Plasmodium falciparum* genomic library (Hayward *et al.* 2000). Arrays are also being created from cDNAs that exist only as open reading frames in sequenced genomes. Chervitz and colleagues (1998) compared the genomes of yeast (6000 genes) and worms (18 000 genes) and found that the regions encoding core functions were conserved and represented a similar number of genes in the two organisms. This parallel approach of comparing the genomes of model organisms is predicted to provide a means of accurately annotating coding sequence and assigning tentative function to the genes within each genome.

Research

Gene expression transcript profiling is expected to provide insights into disease mechanisms, provide clinical markers for disease classification and also identify potential targets for therapeutic interventions. Cancer has been a logical focus for

Table 8.3 The MIAME Checklist

Experiment design
- Type of experiment: for example, is it a comparison of normal vs. diseased tissue, a time course, or is it designed to study the effects of a gene knockout?
- Experimental factors: the parameters or conditions tested, such as time, dose or genetic variation.
- The number of hybridizations performed in the experiment.
- The type of reference used for the hybridizations, if any.
- Hybridization design: if applicable, a description of the comparisons made in each hybridization, whether to a standard reference sample, or between experimental samples. An accompanying diagram or table may be useful.
- Quality control steps taken: for example, replicates or dye swaps.
- URL of any supplemental websites or database accession numbers.

Samples used, extract preparation and labelling
- The origin of the biological sample (for instance, name of the organism, the provider of the sample) and its characteristics: for example, gender, age, developmental stage, strain, or disease state.
- Manipulation of biological samples and protocols used: for example, growth conditions, treatments, separation techniques.
- Protocol for preparing the hybridization extract: for example, the RNA or DNA extraction and purification protocol.
- Labeling protocol(s).
- External controls (spikes).

Hybridization procedures and parameters
- The protocol and conditions used during hybridization, blocking and washing.

Measurement data and specifications
- The quantitations based on the images.
- The set of quantitations from several arrays upon which the authors base their conclusions. While access to images of raw data is not required (although its value is unquestionable), authors should make every effort to provide the following:
  - Type of scanning hardware and software used: this information is appropriate for a materials and methods section.
  - Type of image analysis software used: specifications should be stated in the materials and methods.
  - A description of the measurements produced by the image-analysis software and a description of which measurements were used in the analysis.
  - The complete output of the image analysis *before* data selection and transformation (spot quantitation matrices).
  - Data selection and transformation procedures.
  - Final gene expression data table(s) used by the authors to make their conclusions *after* data selection and transformation (gene expression data matrices).

Array design
- General array design, including the platform type (whether the array is a spotted glass array, an *in situ* synthesized array, etc.); surface and coating specifications (when known – often commercial suppliers do not provide this data); and the availability of the array (the name or make of commercially available arrays).

(Continues)

Table 8.3   (*Continued*)

- For each feature (spot) on the array, its location on the array and the ID of its respective reporter (molecule present on each spot) should be given.
- For each reporter, its type (e.g. cDNA or oligonucleotide) should be given, along with information that characterizes the reporter molecule unambiguously, in the form of appropriate database reference(s) and sequence (if available).
- For commercial arrays: a reference to the manufacturer should be provided, including a catalogue number and references to the manufacturer's website if available.
- For non-commercial arrays, the following details should be provided:

  - The source of the reporter molecules: for example, the cDNA or oligo collection used, with references.
  - The method of reporter preparation.
  - The spotting protocols used, including the array substrate, the spotting buffer, and any post-printing processing, including cross-linking.
  - Any additional treatment performed prior to hybridization.

initial studies using microarray technology because of its clinical importance, the relatively easily accessible material as RNA samples and the many fundamental genetic derangements already identified in its pathogenesis. Researchers studying other disease states are now beginning to make progress using microarray technology. Cardiovascular diseases may not involve the large changes in gene transcription observed in cancerogenic cells but identification of alterations in the gene expression profiles reflecting primary, secondary or treatment related effects will be important in understanding the mechanisms and management of these disease states. Initial studies have identified gene expression changes associated with cardiac hypertrophy (Hwang *et al.* 2000; Friddle *et al.* 2000), myocardial infarction (Stanton *et al.* 2000), human heart failure (Yang *et al.* 2000; Haase *et al.* 2002; Barrans *et al.* 2001; Hwang *et al.* 2002) and primary pulmonary hypertension (Geraci *et al.* 2002). The gene response profiles to beta-blocking agents were examined by Lowes and colleagues (2002) and recently Aronow and colleagues (2001) used microarrays to identify common and distinct gene expression patterns occurring in samples from disease states with grossly similar phenotypes which arise through different mechanisms. The information obtained from these and other studies is enhancing our knowledge of the mechanisms of cardiovascular disease (Liu *et al.* 2001; Paoni and Lowe 2001; Cook *et al.* 2002; Cook and Rosenzweig 2002).

Microarrays are also being used to investigate disease susceptibility genes for complex disease states (Lyons 2002). Karp and colleagues used a combination of microarrays and quantitative trait locus analysis to identify the gene encoding complement factor 5 as a susceptibility locus in a murine experimental allergic asthma model (Karp *et al.* 2000), whilst Eaves and colleagues used a combination of microarrays and congenic strain analysis to study genes linked with diabetes susceptibility in the non-obese diabetic mouse model (Eaves *et al.*

2002). Some of the potential applications for microarrays in immunological studies were described in a recent review by Glynne and colleagues (Glynne *et al.* 2000b).

Pharmaceutical companies are investing heavily in this technology for the development of drug targets. The potential to identify genes that are differentially expressed will have a profound effect on the way new drug targets are identified. Glynne and colleagues used microarrays to identify genes that had altered expression profiles between anergic and activated B cells (Glynne *et al.* 2000a). They were able to identify genes that were involved in self-tolerance in the immune system and with this information were subsequently able to assess the effectiveness of an immunosuppressive drug in recreating this state. This study demonstrates one of the ways in which expression profiling can be incorporated in the drug development. Access to the appropriate cDNAs or oligonucleotide design for the establishment of microarrays for determining these expression profiles will be an important factor as these techniques become more widespread. Many companies have access to very large collections of proprietary ESTs that will provide them with unique material to aid this drug discovery process.

The high density oligonucleotide arrays are of interest to researchers developing techniques for high throughput analysis of SNPs. Each oligonucleotide on the array represents an allele specific probe. The perfectly matched sequences will hybridize more efficiently than the mis-matched probes resulting in a stronger fluorescent signal. The resultant intensity pattern can identify DNA alterations such as polymorphism or mutations, insertions or deletions (Chee *et al.* 1996; Lipshutz *et al.* 1999). A set of short oligonucleotide probes that cover an entire DNA fragment can be arrayed on a chip to enable sequencing by hybridization. Over 100 such arrays have been used to screen SNPs covering more than 2 Mb of genomic DNA (Wang *et al.* 1998). In a similar way researchers have been able to screen thousands of SNPs very rapidly using chip-based resequencing (Hacia *et al.* 1999; Hacia 1999).

Microarray technology is also being explored for the study of food-borne pathogens and bacterial identification (Al-Khaldi *et al.* 2002). Twenty to 70 bp oligonucleotide microarrays are being used to make it possible rapidly to detect and identify numerous pathogens and rapidly to look for the presence of virulence factors and antibiotic resistance genes. The potential automation of this technology means that it could have a role in food processing facilities and food safety monitoring agencies to improve the safety of our food supplies.

Several studies have demonstrated the potential for microarray technology to be implemented in the analysis of microbial community structure, function and dynamics (de Saizieu *et al.* 1998; Lockhart *et al.* 1996; Wu *et al.* 2001; Small *et al.* 2001). The development of microarray technology for the study of environmental samples is in the early stages. There are a number of challenges regarding the specificity and sensitivity of the arrays that are available for this field and the bioinformatics tools that have been developed for the analysis of

complex environmental samples that must be overcome before the full potential of microarrays for microbial ecology studies will be realized (Zhou and Thompson 2002).

Diagnostic

Heller and colleagues (1997) demonstrated the suitability of cDNA microarrays for profiling diseases and identifying disease-related genes when they compared gene expression patterns in tissue samples of rheumatoid arthritis and inflammatory bowel disease (Figure 8.9). There are now many small clinical

(a)

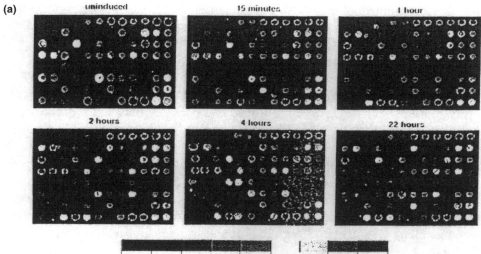

(b)

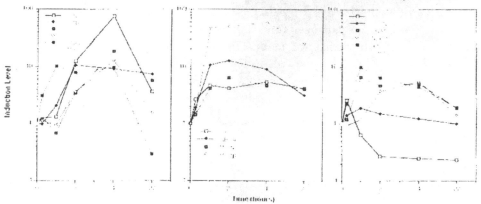

studies that suggest microarrays may contribute valuable insights into disease classification.

One of the more advanced applications where microarrays are predicted to have the potential to aid diagnosis is in the subclassing of malignant lesions. Alizadeh and colleagues demonstrated the classification of B cell malignancies using the gene expression profiles of a collection of cDNAs (Alizadeh *et al.* 2000). They were able to sub-classify histopathologically indistinguishable diffuse large B cell lymphomas into two groups with expression patterns for germinal centre or activated B cells (Figure 8.10). These groupings were thought to be predictive of disease progression (Figure 8.11) (Alizadeh *et al.* 2000; Rosenwald *et al.* 2002). Golub and colleagues demonstrated that expression profiles of clusters of mRNAs were able to distinguish between acute myeloid leukaemia and acute lymphocytic leukaemia (Golub *et al.* 1999). More recently, Alizadeh and Staudt created a 'lymphochip' for the expression profiling of normal and transformed human lymphoid cells (Alizadeh *et al.* 2000; Alizadeh and Staudt 2000). They assembled and arrayed ~18 000 unique lymphoid cDNAs, a large percentage of which were novel ESTs, and were able to detect clusters of mRNAs whose expression patterns could be used to differentiate all of the known leukaemia classes. They also identified expression patterns that distinguished two previously unrecognized subclasses of diffuse large-cell lymphoma that divided these into different clinical regimes (Alizadeh *et al.* 2000; Alizadeh and Staudt 2000).

cDNA microarrays have been used in many types of human cancer to distinguish histopathologically similar tumours and to identify previously unrecognized subclasses (van't Veer *et al.* 2002; Perou *et al.* 2000; Dhanasekaran *et al.* 2001; Bertucci *et al.* 2000; Takahashi *et al.* 2001; Gordon *et al.* 2002; Bittner *et al.* 2000). Wang and colleagues used multiphoton microscopy and cDNA microarrays to identify potentially important indicators for breast cancer invasion and metastasis in nonmetastatic and metastatic cells in culture and within live primary tumours (Wang *et al.* 2002). Their study showed that aligning the *in vivo* cell behaviour with patterns of gene expression could provide new information about

Figure 8.9 Gene expression profiling over a time course following LPS/PMA stimulation of MM6 cells. (A) pseudocolour representation of fluorescent scans corresponding to gene expression levels at each time point. The array is made up of nine *Arabidopsis* control probes and 86 human cDNA probes, the majority of which are genes with known or suspected involvement in inflammation. The colour bars provide a comparative calibration scale between the arrays and are derived from the *Arabidopsis* mRNA samples that are introduced in equal amounts during probe preparation. Fluorescent targets were made by labelling untreated MM6 cells or LPS and PMA treated cells. mRNA was isolated at indicated times after stimulation. (B I–III) The two colour samples were co-hybridized, and microarray scans provided the data for the levels of select transcripts at different time points relative to abundance at time zero. The analysis was performed using normalized data collected from eight-bit images. Reproduced from Heller *et al.* (1997) with permission

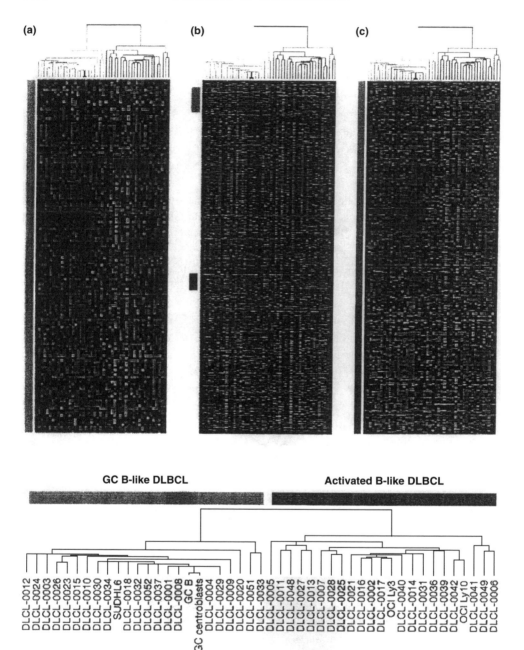

Figure 8.10 Discovery of diffuse large B-cell lymphoma (DLBCL) subtypes by gene expression profiling. The used in this clustering analysis are shown at the bottom. (a) Hierarchical clustering of DLBCL cases (blue and orange) and germinal centre B cells (black) based on the genes previously shown (Alizadeh *et al.* 2000) to identify the germinal B-cell gene expression signature. Two DLBCL subgroups, GC B-like (orange) and activated B-like DLBCL (blue) were defined by this process. (b) Discovery

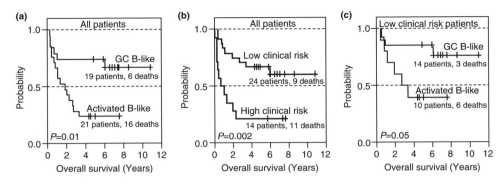

Figure 8.11 Clinically distinct DLBCL subgroups defined by gene expression profiling. (a) Kaplan-Meier plot of overall survival of DLBCL patients grouped on the basis of gene expression profiling. (b) Kaplan-Meier plot of overall survival of DLBCL patients grouped according to the International Prognostic Index (IPI). Low clinical risk patients (IPI score 0–2) and high clinical risk patients (IPI score 3–5) are plotted separately. (c), Kaplan-Meier plot of overall survival of low clinical risk DLBCL patients (IPI score 0–2) grouped on the basis of their gene expression profiles. Reproduced from Alizadeh *et al.* (2000) with permission

the role of the microenvironment of carcinoma cells and the molecular mechanisms driving cell behaviour.

In breast carcinomas (Ginestier *et al.* 2002), urinary bladder cancer (Richter *et al.* 2000) and renal cell carcinomas (Moch *et al.* 1999) the profiles obtained with cDNA microarrays has been coupled with the recently developed tissue microarray technology (Kononen *et al.* 1998) to obtain complementary information about the gene expression profiles and also other distinct information about protein expression. This coupling of techniques may have important implications for the next step in the validation of microarray data and the potential for the introduction of these types of technologies into routine hospital settings (Lakhani and Ashworth 2001; Bertucci *et al.* 2001).

Microarray technology is also being considered for other diagnostic purposes. Neo Gen Screening has pioneered the development of a DNA microarray for newborn screening for use in the identification programme looking for genetic

---

of genes that are selectively expressed in GC B-like DLBCL and activated B-like DLBCL. All genes from microarray experiments described previously (Alizadeh *et al.* 2000), with the exception of the genes in the proliferation, T-cell and lymph-node gene expression signatures, were ordered by hierarchical clustering while maintaining the order of samples determined in (a). Genes selectively expressed in GC B-like DLBCL (orange) and activated B-like DLBCL (blue) are indicated. (c) Hierarchical clustering of the genes selectively expressed in GC B-like DLBCL and activated B-like DLBCL, which was determined from (b). Reproduced from Alizadeh *et al* (2000) with permission

abnormalities leading to diseases such as sickle cell disease, α-1-antitrypsin deficiency and Factor V Leiden (Dobrowolski *et al.* 1999).

The future?

The development of microarray technology has provided an important step forward in our ability to perform experiments to understand the identity and function of the complete genome. The next step is to be able to understand the functional outcomes associated with particular expression profiles and to try to link these with the expression of the mRNA in particular cell or tissue types. The correlation of the gene expression profiles with, first, the corresponding protein expression patterns and then the *in vivo* cell behaviour are huge advances in expanding our knowledge of the molecular mechanisms driving the cell systems that we are studying. Already strategies and software for data mining that allow the correlation of information about gene expression profiles and drug entity structure-activity studies are being developed (Blower *et al.* 2002). This information aims to identify relationships between the structural features, at the molecular level, of new drug compounds and the gene or gene products that predict the activity of the compound against the cell. This development will be a major advance for the refinement of the development of new drugs.

The parallel development of arrays of proteins, peptides, antibodies and other molecules is enabling high-throughput studies in these areas which complement the current areas of intense research. Microarray technology, coupled with the other emerging high-throughput technologies will provide further understanding of the underlying causes of pathology and will also provide new tools for directed drug development and gene therapies to preferentially target genes for disease intervention.

References

Anon (2002) A guide to microarray experiments–an open letter to the scientific journals. *Lancet* **360**: 1019; author reply 1019.

Anon (2002) Microarray standards at last. *Nature* **419**: 323.

Adams, M.D., Kelley, J.M., Gocayne, J.D., Dubnick, M., *et al.* (1991) Complementary DNA sequencing: expressed sequence tags and human genome project. *Science* **252**: 1651–1656.

Adessi, C., Matton, G., Ayala, G., Turcatti, G., *et al.* (2000) Solid phase DNA amplification: characterisation of primer attachment and amplification mechanisms. *Nucleic Acids Res* **28**(20): E87.

Alizadeh, A., Eisen, M., Davis, R., Ma, C., *et al.* (2000) Distinct types of diffuse large B-cell lymphoma identified by gene expression profiling. *Nature* **403**: 503–511.

Alizadeh, A.A. and Staudt, L.M. (2000) Genomic-scale gene expression profiling of normal and malignant immune cells. *Current Opin Immunol* **12**: 219–225.

Al-Khaldi, S.F., Martin, S.A., Rasooly, A. and Evans, J.D. (2002) DNA microarray technology used for studying foodborne pathogens and microbial habitats: minireview. *J AOAC Int* **85**(4): 906–910.

Alwine, J.C., Kemp, D.J. and Stark, G.R. (1977) Method for detection of specific RNAs in agarose gels by transfer to diazobenzyloxymethyl-paper and hybridization with DNA probes. *Proc Natl Acad Sci USAmerica* **74**(12): 5350–5354.

Aronow, B.J., Toyokawa, T., Canning, A., Haghighi, K., *et al.* (2001) Divergent transcriptional responses to independent genetic causes of cardiac hypertrophy. *Physiol Genomics* **6**(1): 19–28.

Ball, C.A., Sherlock, G., Parkinson, H., Rocca-Sera, P., *et al.* (2002) Standards for Microarray Data. *Science* **298**: 539b–.

Barrans, J.D., Stamatiou, D. and Liew, C. (2001) Construction of a human cardiovascular cDNA microarray: portrait of the failing heart. *Biochem Biophys Res Commun* **280**(4): 964–969.

Beier, M. and Hoheisel, J.D. (1999) Versatile derivatisation of solid support media for covalent bonding on DNA-microchips. *Nucleic Acids Res* **27**(9): 1970–1977.

Bertucci, F., Houlgatte, R., Benziane, A., Granjeaud, S., *et al.* (2000) Gene expression profiling of primary breast carcinomas using arrays of candidate genes. *Hum Mol Genet* **9**(20): 2981–2991.

Bertucci, F., Houlgatte, R., Nguyen, C., Viens, P., *et al.* (2001) Gene expression profiling of cancer by use of DNA arrays: how far from the clinic? *Lancet Oncol* **2**(11): 674–682.

Bittner, M., Meltzer, P., Chen, Y., Jiang, Y., *et al.* (2000) Molecular classification of cutaneous malignant melanoma by gene expression profiling. *Nature* **406**: 536–540.

Blower, P.E., Yang, C., Fligner, M.A., Verducci, J.S., *et al.* (2002) Pharmacogenomic analysis: correlating molecular substructure classes with microarray gene expression data. *Pharmacogenomics J* **2**(4): 259–271.

Bowtell, D.D. (1999) Options available – from start to finish – for obtaining expression data by microarray. *Nature Genetics* **21**(1 Suppl): 25–32.

Brazma, A., Hingamp, P., Quackenbush, J., Sherlock, G., *et al.* (2001) Minimum information about a microarray experiment (MIAME) – toward standards for microarray data. *Nat Genet* **29**(4): 365–371.

Brazma, A., Sarkans, U., Robinson, A., Vilo, J., *et al.* (2002) Microarray data representation, annotation and storage. *Adv Biochem Eng Biotechnol* **77**: 113–139.

Brazma, A. and Vilo, J. (2000) Gene expression data analysis. *FEBS Lett* **480**(1): 17–24.

Brazma, A. and Vilo, J. (2001) Gene expression data analysis. *Microbes Infect* **3**(10): 823–829.

Chan, V., Graves, D.J. and McKenzie, S.E. (1995) The biophysics of DNA hybridization with immobilized oligonucleotide probes. *Biophys J* **69**(6): 2243–2255.

Chee, M., Yang, R., Hubbell, E. and Berno, A. (1996) Accessing genetic information with high-density DNA arrays. *Science* **274**: 610–614.

Chen, Y., Dougherty, E. and Bittner, M. (1997) Ratio-based decisions and the quantitative analysis of cDNA microarray images. *J Biomedical Optics* **2**: 364–374.

Chervitz, S., Aravind, L., Sherlock, G., Ball, C., *et al.* (1998) Comparison of the complete protein sets of worm and yeast: orthology and divergence. *Science* **282**: 2022–2028.

Cheung, V.G., Morley, M., Aguilar, F., Massimi, A. *et al.* (1999) Making and reading microarrays. *Nature Genetics* **21**(1 Suppl): 15–19.

Clark, T.A., Sugnet, C.W. and Ares, M., Jr. (2002) Genomewide analysis of mRNA processing in yeast using splicing-specific microarrays. *Science* **296**: 907–910.

Cook, S.A., Matsui, T., Li, L. and Rosenzweig, A. (2002) Transcriptional effects of chronic Akt activation in the heart. *J Biol Chem* **277**(25): 22528–22533.

Cook, S.A. and Rosenzweig, A. (2002) DNA microarrays: implications for cardiovascular medicine. *Circ Res* **91**(7): 559–564.

Coombes, K.R., Highsmith, W.E., Krogmann, T.A., Baggerly, K.A., *et al.* (2002) Identifying and quantifying sources of variation in microarray data using high-density cDNA membrane arrays. *J Comput Biol* **9**(4): 655–669.

de Saizieu, A., Certa, U., Warrington, J., Gray, C., *et al.* (1998) Bacterial transcript imaging by hybridization of total RNA to oligonucleotide arrays. *Nat Biotechnol* **16**(1): 45–48.

Dhanasekaran, S.M., Barrette, T.R., Ghosh, D., Shah, R., *et al.* (2001) Delineation of prognostic biomarkers in prostate cancer. *Nature* **412**: 822–826.

Dobrowolski, S.F., Banas, R.A., Naylor, E.W., Powdrill, T. and Thakkar, D. (1999) DNA microarray technology for neonatal screening. *Acta Paediatrica. Suppl* **88**(432): S61–S64.

Duggan, D.J., Bittner, M., Chen, Y., Meltzer, P. and Trent, J.M. (1999) Expression profiling using cDNA microarrays. *Nature Genetics* **21**(1 Suppl): 10–14.

Eaves, I.A., Wicker, L.S., Ghandour, G., Lyons, P.A., *et al.* (2002) Combining mouse congenic strains and microarray gene expression analyses to study a complex trait: the NOD model of type 1 diabetes. *Genome Res* **12**(2): 232–243.

Eberwine, J., Yeh, H., Miyashiro, K., Cao, Y., *et al.* (1992) Analysis of gene expression in single live neurons. *Proc Natl Acad Sci USA* **89**(7): 3010–3014.

Eisen, M.B. and Brown, P.O. (1999) DNA arrays for analysis of gene expression. *Methods Enzymol* **303**: 179–205.

Eisen, M.B., Spellman, P.T., Brown, P.O. and Botstein, D. (1998) Cluster analysis and display of genome-wide expression patterns. *Proc Natl Acad Sci USA* **95**(25): 14863–14868.

Ekins, R. (1996) Microspot(R), an array based ligand assay – is this the ultimate micro-analytical technology. *J Clin Ligand Assay* **19**(2): 145–156.

Ekins, R. and Chu, F.W. (1999) Microarrays: their origins and applications. *Trends Biotechnol* **17**(6): 217–218.

Ekins, R., Chu, F. and Micallef, J. (1989) High specific activity chemiluminescent and fluorescent markers: their potential application to high sensitivity and 'multi-analyte' immunoassays. *J Biolumin Chemilumin* **4**(1): 59–78.

Ekins, R.P. (1998) Ligand assays: from electrophoresis to miniaturized microarrays. *Clin Chem* **44**(9): 2015–2030.

Ermolaeva, O., Rastogi, M., Pruitt, K.D., Schuler, G.D., *et al.* (1998) Data management and analysis for gene expression arrays. *Nature Genetics* **20**(1): 19–23.

Feldman, A.L., Costouros, N.G., Wang, E., Qian, M., *et al.* (2002) Advantages of mRNA amplification for microarray analysis. *Biotechniques* **33**(4): 906–912, 914.

Franssen-van Hal, N.L., Vorst, O., Kramer, E., Hall, R.D. and Keijer, J. (2002) Factors influencing cDNA microarray hybridization on silylated glass slides. *Anal Biochem* **308**(1): 5–17.

Friddle, C.J., Koga, T., Rubin, E.M. and Bristow, J. (2000) Expression profiling reveals distinct sets of genes altered during induction and regression of cardiac hypertrophy. *Proc Natl Acad Sci USA* **97**(12): 6745–6750.

Geraci, M.W., Hoshikawa, Y., Yeager, M., Golpon, H., *et al.* (2002) Gene expression profiles in pulmonary hypertension. *Chest* **121**(3 Suppl): 104S–105S.

Ginestier, C., Charafe-Jauffret, E., Bertucci, F., Eisinger, F., *et al.* (2002) Distinct and complementary information provided by use of tissue and DNA microarrays in the study of breast tumor markers. *Am J Pathol* **161**(4): 1223–1233.

Glynne, R., Akkaraju, S., Healy, J.I., Rayner, J., *et al.* (2000) How self-tolerance and the immunosuppressant drug FK506 prevent B-cell mitogensis. *Nature* **403**: 672–676.

Glynne, R.J., Ghandour, G. and Goodnow, C.C. (2000) Genomic-scale gene expression analysis of lymphocyte growth, tolerance and malignancy. *Current Opin Immunol* **12**: 210–214.

Golub, T., Slonim, D., Tamayo, P., Huard, C., *et al.* (1999) Molecular classification of cancer: class discovery and class prediction by gene expression monitoring. *Science* **286**: 531–537.

Gordon, G.J., Jensen, R.V., Hsiao, L.L., Gullans, S.R., *et al.* (2002) Translation of microarray data into clinically relevant cancer diagnostic tests using gene expression ratios in lung cancer and mesothelioma. *Cancer Res* **62**(17): 4963–4967.

Haase, D., Lehmann, M.H., Korner, M.M., Korfer, R., *et al.* (2002) Identification and validation of selective upregulation of ventricular myosin light chain type 2 mRNA in idiopathic dilated cardiomyopathy. *Eur J Heart Fail* **4**(1): 23–31.

Hacia, J.G. (1999) Resequencing and mutational analysis using oligonucleotide microarrays. *Nature Genetics* **21**(1 Suppl): 42–7.

Hacia, J.G., Fan, J.B., Ryder, O., Jin, L., *et al.* (1999) Determination of ancestral alleles for human single-nucleotide polymorphisms using high-density oligonucleotide arrays [see comments]. *Nature Genetics* **22**(2): 164–167.

Hay, I.D., Bayer, M.F., Kaplan, M.M., Klee, G.G., *et al.* (1991) American Thyroid Association assessment of current free thyroid hormone and thyrotropin measurements and guidelines for future clinical assays. The Committee on Nomenclature of the American Thyroid Association. *Clin Chem* **37**(11): 2002–2008.

Hayward, R., Derisi, J., Alfadhli, S., Kaslow, D., *et al.* (2000) Shotgun DNA microarrays and stage-specific gene expression in *Plasmodium falciparum* malaria. *Molecular Microbiol* **35**(1): 6–14.

Hegde, P., Qi, R., Abernathy, K., Gay, C., *et al.* (2000) A concise guide to cDNA microarray analysis. *Biotechniques* **29**(3): 548–550, 552–554, 556 passim.

Heller, R.A., Schena, M., Chai, A., Shalon, D., *et al.* (1997) Discovery and analysis of inflammatory disease-related genes using cDNA microarrays. *Proc Natl Acad Sci USA* **94**(6): 2150–2155.

Hess, K.R., Zhang, W., Baggerly, K.A., Stivers, D.N. and Coombes, K.R. (2001) Microarrays: handling the deluge of data and extracting reliable information. *Trends Biotechnol* **19**(11): 463–468.

Hughes, T.R., Mao, M., Jones, A.R., Burchard, J., *et al.* (2001) Expression profiling using microarrays fabricated by an ink-jet oligonucleotide synthesizer. *Nat Biotechnol* **19**(4): 342–347.

Hughes, T.R. and Shoemaker, D.D. (2001) DNA microarrays for expression profiling. *Curr Opin Chem Biol* **5**(1): 21–25.

Hwang, D.M., Dempsey, A.A., Lee, C.Y. and Liew, C.C. (2000) Identification of differentially expressed genes in cardiac hypertrophy by analysis of expressed sequence tags. *Genomics* **66**(1): 1–14.

Hwang, J.J., Allen, P.D., Tseng, G.C., Lam, C.W. *et al.* (2002) Microarray gene expression profiles in dilated and hypertrophic cardiomyopathic end-stage heart failure. *Physiol Genomics* **10**(1): 31–44.

Iyer, V.R., Eisen, M.B., Ross, D.T., Schuler, G. *et al.* (1999) The transcriptional program in the response of human fibroblasts to serum. *Science* **283**: 83–87.

Kafatos, F.C., Jones, C.W. and Efstratiadis, A. (1979) Determination of nucleic acid sequence homologies and relative concentrations by a dot hybridization procedure. *Nucleic Acids Res* **7**(6): 1541–1552.

Karp, C.L., Grupe, A., Schadt, E., Ewart, S.L. *et al.* (2000) Identification of complement factor 5 as a susceptibility locus for experimental allergic asthma. *Nat Immunol* **1**(3): 221–226.

Kononen, J., Bubendorf, L., Kallioniemi, A., Barlund, M. *et al.* (1998) Tissue microarrays for high-throughput molecular profiling of tumor specimens [see comments]. *Nature Med* **4**(7): 844–847.

Kuang, W.W., Thompson, D.A., Hoch, R.V. and Weigel, R.J. (1998) Differential screening and suppression subtractive hybridization identified genes differentially expressed in an estrogen receptor-positive breast carcinoma cell line. *Nucleic Acids Res* **26**(4): 1116–1123.

Lakhani, S.R. and Ashworth, A. (2001) Microarray and histopathological analysis of tumours: the future and the past? *Nat Rev Cancer* **1**(2): 151–157.

Lander, E.S. (1999) Array of hope. *Nature Genetics* **21**(1 Suppl): 3–4.

Lander, E.S., Linton, L.M., Birren, B., Nusbaum, C. *et al.* (2001) Initial sequencing and analysis of the human genome. *Nature* **409**: 860–921.

Lennon, G.G. and Lehrach, H. (1991) Hybridization analyses of arrayed cDNA libraries. *Trends in Genetics* **7**(10): 314–317.

Liang, P. and Pardee, A.B. (1992) Differential display of eukaryotic messenger RNA by means of the polymerase chain reaction. *Science* **257**: 967–971.

Lipshutz, R.J., Fodor, S.P., Gingeras, T.R. and Lockhart, D.J. (1999) High density synthetic oligonucleotide arrays. *Nature Genetics* **21**(1 Suppl): 20–24.

Lipshutz, R.J., Morris, D., Chee, M., Hubbell, E. *et al.* (1995) Using oligonucleotide probe arrays to access genetic diversity. *Biotechniques* **19**(3): 442–447.

Lisitsyn, N. and Wigler, M. (1993) Cloning the differences between two complex genomes. *Science* **259**: 946–951.

Liu, T., Lai, H., Wu, W., Chinn, S. and Wang, P.H. (2001) Developing a strategy to define the effects of insulin-like growth factor-1 on gene expression profile in cardiomyocytes. *Circ Res* **88**(12): 1231–1238.

Lockhart, D.J., Dong, H., Byrne, M.C., Follettie, M.T. *et al.* (1996) Expression monitoring by hybridization to high-density oligonucleotide arrays. *Nature Biotechnol* **14**(13): 1675–1680.

Lowes, B.D., Gilbert, E.M., Abraham, W.T., Minobe, W.A. *et al.* (2002) Myocardial gene expression in dilated cardiomyopathy treated with beta-blocking agents. *N Engl J Med* **346**(18): 1357–1365.

Luo, L., Salunga, R.C., Guo, H., Bittner, A. *et al.* (1999) Gene expression profiles of laser-captured adjacent neuronal subtypes. *Nature Med* **5**(1): 117–122.

Lyons, P.A. (2002) Gene-expression profiling and the genetic dissection of complex disease. *Curr Opin Immunol* **14**(5): 627–630.

Mir, K.U. and Southern, E.M. (1999) Determining the influence of structure on hybridization using oligonucleotide arrays. *Nat Biotechnol* **17**(8): 788–792.

Moch, H., Schraml, P., Bubendorf, L., Mirlacher, M. *et al.* (1999) High-throughput tissue microarray analysis to evaluate genes uncovered by cDNA microarray screening in renal cell carcinoma [see comments]. *Am J Pathol* **154**(4): 981–986.

Nagle, R.B. (2001) New molecular approaches to tissue analysis. *J Histochem Cytochem* **49**(8): 1063–1064.

Nees, M. and Woodworth, C.D. (2001) Microarrays: spotlight on gene function and pharmacogenomics. *Curr Cancer Drug Targets* **1**(2): 155–175.

Ohnmacht, G.A., Wang, E., Mocellin, S., Abati, A. *et al.* (2001) Short-term kinetics of tumor antigen expression in response to vaccination. *J Immunol* **167**(3): 1809–1820.

Ohyama, H., Zhang, X., Kohno, Y., Alevizos, I. *et al.* (2000) Laser capture microdissection-generated target sample for high-density oligonucleotide array hybridization. *Biotechniques* **29**(3): 530–536.

Paoni, N.F. and Lowe, D.G. (2001) Expression profiling techniques for cardiac molecular phenotyping. *Trends Cardiovasc Med* **11**(6): 218–221.

Perou, C.M., Sorlie, T., Eisen, M.B., van de Rijn, M. *et al.* (2000) Molecular portraits of human breast tumours. *Nature* **406**: 747–752.

Peterson, A.W., Heaton, R.J. and Georgiadis, R.M. (2001) The effect of surface probe density on DNA hybridization. *Nucleic Acids Res* **29**(24): 5163–5168.

Phimister, B. (1999) Going global. *Nat Genet* **21**(1 Suppl): 1.

Pietu, G., Alibert, O., Guichard, V., Lamy, B. *et al.* (1996) Novel gene transcripts preferentially expressed in human muscles revealed by quantitative hybridization of a high density cDNA array. *Genome Res* **6**(6): 492–503.

Puskas, L.G., Zvara, A., Hackler, L. Jr., Micsik, T. and van Hummelen, P. (2002) Production of bulk amounts of universal RNA for DNA microarrays. *Biotechniques* **33**(4): 898–900, 902, 904.

Quackenbush, J. (2001) Computational analysis of microarray data. *Nat Rev Genet* **2**(6): 418–427.

Quackenbush, J. (2002) Microarray data normalization and transformation. *Nat Genet* **32 Suppl**: 496–501.

Rajeevan, M.S., Dimulescu, I.M., Unger, E.R. and Vernon, S.D. (1999) Chemiluminescent analysis of gene expression on high-density filter arrays. *J Histochem Cytochem* **47**(3): 337–342.

Richter, A., Schwager, C., Hentze, S., Ansorge, W. *et al.* (2002) Comparison of fluorescent tag DNA labeling methods used for expression analysis by DNA microarrays. *Biotechniques* **33**(3): 620–628, 630.

Richter, J., Wagner, U., Kononen, J., Fijan, A. *et al.* (2000) High-throughput tissue microarray analysis of cyclin E gene amplification and overexpression in urinary bladder cancer. *Am J Pathol* **157**(3): 787–794.

Rosenwald, A., Wright, G., Chan, W.C., Connors, J.M. *et al.* and the Lymphoma/Leukemia Molecular Profiling Project (2002) The use of molecular profiling to predict survival after chemotherapy for diffuse large-B-cell lymphoma. *N Engl J Med* **346**(25): 1937–1947.

Schena, M., Heller, R.A., Theriault, T.P., Konrad, K. *et al.* (1998) Microarrays: biotechnology's discovery platform for functional genomics. *Trends Biotechnol* **16**(7): 301–306.

Schena, M., Shalon, D., Davis, R.W. and Brown, P.O. (1995) Quantitative monitoring of gene expression patterns with a complementary DNA microarray. *Science* **270**: 467–470.

Schena, M., Shalon, D., Heller, R., Chai, A., *et al.* (1996) Parallel human genome analysis: microarray-based expression monitoring of 1000 genes. *Proc Natl Acad Sci USA* **93**(20): 10614–10619.

Scherf, U., Ross, D.T., Waltham, M., Smith, L.H. *et al.* (2000) A gene expression database for the molecular pharmacology of cancer. *Nat Genet* **24**(3): 236–244.

Schuchhardt, J., Beule, D., Malik, A., Wolski, E., *et al.* (2000) Normalization strategies for cDNA microarrays. *Nucleic Acids Res* **28**(10): E47–e47.

Schulze, A. and Downward, J. (2001) Navigating gene expression using microarrays – a technology review. *Nat Cell Biol* **3**(8): E190–E195.

Shchepinov, M.S., Case-Green, S.C. and Southern, E.M. (1997) Steric factors influencing hybridisation of nucleic acids to oligonucleotide arrays. *Nucleic Acids Res* **25**(6): 1155–1161.

Sherlock, G. (2000) Analysis of large-scale gene expression data. *Current Opin Immunol* **12**: 201–205.

Shoemaker, D.D., Schadt, E.E., Armour, C.D., He, Y.D. *et al.* (2001) Experimental annotation of the human genome using microarray technology. *Nature* **409**: 922–927.

Singh-Gasson, S., Green, R.D., Yue, Y., Nelson, C. *et al.* (1999) Maskless fabrication of light-directed oligonucleotide microarrays using a digital micromirror array. *Nature Biotechnol* **17**(10): 974–978.

Small, J., Call, D.R., Brockman, F.J., Straub, T.M. and Chandler, D.P. (2001) Direct detection of 16S rRNA in soil extracts by using oligonucleotide microarrays. *Appl Environ Microbiol* **67**(10): 4708–4716.

Southern, E., Mir, K. and Shchepinov, M. (1999) Molecular interactions on microarrays. *Nature Genetics* **21**(1 Suppl): 5–9.

Southern, E.M. (1975) Detection of specific sequences among DNA fragments separated by gel electrophoresis. *J Molecular Biol* **98**(3): 503–517.

Spellman, P.T., Sherlock, G., Zhang, M.Q., Iyer, V.R. *et al.* (1998) Comprehensive identification of cell cycle-regulated genes of the yeast *Saccharomyces cerevisiae* by microarray hybridization. *Molecular Biol Cell* **9**(12): 3273–3297.

Stanton, L.W., Garrard, L.J., Damm, D., Garrick, B.L. *et al.* (2000) Altered patterns of gene expression in response to myocardial infarction. *Circ Res* **86**(9): 939–945.

Sterrenburg, E., Turk, R., Boer, J.M., van Ommen, G.B. and den Dunnen, J.T. (2002) A common reference for cDNA microarray hybridizations. *Nucleic Acids Res* **30**(21): e116.

Stillman, B.A. and Tonkinson, J.L. (2001) Expression microarray hybridization kinetics depend on length of the immobilized DNA but are independent of immobilization substrate. *Anal Biochem* **295**(2): 149–157.

Takahashi, M., Rhodes, D.R., Furge, K.A., Kanayama, H. *et al.* (2001) Gene expression profiling of clear cell renal cell carcinoma: gene identification and prognostic classification. *Proc Natl Acad Sci USA* **98**(17): 9754–9759.

Tefferi, A., Bolander, M.E., Ansell, S.M., Wieben, E.D. and Spelsberg, T.C. (2002) Primer on medical genomics. Part III: Microarray experiments and data analysis. *Mayo Clin Proc* **77**(9): 927–940.

van't Veer, L.J., Dai, H., van de Vijver, M.J., He, Y.D. *et al.* (2002) Gene expression profiling predicts clinical outcome of breast cancer. *Nature* **415**: 530–536.

Venter, J.C., Adams, M.D., Myers, E.W., Li, P.W. *et al.* (2001) The sequence of the human genome. *Science* **291**: 1304–1351.

Wang, D.G., Fan, J.B., Siao, C.J., Berno, A. *et al.* (1998) Large-scale identification, mapping, and genotyping of single-nucleotide polymorphisms in the human genome. *Science* **280**: 1077–1082.

Wang, W., Wyckoff, J.B., Frohlich, V.C., Oleynikov, Y. *et al.* (2002) Single cell behavior in metastatic primary mammary tumors correlated with gene expression patterns revealed by molecular profiling. *Cancer Res* **62**(21): 6278–6288.

Warrington, J.A., Nair, A., Mahadevappa, M. and Tsyganskaya, M. (2000) Comparison of human adult and fetal expression and identification of 535 housekeeping/maintenance genes. *Physiol Genomics* **2**(3): 143–147.

Wildsmith, S.E., Archer, G.E., Winkley, A.J., Lane, P.W. and Bugelski, P.J. (2001) Maximization of signal derived from cDNA microarrays. *Biotechniques* **30**(1): 202–206, 208.

Wildsmith, S.E. and Elcock, F.J. (2001) Microarrays under the microscope. *Mol Pathol* **54**(1): 8–16.

Wu, L., Thompson, D.K., Li, G., Hurt, R.A., Tiedje, J.M. and Zhou, J. (2001) Development and evaluation of functional gene arrays for detection of selected genes in the environment. *Appl Environ Microbiol* **67**(12): 5780–5790.

Yang, J., Moravec, C.S., Sussman, M.A., DiPaola, N.R. *et al.* (2000) Decreased SLIM1 expression and increased gelsolin expression in failing human hearts measured by high-density oligonucleotide arrays. *Circulation* **102**(25): 3046–3052.

Zhao, N., Hashida, H., Takahashi, N., Misumi, Y. and Sakaki, Y. (1995) High-density cDNA filter analysis: a novel approach for large-scale, quantitative analysis of gene expression. *Gene* **156**(2): 207–213.

Zhou, J. and Thompson, D.K. (2002) Challenges in applying microarrays to environmental studies. *Curr Opin Biotechnol* **13**(3): 204–207.

9

Overview of Differential Gene Expression by High-throughput Analysis

Kin-Ying To

Introduction

By comparing the individual mRNAs from different genotypes, developmental stages or growth conditions, genes can be identified that are differentially expressed and hence may have specific metabolic or morphogenetic functions. In addition, the study of differences in gene-expression patterns is also a powerful tool in understanding mechanisms of differentiation and development. In recent years, a variety of techniques have been developed to analyse differential gene expression, including comparative expressed sequence tag sequencing, differential display, cDNA-amplified fragment length polymorphism, subtractive hybridization, representational difference analysis, cDNA or oligonucleotide microarrays, and serial analysis of gene expression. These methods are typically used to identify genes that are critical for a developmental process, to identify genes that mediate cellular responses to a variety of chemical or physical stimuli, or to understand the molecular events resulting from mutations in particular genes (Carulli *et al.* 1998; Kozian and Kirschbaum 1999; To 2000; Kuhn 2001; Donson *et al.* 2002). This chapter is an overview of current methods for transcription imaging, with an emphasis on technologies that have been developed during the past few years and allowed the high-throughput analysis of mRNA expression profiles. In addition,

Molecular Analysis and Genome Discovery edited by Ralph Rapley and Stuart Harbron
© 2004 John Wiley & Sons, Ltd   ISBN 0 471 49847 5 (cased)   ISBN 0 471 49919 6 (pbk)

microarray-based plant functional genomic studies in my laboratory are used as examples to introduce the principle and application of DNA microarrays.

Differential display

Since the development of the differential display (DD) process in 1992 (Liang and Pardee 1992), the application of this process has expanded into nearly all areas of biological science. The concept of differential display is to use a limited number of arbitrary primers in combination with anchored oligo(dT) primers to amplify and visualize most of the mRNA in a cell. A schematic representation of mRNA differential display is shown in Figure 9.1. This technique is a four-step process: (1) Reverse transcription of mRNA with anchored oligo(dT)NG primers; (2) PCR amplification of cDNA with arbitrary primers; (3) Resolution of amplified cDNA by polyacrylamide gel electrophoresis (PAGE); and (4) Isolation of resolved fragments from the gel, followed by cloning and sequencing analysis. Using this technique, differentially expressed genes under several conditions (e.g. light-grown seedlings versus dark-grown seedlings, or green leaves versus yellow leaves) of rice plants have been identified by the presence or absence of bands on the gel (To 2000). These darkness-induced or senescence-associated genes may be useful in understanding the mechanisms of plant growth and development. The most critical point that should be considered when applying differential display is the source and purity of RNA. Poorly defined maintenance conditions for the cells and tissues might render further work difficult. Since 1992, there has been developed variations based on classical differential display (Matz and Lukyanov 1998).

Limitations of the classical differential display approach and associated methods are a high percentage of false-positive bands, lack of sensitivity and difficulties with reproducibility. In conclusion, differential display methods are relatively low in cost and simple, and are particularly good where the availability of RNA is limited. However, they are not very accurate in quantitatively profiling global levels of gene expression, as illustrated by the number of false-positives generated.

cDNA-Amplified fragment length polymorphism

To overcome the problems associated with differential display, cDNA-amplified fragment length polymorphism (cDNA-AFLP) was developed using stringent PCR conditions afforded by the ligation of adapters to restriction fragments and the use of specific primer sets (Vos *et al.* 1995; Bachem *et al.* 1996). The basic protocol of cDNA-AFLP (Figure 9.2) can be divided into seven steps:

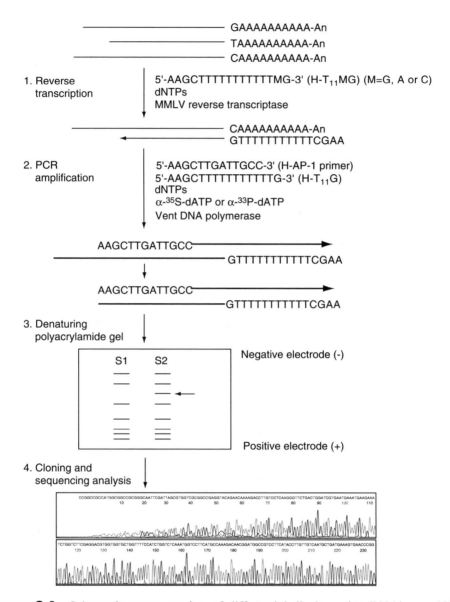

Figure 9.1 Schematic representation of differential display using RNAimage (Gen-Hunter, USA). Three one-base anchored oligo(dT) primers with *Hin*dIII restriction site (5′-AAGCTT-3′) were used in combination with a series of arbitrary 13 mers containing a *Hin*dIII restriction site by reverse transcription and amplify mRNA from a cell in the presence of isotope-labelled α-^{35}S-dATP or α-^{33}P-dATP. After amplification, the products were separated by electrophoresis on 6% denaturing polyacrylamide gel. The dried gel was then followed by autoradiography on X-Omat AK file (Kodak, USA). An arrow indicates the DNA fragment which is differentially expressed in sample S2 but not sample S1. The identified fragment was then isolated from the gel, and then dissolved in buffer. After brief purification, DNA fragment was cloned into vector and then sequenced

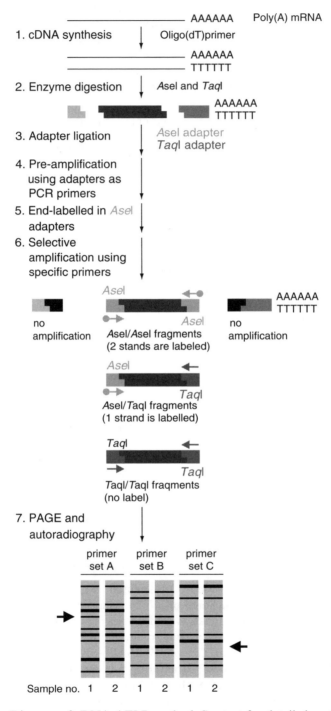

Figure 9.2 Diagram of cDNA-AFLP method. See text for detailed explanation

(1) synthesis of cDNA from total RNA or poly(A) mRNA of eukaryotic cells using a poly(dT) oligonucleotide; (2) production of primary template by restriction digestion with *Ase*I and *Taq*I, which recognize six and four basepairs (bp), respectively. Complete digestion of cDNA produces five different types of molecules: 5′-end fragments, *Ase*I/*Ase*I fragments (relatively rare due to its longer recognition sequence), *Ase*I/*Taq*I fragments, *Taq*I/*Taq*I fragments (relatively abundant due to its shorter recognition sequence), and 3′-end fragments; (3) following digestion, double-stranded *Taq*I and *Ase*I adapters are ligated to the restriction fragments, generating templates for amplification; (4) pre-amplification is carried out using primers complementary to the two adapter sequences; (5) the primer complementary to *Ase*I anchors was end-labelled using γ-^{32}P-ATP and polynucleotide kinase; (6) selective amplification is then followed using primers bearing at their 3′ end two additional nucleotides which extend into the sequence of the restriction fragments, allowing only a subpopulation to be amplified; and (7) the final PCR products are separated on polyacrylamide gel and visualized by autoradiography. Most of the bands represent *Ase*I/*Taq*I fragments because *Ase*I/*Ase*I fragments are rare and *Taq*I/*Taq*I fragments are not visible on the X-ray film. RNA from different samples (e.g. sample 1 and sample 2) will produce different cDNA-AFLP banding patterns, which allow differentially expressed cDNAs to be identified. A modified method employing only one restriction enzyme (*Taq*I) has also been developed (Habu *et al.* 1997). In addition, there are a number of other similar methods to cDNA-AFLP (Matz and Lukyanov 1998). In contrast to differential display, cDNA-AFLP allows for a systematic survey of the organism's transcriptome through the use of selective fragment amplification.

The cDNA-AFLP technique is an attractive alternative to the conventional differential display and differential screening of cDNA libraries. In cDNA-AFLP, universal adapters are added to DNA fragments produced by restriction of cDNA with one or two restriction enzymes. The method needs only minute amounts of RNA due to the pre-amplification step performed with non-selective primers. Because stringent hybridization conditions are used in the amplification reactions, mismatched priming events are observed only in cases where transcript levels are extremely high. This results in cDNA-AFLP banding patterns being highly reproducible and almost free of false positives. Furthermore, no prior sequence information requirement and genome-wide expression analysis are attractive advantages of this high-throughput transcript technology. A significant disadvantage of the cDNA-AFLP method is the requirement for comprehensive reference database. In addition, isolating, cloning and sequencing the identified bands and maintaining these bacterial clones are time-consuming and laboring work.

Representational difference analysis

Representational difference analysis (RDA) was originally developed to identify differences between two complex genomes (Lisitsyn *et al.* 1993) and was adapted to analyse differential gene expression by taking advantages of both subtractive hybridization and PCR (Lisitsyn *et al.* 1993; Hubank and Schatz 1994). In the first step, mRNA derived from two different populations, the tester and the driver (control), is reverse transcribed; the tester cDNA represents the cDNA population in which differential gene expression is expected to occur. Following digestion with a frequently cutting restriction endonuclease, adapters are ligated to both ends of the cDNA. A PCR step then generates the initial representation of the digest, and a new adapter is ligated to the ends of the tester cDNA. The tester and driver cDNAs are then mixed in a 1 : 100 ratio with an excess of driver cDNA in order to promote hybridization between single-stranded cDNA common in both tester and driver cDNA pools. Following hybridization of the cDNAs, a PCR exponentially amplifies only those homoduplexes generated by the tester cDNA, via the priming sites on both ends of the double-stranded cDNA.

The major advantage of RDA is the specific amplification of fragments exclusively present in one cDNA pool, owing to an enrichment of rarely expressed tester sequences. RDA can be used to provide a picture of mRNA composition of cells, by displaying subsets of mRNAs as short cDNA bands. This approach is similar to two-dimensional protein gels, for example, in observing alterations in gene expression. Secondly, these cDNAs can be quickly sequenced; thereby a sequence unique for each mRNA can readily be obtained and compared with sequences in public databases. Thirdly, individual bands can readily be cloned and then used as probes for northern or Southern blotting and to isolate genes from genomic or cDNA libraries. A comprehensive set of these genes obtained from a display are useful for genetic applications and for preparing antibodies via corresponding amino acid sequences. However, this technique is unlikely to identify differences due to point mutations, very small deletions or insertions, fragments from the ends of transcripts, or fragments which lack appropriate enzyme sites.

Subtractive hybridization

Differential screening or subtractive hybridization (SH) methods have been used for many years to compare two populations of mRNA and obtain clones representing mRNAs that are expressed in one population but not in the other. There are numerous protocols for SH but the principle remains the same. Usually cDNA called 'tester' that contains differentially expressed genes of interest is hybridized with an excess of mRNA called 'driver' that does not contain these genes. Consequently, unhybridized and differentially expressed cDNAs are obtained that are present in the tester cDNA, but absent from the driver mRNA. A modified method, termed suppression subtractive hybridization (SSH) has been developed for generating a testis-specific cDNA library and by using the subtracted cDNA mixture as a hybridization probe to identify homologous sequences in human Y chromosome cosmid library (Diatchenko *et al.* 1996). SSH is a PCR-based cDNA subtraction method, and is used selectively to amplify target cDNA fragments (differentially expressed) and simultaneously suppress non-target DNA amplification. The method is based on the suppression PCR effect previously described by Clontech Laboratories (Palo Alto, CA, USA): long inverted terminal repeats when attached to DNA fragments can selectively suppress amplification of undesirable sequences in PCR procedures. Considering their handling and efficiency, differential mRNA display, RDA and subtractive hybridization are the most suitable methods for analysing differential gene expression for small laboratories.

Expressed sequence tags

The concept of expressed sequence tags (ESTs) was first proposed by Adams *et al.* in 1991. The basic idea is simple: create cDNA libraries from tissues of interest, randomly pick up clones from these libraries, and then perform a single sequencing reaction from a large number of clones. Each sequencing reaction generates 300–500 nucleotides or so of sequences that represents a unique sequence tag for a particular transcript. To generate a meaningful EST database, the quality of the cDNA library is extremely important and may be monitored by following parameters: (1) the titre of the unamplified library should be higher than 1.0×10^5 plaque forming unit (pfu) altogether; (2) the percentage of recombinant clones, in most case it can be indicated as transparent plaques on medium containing IPTG (isopropyl β-D-thiogalactopyranoside) and X-gal (5-bromo-4-chloro-3-indolyl β-D-galactopyranoside), should be higher than 80%; (3) the average length of cDNA inserts, which can be simply examined by PCR amplification using the obtained phage library as DNA template and flanking sequences of multiple cloning sites in cloning vector as PCR primers; (4) randomly select up to 50 plaques, convert into smaller size of plasmid form, and isolate the plasmid DNA. Enzyme digestion is carried out to study the length of cDNA insert from individual clone; and (5) most importantly, sequencing reaction should be performed from both ends of each clone from randomly selected samples until the clone is completely sequenced. Sequence analysis is carried out by several programs (GCG, Omiga and other commercial available softwares) to determine whether the examined clone contains partial or full-length cDNA sequence, and to compare similarity among these clones. If the cDNA library contains high percentages of small inserts (e.g. less than 0.5 kb), abundant identical/homologous clones, or most of them contain only partial cDNA sequence, suggesting the library is not a representative library and should be reconstructed, even though the previously mentioned parameters seem to be good.

An EST sequencing project is technically simple to execute; however, to generate meaningful amounts of EST database, it requires not only a high-quality cDNA library, it also requires automatic robotic colony-picking system, large-scale PCR amplification system, DNA sequencing facility and bioinformatics protocols for analysing huge of sequences and managing massive bacterial clones as well as DNA. It should be mentioned that EST approach does not seem to be the right tool for medium-size laboratories.

EST sequencing can be accomplished using normalized or non-normalized cDNA library. A normalized cDNA library is one in which each transcript is represented in more or less equal numbers (Patanjali *et al.* 1991; Soares *et al.* 1994). The advantage of using normalized cDNA libraries is that redundant sequencing of highly expressed genes is minimized, and the potential for identification of rare transcripts is maximized (Bonaldo *et al.* 1996). An advantage of non-normalized and non-amplified libraries is that the transcript abundance of the original cell or tissue is accurately reflected in the frequency of clones in the library. Non-normalized libraries can be used for an EST project to identify highly expressed, unknown genes and to compare the expression of highly expressed genes in different cell or tissue samples. The accumulation of numerous ESTs has led to the creation of a new databank, namely dbEST (Boguski *et al.* 1993), which contains 15 879 985 sequences from a number of different organisms at the date of 7 March 2003 (Figure 9.3). Since the EST database grows very fast, readers may visit the web site (http://www.ncbi.nlm.nih.gov/dbEST/) for current status.

Information extracted from EST programs have been extensively used in identification of tissue-specific gene expression patterns, construction of an RFLP (restriction fragment length polymorphism) linkage map and physical maps of the chromosomes, investigations of the mechanisms of expression of various isozymes and family genes, verification of genomic sequences, and so on. In higher plants, large-scale EST projects have been carried out including *Arabidopsis*, rice, tomato, maize, wheat, barley, soybean, sorghum, *Brassica*, cherry and *Vigna unguiculata* (Richmond and Somerville 2000). Moreover, since the complete sequence of the *Arabidopsis* genome is available (for detail please visit the website of The Arabidopsis Information Resource at http://www.arabidopsis.org/), comparative study between *Arabidopsis* genome and important crops such as wheat ESTs (Clarke *et al.* 2003) and maize ESTs (Brendel *et al.* 2002) have been performed. Most recently, draft sequences for two widely cultivated subspecies of rice (namely *Oryza sativa* L. ssp. *Indica* and *O. sativa* L. ssp. *Japonica*) have been reported (Goff *et al.* 2002; Yu *et al.* 2002), it is expected that more and more comparative studies will be carried out between complete genome sequences of *Arabidopsis* or rice and other plant or even other organism sequences.

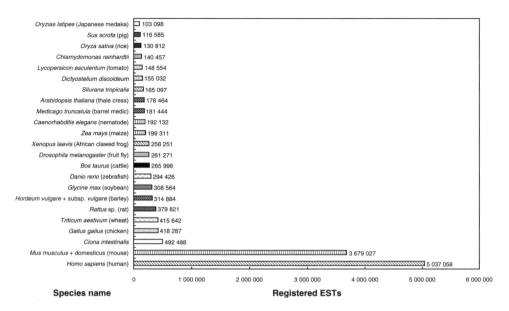

Figure 9.3 Number of ESTs registered in NCBI GenBank dbEST as of 7 March 2003 for various organisms including human, mouse, *Arabidopsis*, tomato, rice and so on. Organisms containing over 100 000 EST entries are shown in the figure. The total number of EST entries in the database at that time was 15 879 985

Serial analysis of gene expression

Serial analysis of gene expression (SAGE) is a sequence-based approach that is essentially an accelerated version of EST sequencing, allowing a rapid analysis of thousands of transcripts (Velculescu *et al.* 1995). In this method, unique and very short sequence tags (9 to 14 bp) which contain sufficient information for each transcript are first generated in the cell or tissue of interest, followed by ligation of sequence tags to obtain concatemers that can be cloned and sequenced, and then compared with data bases to determine differences in expression of genes that have been identified by the tags. A public database is available for analysing SAGE profiles among various organisms (Lash *et al.* 2000).

A schematic diagram of SAGE is shown in Figure 9.4. To generate sequence tags, mRNA is generated by reverse transcription with a biotinylated oligo(dT) primer and then digested with a frequent-cutting restriction enzyme (e.g. *Nla*III restricts 4-bp recognition sequence) called an anchoring enzyme. The 3' portion of the cDNA is further purified through streptavidin beads and is split into two fractions, which are ligated with adapters A and B. These adapters contain a recognition site for a type II restriction enzyme, which cleaves DNA at a defined distance from the recognition site. One example of this type of restriction enzyme is *Fok*I, which recognizes and restricts the following DNA sequence:

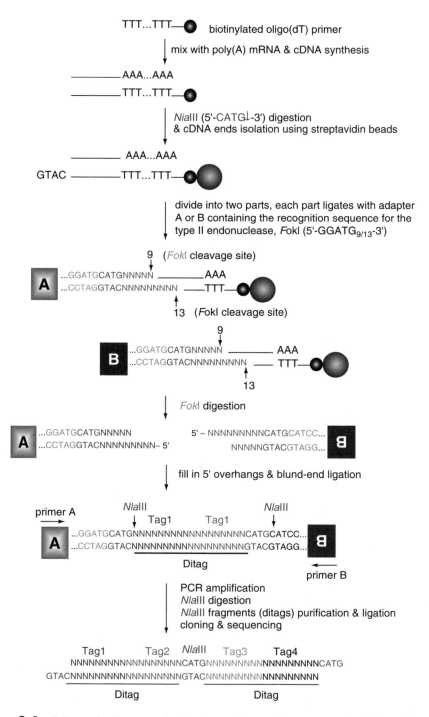

Figure 9.4 Schematic diagram of SAGE experiment. See text for detailed explanation

$$5'\text{-GGATG(N)}_9^{\blacktriangledown}\text{-}3'$$

$$3'\text{-CCTAC(N)}_{13_\blacktriangle}\text{-}5'$$

After digestion with the relevant type II restriction enzyme (named the tagging enzyme) and elution of the digested and unbound DNA portion, the eluted restriction fragments carrying adapter sequences are first filled in their 5′ overhangs and then subjected to ligation. PCR amplification is carried out using primers specific to adapters A and B. Following amplification, the adapter sequences A and B are removed with the anchoring enzyme *Nla*III. The sticky ends thus generated enable the DNA fragments to form concatemers, which can then be cloned into a vector. The structure of the concatemers has a typical pattern: between each anchoring site (e.g. *Nla*III restriction site), a so-called 'ditag' contains the sequence information of two independent cDNA tags.

Transcription profiles are created by sequencing each SAGE library. Since each sequencing reaction yields information for 20 or more genes, it is possible to generate data points for tens of thousands of transcripts in a modest sequencing effort. The relative abundance of each gene is determined by counting or clustering sequence tags. For most genes this short sequence tag is sufficient to provide a unique identifier by standard database searches. For a previously undescribed gene, the SAGE tag can be used to obtain a cDNA clone by PCR or hybridization-based methods.

The advantage of SAGE over many other methods is that SAGE can provide both quantitative and qualitative data about gene expression, and is able to accumulate and compare SAGE tag data from a variety of organisms. The disadvantages are related to the technical difficulties in generating good SAGE libraries and in analysing the data. So far, only a limited number of organisms including *Arabidopsis*, cow, human, barrel medic, house mouse and Norway rat have effectively used the SAGE analysis (please visit the SAGE website at *http:// www.ncbi.nlm.nih.gov/SAGE/* for current status).

DNA microarrays and plant functional genomics

Entering the 21st century, two research areas in biosciences have quickly evolved into new technology systems: (1) DNA microarrays, which are currently fabricated and assayed by two main approaches involving *in situ* synthesis of oligonucleotides (namely oligonucleotide microarrays) or deposition of amplified DNA fragments (namely cDNA microarrays) on solid surfaces; and (2) bioinformatic analysis for functional genomics, proteomics and metabolomics. The Human Genome Project has already generated a large body of molecular genetics, DNA sequence and cell biological information for 'Human Biology' studies (for details please visit the website at *http://www.genome.gov/*). The complete sequence of the human genome was finished in 2001 (Venter *et al*. 2001). In a similar approach, plant biologists have focused on plant genome projects of *Arabidopsis* (Arabidopsis Genome Initiative 2000), a small flowering plant widely used as a model organism in plant biology, and rice (Goff *et al*. 2002; Yu *et al*. 2002), the smallest genome among economically important crop plants. Up to March 2003, complete genomes including three major domains of life: bacteria (119 species), archaea (16 species) and eukaryotae (nine species), as well as 1042 viruses and 428 organelles are available on public databases (*http://www.ncbi.nih.gov/entrez/query.fcgi?db = Genome*). By observing the fast-growing database with increasing numbers of 'cloned' but unknown functional genes, biological scientists have soon realized that there is an even bigger task lying ahead of us once the genome projects are finished. We now recognize that it is essential to determine systematically and efficiently what roles many of these sequenced genes play, and how, where and when do they function and interact with each other. Based on this urgent need and consideration, various high throughput technology systems and better bioinformatic softwares have already been developed in order to meet this challenge. More importantly, worldwide competition on 'Functional Genomics and Bioinformatics' experimental research systems apparently has just started in late 1999. Among them, DNA microarray approach has been demonstrated to be the key technology system for 'functional genomics' study (see Chapter 8, and for reviews see Marshall and Hodgson 1998; Ramsay 1998; Brown and Botstein 1999; Duggan *et al*. 1999; Khan *et al*. 1999; Kurian *et al*. 1999; Zhu and Wang 2000; Aharoni and Vorst 2002; Finkelstein *et al*. 2002).

The use of DNA microarray for comprehensive RNA expression analysis has caused a great deal of interest recently. Gene expression monitoring by microarray analysis was first described using radioactive targets hybridized onto filter-immobilized cDNA clones. Large-scale cDNA microarrays were used by Drmanac and Drmanac (1994) and Drmanac *et al*. (1996), who have produced DNA microarrays containing up to 31104 cDNA clones which were PCR-amplified and robotically spotted onto nylon membranes for gene expression and discovery experiments. Since then, technical developments including increased sensitivity, array generation and data interpretation make microarray-based expression analysis a powerful tool. Among them, DNA microarrays printed on glass and hybridized with fluorescently labelled cDNA are a significant improvement on the filter-immobilized DNA arrays. The technology was first described by Schena *et al*. (1995), who printed 48 genes of *Arabidopsis thaliana* onto glass slides and measured differential expression of genes between two different tissues – root and leaf. Fluorescent targets were made from each tissue by reverse transcription of mRNA using distinct fluorochromes. By measuring the intensity ratio for each printed gene they are able to show widespread differences in gene expression between these two samples. The two-colour fluorescence detection scheme has the advantage over radioactively labelled targets of allowing rapid and simultaneous differential expression analysis of independent biological samples. In addition, the use of ratio measurements compensates for probe-to-probe variations of intensity due to DNA concentration and hybridization.

One of the research interest in my laboratory is tissue-specific expression in tomato, an important crop worldwide. To identify gene expression patterns in tomato plants, leaf and root tissues are harvested from different growth stages of rapid-grown young plants and mature plants with flowers. mRNA is isolated from these samples and then labelled with different fluorescent dyes (Figure 9.5). To set up cDNA microarray analysis, approximately 12 000 tomato cDNA clones including approximately 6000 unique clones from four tomato root libraries (mineral deficiency, pre-anthesis stage, post-anthesis stage and germinating seedlings) purchased from Clemson University Genomics Institute (GUGI) and approximately 6000 subtractive or normal cDNA clones from individual laboratories of Crop Plant Improvement (CPI) Group within our Institute (Institute of BioAgricultural Sciences, Academia Sinica, Taipei, Taiwan), were PCR amplified and purified, and then printed onto a glass slide by arrayer. Typical instrumentation for microarray analysis has been set up in our Core Facility Laboratory (Figure 9.6). Microarray experiments including labelling and hybridization were carried out according to the standard protocols available on the Arabidopsis Information Resource (*http://www.arabidopsis.org/*). After hybridization, laser excitation of the incorporation targets yields an emission with a characteristic spectrum, which is measured using a scanning confocal laser microscope. Monochrome images from the scanner are imported into software in which the images are pseudo-coloured and merged. The relative abundance of mRNA from the same tissue (e.g. root) in young plant versus mature plant is reflected by the ratio of green/red as measured by the fluorescence emitted from the corresponding array element. Image analysis software (e.g. GeneSpring) is used to determine fluorescence intensities that allow the quantitative comparison between the two stages of root development for all genes on the array (Figure 9.5). Currently, we have isolated some tissue-specific genes using a cDNA microarray approach, and molecular characterization will be carry out to address the functions of these genes.

As mentioned, DNA microarrays can also be assembled from synthetic oligonucleotides (Lockhart *et al.* 1996; Lipshutz *et al.* 1999; Lockhart and Winzeler 2000). These arrays are referred as oligonucleotide microarrays (see Chapter 7). Affymetrix, the most famous manufacturer for making gene chips, fabricates oligonucleotide arrays on the chip (GeneChip) using photolithography (Lockhart *et al.* 1996; Warrington *et al.* 2000). A mercury lamp is shone through a photolithographic mask onto the chip surface, which removes a photoactive group, resulting in a 5′ hydroxyl group capable of reacting with another nucleoside. The mask therefore predetermines which nucleotides are activated. Successive rounds of deprotection and chemistry result in oligonucleotides up to 30 bases in length.

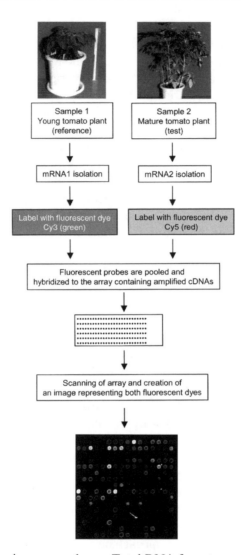

Figure 9.5 cDNA microarray schema. Total RNA from two samples to be compared (e.g. one sample is referred as reference and another sample is referred as test) are labelled with fluorescent dyes Cy3 (green) and Cy5 (red), respectively, by reverse transcription. The fluorescent probes are pooled and allowed to hybridize under stringent conditions to the array containing PCR-amplified fragments for genes of interest and printing on coated glass microscope slide or nylon filter by a computer-controlled, high-speed arrayer. After hybridization, highly sensitive laser scanner is used to detect the expression signal in each spot on the slide. Monochrome images from the scanner are imported into software in which the images pseudo-coloured and merged. This image represents the relative expression in the two samples we examined (e.g. spot with red colour represents relative high expression in sample 1, while spot with green colour represents relative high expression in sample 2). A third colour of yellow is introduced to represent a similar expression level between the two samples

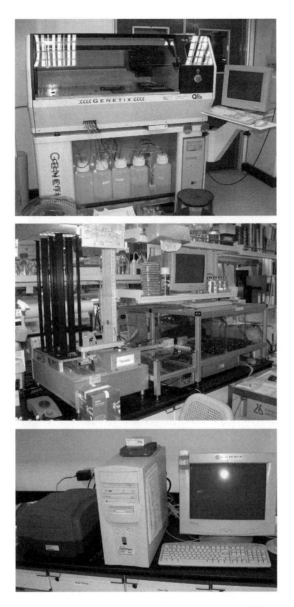

Figure 9.6 DNA microarray system facility at the Institute of BioAgricultural Sciences of Academia Sinica in Taiwan. Top panel: Genetix Q-Pix Colony Picker. The major function of this picker is picking the recombinant colonies (those white clones in medium containing IPTG and X-gal) from a large plate (22.2 cm × 22.2 cm) onto 384-well or 96-well culture plates. In addition to colony picking, this picker is also used for gridding, replicating, rearraying and microarraying. Middle panel: Cartesian SynQUAD Arrayer. This arrayer is used for printing DNA onto slides or nylon filters. Bottom panel: Axon GeneRix 4000B scanner. This scanner is used for microarray data mining and data analysis

Apparently, a disadvantage of oligonucleotide microarrays related to cDNA microarrays is the limitation of the technology to genes of known sequences, only limited genome-wide genechip arrays for *Arabidopsis thaliana, Caenorhabditis elegans, Escherichia coli, Drosophila melanogaster, Pseudomonas aeruginosa*, human, mouse, rat and yeast are now available (for details please visit the website at http://www.affymetrix.com/). For the commercially available *Arabidopsis* ATH1 GeneChip (Affymetrix), each slide contains more than 22 500 probe sets representing approximately 24 000 genes (all open reading frames within *Arabidopsis* genome). For each gene, 11 probe pairs were designed. One set of 11 consisted of 'mismatch oligonucleotides' that were identical to a 'perfect match' set except for the 13th nucleotide in each 25-mer. These mismatched oligonucleotides are used to assess cross-hybridization and background signals. Affymetrix GeneChips are expensive to make, not least because of the need to manufacture the glass masks. Recently, there are other promising oligonucleotide-based technologies, including those that array 5'-terminally modified oligonucleotides, unmodified oligonucleotides and phosphoamidities for the *in situ* synthesis of oligonucleotides.

For oligonucleotide microarray hybridization, poly(A) mRNA from samples are isolated and then transcribed into double-stranded cDNA by reverse transcription. The cDNA is then copied to antisense RNA (cRNA) by an *in vitro* transcription reaction performed in the presence of biotin-labelled ribonucleotide triphosphate (biotin-UTP or biotin-CTP). Fragmented cRNA (50 to 100 nucleotides) is used for hybridization. After a brief washing step to remove unhybridized cRNA, the microarrays are stained by streptavidin phycoerythrin and scanned. Detailed protocols for Affymetrix GeneChip microarray expression analysis are available on the company's website.

Currently, we are employing oligonucleotide microarray and bioinformatics approaches to conduct a functional study of plant leaf senescence under high temperature. Like many other developmental process, leaf senescence is a genetically controlled programme regulated by a variety of autonomous factors (e.g. age, reproductive development and hormone levels) and environmental factors (e.g. extremes of temperature, drought, ozone, nutrient deficiency, pathogen infection, wounding and shading). Among the various environmental factors, high temperature or thermal stress is one of the important limiting factors on crop production worldwide, especially in tropical and subtropical areas such as Taiwan. However, no report can be found in the research area of heat-induced senescence in plant leaves. In parallel to screening heat-induced senescent mutants, it is important to understand the differential gene expression pattern in senescent leaves during heat treatment. To isolate and characterize senescence-associated genes (SAGs) during leaf senescence under high temperature, surface-sterilized *Arabidopsis* seeds were germinated at normal temperature of 24°C growth chamber with light/dark cycle. After 10 days, partial seedlings were continued to grow on the same growth chamber (24°C), while partial seedlings were transferred to an another growth chamber of 35°C with light/dark cycle for a period of 5 days (S1 sample), 10 days (S2 sample) and 15 days (S3 sample) (Figure 9.7). All seedlings were healthy grown, cotyledons and leaves were green, roots were normally elongated during the 10 days at normal conditions (C sample). However, under a period of continuous high-temperature incubation, growth of young seedlings was inhibited, and cotyledons and leaves turned yellow, and roots stopped elongating. These phenotypic damages were severely increased as the duration at high temperature was prolonged. Figure 9.8 shows the typical growth patterns of seedlings grown on normal temperature (labelled as C) and 15 days under high temperature of 35°C (labelled as S3). Total RNA was isolated from each sample and *Arabidopsis* GeneChip microarray expression analysis was conducted (Affymetrix). As shown in Figure 9.9, abundant up-regulated and down-regulated genes (each white spot represents one up- or down-regulated gene in each panel) are found at different stages of S1, S2 and S3 as compared with the control C sample (no high-temperature treatment). To our surprise, the total

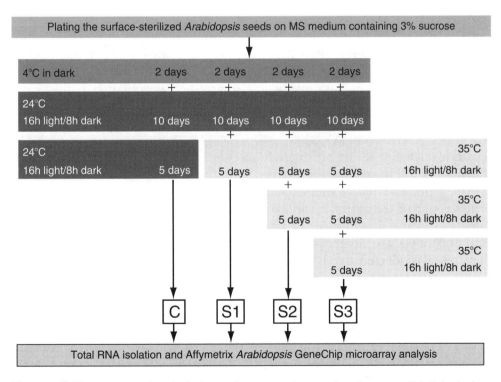

Figure 9.7 Strategy for isolation of senescence-associated genes (SAGs) during leaf senescence under high temperature in *Arabidopsis* by oligonucleotide microarray approach

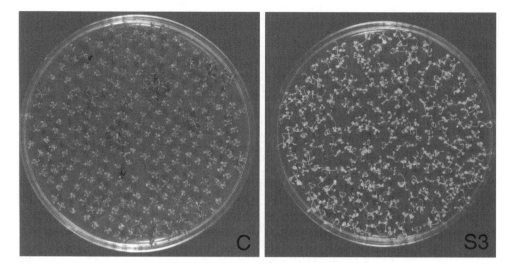

Figure 9.8 Typical growth patterns of *Arabidopsis* seedlings which grew on normal temperature of 24°C (left panel) and continuous high temperature of 35°C (right panel)

number of either up-regulated genes or down-regulated genes is significantly increased as the duration of high-temperature treatment increased (Figure 9.9), suggesting enzymes in both synthetic pathway and degradation pathway are actively involved under unfavourable conditions. We further cross-examined the expression profiles of up-regulated genes or down-regulated genes during different stages of senescence (S1, S2 and S3); a total of 106 up-regulated genes and 95 down-regulated genes are present in S1, S2 and S3 samples (Figure 9.10). Part of these genes will be selected for further studies.

The ability to monitor simultaneously the expression pattern of a large set of genes is one of the major advantage for microarray-based approach. In higher plants including *Arabidopsis*, strawberry, *Petunia*, maize, lima bean, rice, *Vicia* spp. and ice plant, genome-wide expression analysis for different research projects have been conducted by cDNA or oligonucleotide microarrays (Aharoni and Vorst 2002). Based on huge data generated by microarray projects, readers may want to visit the collections for special gene expression databases emphasizing on plant functional genomics (Cartinhour 1997; Reiser *et al.* 2002). While there are many advantages to the cDNA or oligonucleotide microarray approach to analysing differential gene expression, disadvantages include the expense for both set-up instruments and experimental reagents, the requirement of a good set of clones or oligonucleotides for known genes to array, and the relatively large amount of RNA that is required to prepare the probes. The large RNA requirement may make it difficult to analyse small samples.

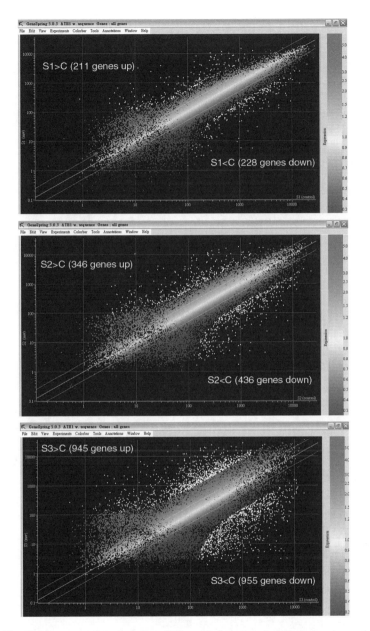

Figure 9.9 The scatter plot showing differential expression patterns in each senescent stage of S1 (top panel), S2 (middle panel) and S3 (bottle panel) by Affymetrix *Arabidopsis* GeneChip microarray expression analysis. Different colour spots (e.g. red, yellow or blue) in each panel represent relative expression levels under the specific stage, and the total number of spots in each panel should be approximately 24 000 (total gene number in the *Arabidopsis* GeneChip). White colour spots are introduced to represent those up-regulated and down-regulated genes, as indicated in each panel. Microarray data analysis was conducted by GeneSpring (version 5.0.3) software

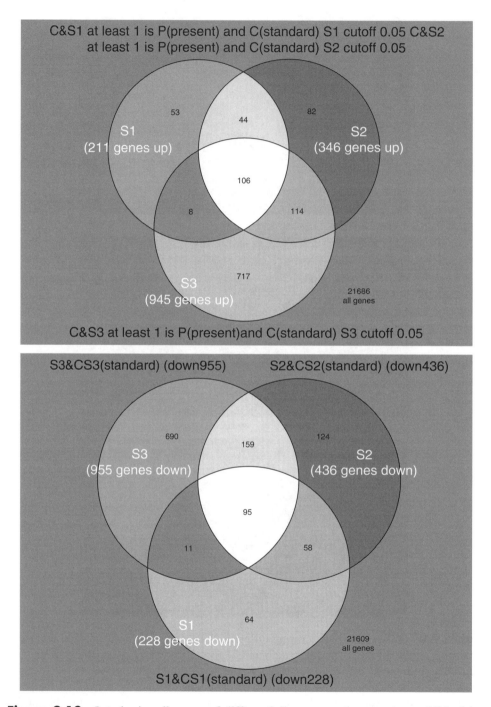

Figure 9.10 Interloping diagram of differentially expressed genes that exhibited increased expression in different stages of senescence (upper panel) or decreased expression in different stages of senescence (lower panel) as compared with the seedlings grown on normal temperature. The threshold for increased or decreased expression was at least

Discussion

In the past few years, a variety of methods have been developed for high through-put analysis of differential gene expression. Each of these methods has a number of unique advantages, such as simplicity (differential display, subtractive hybrid-ization and RDA) or the range (SAGE) of analysis. By contrast, there are also some limitations, including the unidirectional analysis of RDA or the high cost of DNA microarrays. Most of the discussed techniques such as Northern blot analysis, subtractive hybridization or differential plaque hybridization are labour and material intensive, unidirectional, serial or not well suited to the detection of rare mRNA species. However, they remain indispensable tools for sizing mRNAs and full-length cloning. The introduction of differential display provides a fast and technically simple method for the identification of differentially expressed mRNAs. This non-selective approach, which offers the possibility of comparing gene expression in a parallel and bidirectional fashion, is one of the most com-monly used methods for analysis of gene expression. Sequence-based approaches such as SAGE or oligonucleotide arrays are excellent tools for high-throughput screenings of expression profiles but, once differentially expressed mRNAs are detected, the corresponding cDNAs must be cloned. Other promising sequence-based technologies such as massively parallel signature sequencing (MPSS) and mass spectrometry sequencing remain to be evaluated by the scientific commu-nity. It is highly desirable to have an ideal tool, which would allow the unambigu-ous identification of differentially expressed genes in a simple and parallel manner and be suited to the identification of low-abundance mRNA species in a single experiment. It should be emphasized that the high throughput resulting from these methods derives from the number of samples versus the number of genes which can be reasonably analysed. Some methods, such as SAGE and ESTs, could in theory analyse all expressed genes, but on a limited number of samples. Other methods, such as microarrays, may be used to analyse a subset of genes on a large number of samples.

Molecular approaches are still being developed for rapid and efficient identifi-cation of gene-expression patterns. In addition to profiling the mRNA (the tran-scriptome) as mentioned in this chapter, high-throughput techniques for measuring protein (the proteome), metabolite (the metabolome) and phenotype (the phenome) are also developed for plant functional genomic study (Finhn 2002; Kersten *et al.* 2002; Oliver *et al.* 2002; Roberts 2002). For the time being, we will have to combine the currently available tools in an intelligent way. Based on experience, equipment and cost, one is left to decide which technology best suits the needs of the laboratory and project. The combination of classical and modern

Figure 9.10 (continued) four-fold. The number in the overlapping white area in each panel indicates the number of common genes that exhibited up regulation (upper panel) or down regulation (lower panel) within these three different stages (S1, S2 and S3)

molecular tools provide powerful molecular-biological approaches for the discovery of new genes, facilitate our understanding of development and differentiation, and provide a new strategy for biotechnology at the molecular level.

References

Adams, M.D., Kelley, J.M., Gocayne J.D., Dubnick, M., *et al.* (1991) Complementary DNA sequencing: expressed seguence tags and human genome project. *Science* **252**: 1651–1656.

Aharoni, A. and Vorst, O. (2002) DNA microarrays for functional plant genomics. *Plant Mol Biol* **48**: 99–118.

Arabidopsis Genome Initiative (2000) Analysis of the genome sequence of the flowering plant *Arabidopsis thaliana*. *Nature* **408**: 796–815.

Bachem, C.W.B., van der Hoeven, R.S., de Bruijn, S.M., Vreugdenhil, D., *et al.* (1996) Visualization of differential gene expression using a novel method of RNA fingerprinting based on AFLP: analysis of gene expression during potato tuber development. *Plant J.* **9**: 745–753.

Boguski, M.S., Lowe, T.M. and Tolstoshev, C.M. (1993) dbEST-database for 'expressed sequence tags'. *Nature Genet* **4**: 332–333.

Bonaldo, M.F., Lennon, G. and Soares, M.B. (1996) Normalization and subtraction: two approaches to facilitate gene discovery. *Genome Res* **6**: 791–806.

Brendel, V., Kurtz, S. and Walbot, V. (2002) Comparative genomics of *Arabidopsis* and maize: prospects and limitations. *Genome Biol* **3**: reviews 1005.1–1005.6.

Brown, P.O. and Botstein, D. (1999) Exploring the new world of the genome with DNA microarrays. *Nature Genet Suppl* **21**: 33–37.

Cartinhour, S.W. (1997) Public informatics resources for rice and other grasses. *Plant Mol Biol* **35**: 241–251.

Carulli, J.P., Artinger, M., Swain, P.M., Root, C.D., *et al.* (1998) High throughput analysis of differential gene expression. *J Cellular Biochem Suppl* **30/31**: 286–296.

Clarke, B., Lambrecht, M. and Rhee, S.Y. (2003) *Arabidopsis* genomic information for interpreting wheat EST sequences. *Funct Integr Genomics* **3**: 33–38.

Diatchenko, L., Lau, Y.F.C., Campbell, A.P., Chenchik, A., *et al.* (1996) Suppression subtractive hybridization: a method for generating differentially regulated or tissue-specific cDNA probes and libraries. *Proc Natl Acad Sci USA* **93**: 6025–6030.

Donson, J., Fang, Y., Espiritu-Santo, G., Xing, W., *et al.* (2002) Comprehensive gene expression analysis by transcript profiling. *Plant Mol Biol* **48**: 75–97.

Drmanac, S. and Drmanac, R. (1994) Processing of a cDNA and genomic kilobase–size clones for massive screening, mapping and sequencing by hybridization. *BioTechniques* **17**: 328–329, 332–336.

Drmanac, S., Stavropoulos, N.A., Labat, I., Vanau, J., *et al.* (1996) Gene-representing cDNA clusters defined by hybridization of 57 419 clones from infant brain libraries with short oligonucleotide probes. *Genomics* **37**: 29–40.

Duggan, D.J., Bittner, M., Chen,Y., Meltzer, P. and Trent, J.M. (1999) Expression profiling using cDNA microarrays. *Nature Genet Suppl* **21**: 10–14.

Finhn, O. (2002) Metabolomics – the link between genotypes and phenotypes. *Plant Mol Biol* **48**: 155–171.

Finkelstein, D., Ewing, R., Gollub, J., Sterky, F., *et al.* (2002) Microarray data quality analysis: lessons from the AFGC project. *Plant Mol Biol* **48**: 119–131.

Goff, S.A., Ricke, D., Lan, T.H., Presting, G., *et al.* (2002) A draft sequence of the rice genome (*Oryza Sativa* L. ssp. *japonica*). *Science* **296**: 92–100.

Habu, Y., Fukada-Tanaka, S., Hisatomi, Y. and Iida, S. (1997) Amplified restriction fragment length polymorphism-based mRNA fingerprinting using a single restriction enzyme that recognizes a 4-bp sequence. *Biochem Biophys Res Comm* **234**: 516–521.

Hubank, M. and Schatz, D.G. (1994) Identifying differences in mRNA expression by representational difference analysis of cDNA. *Nucl Acids Res* **22**: 5640–5648.

Kersten, B., Bürkle, L., Kuhn, E.J., Giavalisco, P., *et al.* (2002) Large-scale plant proteomics. *Plant Mol Biol* **48**: 133–141.

Khan, J., Saol, L.H., Bittner, M.L., Chen, Y., *et al.* (1999) Expression profiling in cancer using cDNA microarrays. *Electrophoresis* **20**: 223–229.

Kozian, D.H. and Kirschbaum, B.J. (1999) Comparative gene-expression analysis. *Trends Biotechnol* **17**: 73–78.

Kuhn, E. (2001) From library screening to microarray technology: strategies to determine gene expression profiles and to identify differentially regulated genes in plants. *Ann Bot* **87**: 139–155.

Kurian, K.M., Watson, C.J. and Wyllie, A.H. (1999) DNA chip technology. *J Pathol* **187**: 267–271.

Lash, A.E., Tolstoshev, C.M., Wagner, L., Schuler, G.D., *et al.* (2000) SAGEmap: a public gene expression resource. *Genome Res* **10**: 1051–1060.

Liang, P. and Pardee, A.B. (1992) Differential display of eukaryotic messager RNA by means of the polymerase chain reaction. *Science* **257**: 967–971.

Lipshutz, R.J., Fodor, S.P., Gingeras, T.R. and Lockhart, D.J. (1999) High density synthetic oligonucleotide arrays. *Nature Genet Suppl* **21**: 20–24.

Lisitsyn, N., Lisitsyn, N. and Wigler, M. (1993) Cloning the differences between two complex genomes. *Science* **259**: 946–951.

Lockhart, D.J. and Winzeler, E.A. (2000). Genomic, gene expression and DNA arrays. *Nature* **405**: 827–836.

Lockhart, D.J., Dong, H., Byrne, M.C., Follettie, M.T., *et al.* (1996). Expression monitoring by hybridization to high-density oligonucleotide arrays. *Nature Biotechnol.* **13**: 1675–1680.

Marshall, A. and Hodgson, J. (1998) DNA chips: an array of possibilities. *Nature Biotechnol* **16**: 27–31.

Matz, M.V. and Lukyanov, S.A. (1998) Different strategies of differential display: areas of application. *Nucl Acids Res* **26**: 5537–5543.

Oliver, D.J., Nikolau, B. and Wurtele, E.S. (2002). Functional genomics: high-throughput mRNA, protein, and metabolite analysis. *Metabolic Engineering* **4**: 98–106.

Patanjali, S.R., Parimoo, S. and Weissman, S.M. (1991) Construction of a uniform-abundance (normalized) cDNA library. *Proc Natl Acad Sci USA* **88**: 1943–1947.

Ramsay, G. (1998) DNA chips: state-of-the art. *Nature Biotechnol* **16**: 40–44.

Reiser, L., Mueller, L.A. and Rhee, S.Y. (2002) Surviving in a sea of data: a survey of plant genome data resources and issues in building data management systems. *Plant Mol Biol* **48**: 59–74.

Richmond, T. and Somerville, S. (2000) Chasing the dream: plant EST microarrays. *Current Opin Plant Biol* **3**: 108–116.

Roberts, J.K.M. (2002) Proteomics and a future generation of plant molecular biologists. *Plant Mol Biol* **48**: 143–154.

Schena, M., Shalon, D., Davis, R.W. and Brown, P.O. (1995) Quantitative monitoring of gene expression patterns with a complementary DNA microarray. *Science* **270**: 460–470.

Soares, M.B., Bonaldo, M.F., Jelene, P., Su, L., *et al.* (1994) Construction and characterization of a normalize cDNA library. *Proc Natl Acad Sci USA* **91**: 9228–9232.

To, K.Y. (2000) Identification of differential gene expression by high throughput analysis. *Comb Chem High Throughput Screening* **3**: 235–241.

Velculescu, V.E., Zhang, L., Vogelstein, B. and Kinzler, K.W. (1995) Serial analysis of gene expression. *Science* **270**: 484–488.

Venter, J.C., Adams, M.D., Myers, E.W., Li, P.W., *et al.* (2001) The sequence of the human genome. *Science* **291**: 1304–1351.

Vos, P., Hogers, R., Bleeker, M., Reijans, M., *et al.* (1995) AFLP: a new technique for DNA fingerprinting. *Nucl Acids Res* **23**: 4407–4414.

Warrington, J.A., Dee, S. and Trulson, M. (2000) Large-scale genomic analysis using Affymetrix GeneChip probe arrays. In: Schena, M. (Ed.) *Microarray Biochip Technology*, Eaton Publishing, Natick, MA, USA, pp. 119–148.

Yu, J., Hu, S., Wang, J., Wong, G. K.S., *et al.* (2002) A draft sequence of the rice genome (*Oryza sativa* L. ssp. *indica*). *Science* **296**: 79–92.

Zhu, T. and Wang, X. (2000) Large-scale profiling of the *Arabidopsis* transcriptome. *Plant Physiol* **124**: 1472–1476.

Further reading

Brown, T.A. (1999) *Genomes*. BIOS Scientific Publisher Press, Oxford.

Schena, M. (1999) *DNA Microarrays*. Oxford University Press, New York.

The Arabidopsis Book. Online publication at *http://www.aspb.org/publications/arabidopsis/toc.cfm/*

Wilson, Z.A. (2000) *Arabidopsis*. Oxford University Press, New York.

10

Aptamers: Powerful Molecular Tools for Therapeutics and Diagnostics

Eva Baldrich Rubio, Mònica Campàs i Homs and **Ciara K. O'Sullivan**

Introduction

In 1990, within months of each other, the laboratories of G.F. Joyce (La Jolla, USA) (Robertson and Joyce 1990), J.W. Szostak (Boston, USA) (Ellington and Szostak 1990) and L. Gold (Bolder, USA) (Tuerk and Gold 1990) independently reported on the development of an *in vitro* selection and amplification technique, which has allowed the discovery of specific nucleic acid sequences that bind a wide array of non-nucleic acid targets with high affinity and specificity.

The technique by which these oligonucleotide ligands are obtained was coined as SELEX – systematic evolution of ligands by exponential enrichment – and the resulting oligonucleotides are referred to as aptamers, derived from the Latin *aptus*, meaning, 'to fit' (Jayasena 1999). The last decade has seen the selection of an extensive range of aptamers with abilities to bind to small molecules, peptides, proteins, cells, etc. with selectivity, specificity and affinity equal and often superior to those of antibodies (Hesselberth *et al.* 2000; Haller and Sarnow 1997; Gebhardt *et al.* 2000; Wilson *et al.* 2001; Kawakami *et al.* 2000). Aptamers possess the ability to discriminate targets on the basis of subtle structural differences such

Molecular Analysis and Genome Discovery edited by Ralph Rapley and Stuart Harbron
© 2004 John Wiley & Sons, Ltd   ISBN 0 471 49847 5 (cased)   ISBN 0 471 49919 6 (pbk)

as the presence or absence of a methyl or hydroxy group or the D vs. L-enantiomeric configuration of the target (Geiger *et al.* 1996).

The advantages of aptamers over alternative approaches include the relatively simple techniques and apparatus required for their isolation, the number of alternative molecules that can be screened and their chemical simplicity (James 2001). As aptamers can be evolved to bind tightly and specifically to almost any protein, their potential for use in analytical devices, proteomic microarrays and as therapeutic agents is immense. Despite the fact that aptamer technology has not yet been fully embraced by the scientific community (possibly due to extensive patents that exist in the area), automated aptamer selection and increasing commercial availability will facilitate increased accessibility and application.

Aptamer selection

SELEX

The SELEX process is a technique for screening very large combinatorial libraries of oligonucleotides by an iterative process of *in vitro* selection and amplification. Nucleic acid libraries are easily obtained via combinatorial chemistry synthesis. Each sequence synthesized represents a linear oligomer of unique sequence and the molecular diversity is dependent on the number of randomized nucleotide positions. Typically libraries of at least $10^{13} - 10^{18}$ independent sequences are used, with a variable region of 30 bases flanked by primers of 20–25 bases normally being employed (Figure 10.1).

In the process, which is depicted in Figure 10.2, a random sequence oligonucleotide library is incubated with a target of interest, in a buffer of choice, at a given temperature. In the initial cycles of selection, a tiny percentage (0.1–0.5

Library of random oligonucleotides

| Fixed | Variable (n) | Fixed |

$n=10 \quad s=10^6$
$n=20 \quad s=10^{12}$
$n=30 \quad s=10^{18}$
$n=50 \quad s=10^{30}$

Figure 10.1 Representation of library used for aptamer selection. Libraries of at least $10^{13} - 10^{18}$ independent sequences are used, with a variable region of 30 bases flanked by primers of 20–25 bases typically being employed

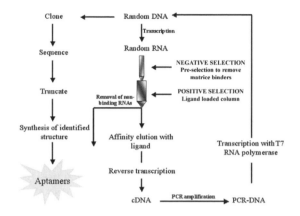

Figure 10.2 Schematic indicating steps employed in SELEX process

per cent) of individual sequences interact with the target. These sequences are separated from the rest of the library by using techniques such as affinity chromatography, magnetic separation or filter binding (Burke and Gold 1997; Holeman *et al.* 1998). This isolated population of sequences is then amplified to obtain an enriched library to be used for the proceeding selection/amplification cycle. The efficiency of enrichment of high-affinity binders is governed by the stringency of selection of each round. To this end, negative selection (removal of aptamers that bind ligand support) and counter selection (removal of aptamers that bind to structures similar to that of the target) are employed. The length of incubation time of library with target used is manipulated to yield aptamers of desired kinetics. Upon achieving affinity saturation, typically after 8–15 cycles of selection/amplification, the enriched library is cloned and sequenced and individual sequences investigated for their ability to bind to the target (e.g. using surface plasmon resonance or fluorescence correlation spectroscopy) (Jeong *et al.* 2001; Schürer *et al.* 2001). Aptamers can then be truncated as desired, eliminating the fixed primer sequences as well as those identified not to be part of the consensus motif (i.e. sequence required for binding). When the desired sequence has been identified, the aptamer can be produced in sizeable quantities by chemical synthesis (Famulok 1994).

PhotoSELEX

PhotoSELEX aims at producing aptamers of even higher specificity and sensitivity using a modified form of SELEX whose selection criteria is the ability to form a photo-induced covalent bond to the target molecule. In PhotoSELEX, a modified nucleotide activated by absorption of light is incorporated in place of a native base in either RNA or in ssDNA random oligonucleotide libraries (Golden *et al.* 2000). An example of this modified oligonucleotide is that of 5-bromo-2′-deoxyuridine (BrdU) (Meisenheimer and Koch 1997) that in the form of 5-Br-dUTP is used as a substituent for TTP whilst synthesizing the RNA/ DNA combinatorial library. The bromouracil chromophore absorbs UV light in the 310 nm range where native chromophores of nucleic acids and proteins absorb very weakly. When excited at 310 nm, the resulting excited singlet state intersystem crosses to the lowest triplet state that specifically cross-links with aromatic and sulphur-bearing amino acid residues of a protein target in suitable proximity (Ito *et al.* 1980; Swanson *et al.* 1981; Dietz and Koch 1987; 1989). Importantly, excited bromouracil in DNA is relatively unreactive in the absence of a proximal reactive amino acid or nucleotide residue (Willis *et al.* 1994; Cook and Greenburg 1996), further increasing the selectivity of the cross-linking reaction.

It is this phenomenon that is exploited in PhotoSELEX. The target is immobilized, for example onto a magnetic bead and introduced into the oligonucleotide library, where aptamers will bind the target. The unbound oligonucleotides are separated using a mild wash and the magnetic beads with bound target-aptamer complexes localized via magnetic attraction. The isolated aptamer-target complex is then exposed to 310 nm laser light and photocross-linking of the complexes occurs. Subsequently, the complex is subjected to denaturing conditions and a harsh wash removes all non-cross-linked aptamers from solution. Denatured protein is digested with protease, releasing the cross-linked photoaptamers, which are then amplified by PCR. The cycle is repeated until the population of target positive photoaptamers is enriched and they are then cloned and sequenced as in the case of routine SELEX.

Automated selection

Automated platforms for aptamer selection will contribute considerably to their accessibility and subsequent use as analytical tools. A robotic platform based on the use of an augmented Beckman Biomek 2000 pipetting robot (Cox *et al.* 1998) that integrates a PCR thermal cycler, a magnetic bead separator, reagent trays and pipette tip station was used to perform a model selection for poly-T binding RNA molecules. This automated selection has been used to generate nucleic acid aptamers against lysosyme in a matter of days rather than weeks or months, currently having the ability to carry out eight selections in parallel, completing 12 rounds in selection in 2 days (Cox and Ellingta 2001). A recent report has outlined initial work on the automation of aptamer selection on silicon chips, which requires miniscule quantities of reagents and could facilitate the attainment of aptamers, in a fraction of the time currently required, without the requirement of sophisticated apparatus (Chambers and Valdes 2002). Selection of lysozyme specific aptamers has been demonstrated in a modular-manual fashion on a silicon chip of 4 cm^2 area and although the selected aptamers were not as robust as those selected using the robotic workstation, the concept has been demonstrated and work is on-going to achieve a fully-automated on-chip selection of aptamers.

Aptamers for therapeutics

The development in applying aptamers for therapeutics has been extensively reviewed and the reader is directed to the excellent overviews given by Brody and Gold (2000) and Osbourne *et al.* (1997). The aim of this section is thus not to give a complete outline of the application of aptamers for therapeutics, rather to focus on the state-of-the-art developments, addressing recent reports regarding improvement of aptamer stability and their resistance to nuclease attack as well as the introduction of intramers and spiegelmers as powerful therapeutic tools.

Aptamer Stability

One crucial problem with the use of either DNA or RNA aptamers is their stability, particularly with reference to nuclease attack. Various routes to stabilize the aptamers have been explored, and most success has been met in the shape of spiegelmers and via the chemical modification of either the oligonucleotide backbone or of the 2' position of the pyrimidine moiety and an overview of the approaches investigated to date is given within this section.

Chemical modification of aptamers

The hydroxyl at the 2'-position is reactive, particularly at higher than neutral pH, and will attack the neighbouring phosphiodiester bond to produce a cyclic 2',3'-phosphate, thereby breaking the nucleic acid backbone. This reaction that is catalysed by many transition metal ions as well as a range of ribonucleases present in biological samples, that result in a very short aptamer lifetime (<10 min) (James 2000; Kusser 2000). Reports have been published detailing approaches to make oligonucleotide sequences resistant to attack by modifying the oligonucleotide backbone. However, most of these modifications result in structures not recognized by the enzymes used in the SELEX process. Nevertheless, by using α-thio-substituted deoxynucleoside triphosphates (cNTPs), a phosphorothioate DNA library was successfully screened for aptamers against the transcription factor NF-IL5 (Ruckman *et al.* 1998) and some progress has been made toward the use of analogous 'thio-RNA' aptamers (King *et al.* 1998; Jhaveri *et al.* 1998).

Modification of the 2' positions of pyrimidine nucleotides with amino/fluoro groups has, however, been demonstrated to increase dramatically the half-life of aptamers in biological fluids. These modifications were pioneered by the Eckstein group (Pieken *et al.* 1991; Williams *et al.* 1991) and are compatible with T7 RNA polymerase for efficient *in vitro* transcription (Padilla and Sousa 1999; Kujau and Wölfl 1997; Kubik *et al.* 1997). When compared, 2'-deoxy and 2'-amino chemistries have been found to produce aptamers with similiar affinity (Wiegand *et al.*

1996) but 2'-fluoro chemistry produces aptamers with greater thermal stability and greater affinity (Pagratis *et al.* 1997).

Spiegelmers

Spiegelmers (whose name was coined from the German word for mirror: *Spiegel*) offer an interesting alternate route to producing nuclease resistant aptamers. Spiegelmers are mirror-image of aptamers, composed of L-ribose or L-2'-deoxyribose units, that bind specifically to a target, as do aptamers. The chiral inversion confers high biological stability and long life (Klussmann *et al.* 1996; Nolte *et al.* 1996). Spiegelmers are produced using a 'mirror-image' SELEX procedure. Initially, D-oligonucleotides are selected from DNA/RNA libraries – composed of 10^{15} different molecules – against synthetic enantiomers of a chosen target. It is accepted that if a D-oligomer (aptamer) binds a D-peptide, its L-oligomer (spiegelmer) will bind the enantiomer L-peptide. The mirror images of the selected D-oligomers can then be synthesized and their binding ability tested against the natural target (Klussmann *et al.* 1996; Nolte *et al.* 1996). It has been demonstrated that spiegelmers show the same high affinity for targets as their aptamer analogues (Leva *et al.* 2002).

Spiegelmers and mirror image nucleotides have never been found in nature and it is assumed that they are not naturally produced, and thus, nucleases do not efficiently recognize or metabolize them, a property that truly sets spiegelmers apart. Additionally, as they are chemically and structurally similar to D-based nucleic acids, they induce minimal immunogenic response, not being neutralized by the immune system, and induce no toxicity (Sooter and Ellington 2002).

The high selectivity of spiegelmers coupled with their nuclease resistance and non-immunogenic properties makes them ideal candidates for *in vivo* and *in vitro* diagnostics. Currently they are being explored as powerful potential candidates as targeted drug delivery by the start-up company Noxxon AG (Germany), the spiegelmers obviously possessing significant advantages (Wlotzka *et al.* 2002; Sooter and Ellington 2002, Leva *et al.* 2002).

Intramers

Intramers are aptamers that are expressed inside cells using their own transcription machinery and retaining the structure and function for which they have been selected *in vitro*. This means they are intracellularly applied and can be used to target a certain protein in its natural environment. As nucleic acids are natural components of cells, intramers produce no toxic effect and induce lower immunogenic response than foreign proteins. This confers on them a great potential as rapidly generated intracellular inhibitors of biomolecules (being able to inhibit signal transduction, transcription, translation, cell growth and viral infection), as a new tool to study protein functions/interactions *in vivo*, and in general in functional genomics, proteomics and drug discovery (Burke and Nickens 2002; Famulok *et al.* 2000; 2001).

The expression of intramers inside cells has been approached in different ways. Intramers have been cloned in vaccinia recombinant viruses (Blind *et al.* 1999). As vaccinia replicates in the cytoplasm, such intramers are directly produced in this cellular compartment without needing transport from the nucleus. When more general delivery is required, intramers have been expressed under retroviral or polymerase III-controlled promoters, and more recently, cloned in expression vectors based in endogenous tRNA^{met} or U6 snRNA fused to the polymerase III-controlled promoters, giving the construct higher stability and expression levels (Good *et al.* 1997; Paul *et al.* 2002). Introduction in cells has been attempted by infection using recombinant viruses (Blind *et al.* 1999; Mayer *et al.* 2001), or via liposome transfection (Konopka *et al.* 1998; Paul *et al.* 2002) and if necessary, specific signals can be added to the construct to induce localized expression (Paul *et al.* 2003)

The clearest application of intramers is to inactivate or modulate a gene product, knocking out or changing protein activity, without altering the genetic material. In this sense, they have been successfully expressed in cell lines to modulate biological function of cytoplasmic membrane protein domains (Blind *et al.* 1999) and to study the *in vivo* function of cytoplasmic proteins (Mayer *et al.* 2001). It is important to point out that antibodies lose their function in the cytoplasm, presumably due to incorrect folding in the reductive intracellular environment, a phenomenon not observed with intramers. Good results have also been obtained in anti-HIV therapy and HIV production could be inhibited in cell culture by blocking viral protein expression in the cell nucleus (Good *et al.* 1997; Konopka *et al.* 1998; Paul *et al.* 2002). However, the most promising result has been the *in vivo* expression in transfected *Drosophila melanogaster*, under an inducible promotor, showing partial inhibition of a nuclear protein (B52, a regulator of RNA splicing) and affecting phenotype increasing lethality by 50 per cent (Shi *et al.* 1999).

Peptide aptamers

Peptide aptamers appear as new combinatorial biological reagents designed to bind with high specificity to intracellular target proteins and inhibit their function or interactions (Colas *et al.* 1996).

Peptide aptamers are selected among millions of molecules present in combinatorial peptide libraries (Colas *et al.* 1996). They consist of variable peptide loops of about 20 random amino acids, attached to the surface of a protein scaffold (usually *E. coli* thioredoxin – a small, soluble, easily purified and expressed protein), and fused to a yeast/mammalian protein. Such peptide libraries are cloned in plasmids and expressed in two-hybrid yeast or mammalian systems. The peptides are long enough to fold into many different patterns of shape and charge, but due to the anchoring within the scaffold protein, which gives structural constraint and increases affinity to levels comparable to antibodies, the folding of every peptide is more restricted and reproducible than linear peptides. This facilitates the ability to discover, by selection, peptide aptamers with the ability to bind any target protein or gene. Interestingly, the selection can also be performed *in vivo* by using yeast colonies expressing the target protein fused to a reporter protein (e.g. Lex A, LacZ; Colas *et al.* 2000), assisting the target protein to assume a 3D structure similar to the native one, and to facilitating checking the potential toxicity of the peptide aptamer.

Peptide aptamers have to date been applied to therapeutics, where the goal is to select peptide aptamers whose binding to the target protein induces inactivation of the biological activity of the latter by inhibiting the usual protein–protein interactions. Not only has the binding strength of these selected aptamers been observed to be comparable to that of antibodies but recognition has also been shown to be highly specific, some peptide aptamers being able to discriminate between members of protein families (Colas *et al.* 1996; James 2001), allelic forms of proteins (Xu and Luo, 2002; Colas *et al.* 2000), and even showing substrate specific inhibition (Cohen *et al.* 1998). It is well established that many of the proteins and transcription factors involved in cell growth and cancer are regulated by their cellular localization. Peptide aptamers have been successfully used in simple *in vivo* systems and delivered to the desired cellular compartments by attaching them to transporter proteins (Colas *et al.* 2000; Kau and Silver 2003).

In summary, the study and application of peptide aptamers is opening doors in a diverse range of potential application areas. Peptide aptamers can be used to bind and inactivate genes coding proteins under study without mutating them, thus facilitating the study of protein pattern expression. They can also powerfully be used to detect cellular binding partners for a given target protein (Klevenz *et al.* 2002). In this sense, they represent a useful method for studying and manipulating protein function *in vivo* (Kolonin and Finley 2000; Colas *et al.* 2000; Geyer *et al.* 1999; Cohen *et al.* 1998).

Successful selection of peptide aptamers with the ability to inhibit cell-cycle progression (Colas *et al.* 2000; Kau and Silver 2003) suggests new ways of studying cancer development and cell cycle regulation (Xu and Luo 2002; Célis and Gromov 2003). Eventually peptide aptamers could be used to target and disrupt protein–protein interactions related to cancer proliferation. A related application field is validation of novel therapeutic targets and development of new therapeutic molecules against these targets (Schmidt *et al.* 2002), with one of many potential applications of peptide aptamers being as specific antiviral agents (Butz *et al.* 2001).

Aptamers in analysis

Aptamers hold much promise as molecular recognition tools for incorporation in analytical devices, such as biosensors, affinity probe capillary electrophoresis, capillary electrochromatography, affinity chromatography and flow cytometry and this section details the research efforts that have been made to exploit aptamers for these analytical applications.

Aptamers in affinity-based assays

The advantages of aptamers over antibodies outlined in Table 10.1 position aptamers as potential candidates for replacing antibodies as biorecognition tools in diagnostics and, despite the fact that this use of aptamers is still in its infancy, the results obtained to date indicate a promising future. Added to the advantages detailed, aptamers are extremely flexible and a range of diverse assay formats exploiting aptamers can be imagined. Some of the potential assay permutations, where mixed antibody–aptamer ELONAs as well as pure aptamer ELONAs, are depicted in Figure 10.3. Moreover, the use of molecular aptamer beacons (detailed later) could contribute significantly to the bioassay arena, facilitating a rapid, one-step assay for target detection.

Table 10.1 Advantages of aptamers over antibodies for analysis using enzyme linked oligonucleotide assays and aptasensors

Antibodies	Aptamers
Limitations against target representing constituents of the body and toxic substances	Toxins as well as molecules that do not elicit good immune response can be used to generate high affinity aptamers
Kinetic parameters of Ab-Ag interactions cannot be changed on demand	Kinetic parameters such as on/off rates can be changed on demand
Antibodies have limited shelf life and are sensitive to temperature and may undergo denaturation	Denatured aptamers can be regenerated within minutes, aptamers are stable to long-term storage and can be transported at ambient temperature
Identification of antibodies that recognize targets under conditions other than physiological is not feasible	Selection conditions can be manipulated to obtain aptamers with properties desirable for *in vitro* assay e.g. non-physiological buffer/T
Antibodies often suffer from batch to batch variation	Aptamers are produced by chemical synthesis resulting in little or no batch to batch variation
Requires the use of animals	Aptamers are identified through an *in vitro* process not requiring animals
Labelling of antibodies can cause loss in affinity	Reporter molecules can be attached to aptamers at precise locations not involved in binding

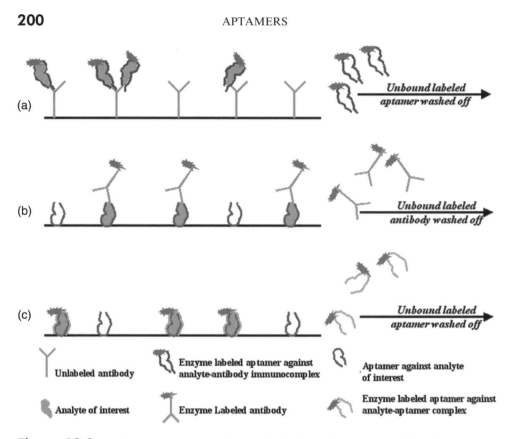

Figure 10.3 Aptamers can be used in affinity based assays as either detector or signalling probes, or as both. As is depicted in the schematic, aptamers can be flexibly used to detect antibody–analyte couples, analyte, or aptamer–analyte couples

One of the examples of the use of aptamers is the sandwich assay developed by Drolet *et al.* (1996), which used a capture monoclonal antibody specific for the vesicular endothelial growth factor (VEGF) to detect VEGF in serum. Subsequent to sample addition, a fluorescein-labelled RNA-based aptamer to VEGF was added and the aptamer-VEGF couple was recognized by an enzyme-labelled anti-fluorescein antibody.

Kawazoe *et al.* (1997) have described an assay for the detection of Reactive Green 19 (RG19), labelling a DNA aptamer specific for this target molecule with fluorescein. Using circular dichroism, the team demonstrated that the fluorescein moiety did not significantly affect the conformation of the aptamer, and consequently, its binding affinity. They used this modified aptamer to semi-quantify the amount of RG19 immobilized on a gel, detecting 2 µmol of the target analyte.

Aptamers have also been used in assays in combination with beads or cells. Davis *et al.* (1996) attached a fluorescein moiety to an anti-human neutrophil elastase (HNE) aptamer and incubated the fluorescent aptamer with HNE-coated beads. The fluorescence emitted by the complexes was measured by flow cytometry. The same authors (Davis *et al.* 1998) went on to use fluorescein- and phycoer-ythrin-labelled aptamers against human CD4 expressed on cell surfaces. The results clearly indicated that the aptamers specifically stained T cells that express CD4 in a heterogeneous mixture of human peripheral blood mononuclear cells.

Exploiting the properties of magnetics beads, Rye and Nustad (2001) developed a hybrid immunobead assay based on a 5'-biotinylated DNA thrombin aptamer and an anti-thrombin antibody. The authors linked sheep anti-mouse IgG to immunomagnetic beads and afterwards incubated the pre-modified beads with IgG anti-thrombin monoclonal antibody, EST-7. The resulting EST-7-coated magnetic beads were incubated with a pre-incubated mixture of thrombin and biotinylated thrombin aptamer. Finally, europium(Eu)-labelled streptavidin was added, followed by an enhancing solution, and the fluorescence measured. With this immunobead assay, it was demonstrated that the thrombin DNA aptamer was able to bind to the target analyte under stringent conditions and at physiological concentrations.

In a similar manner, Bruno and Kiel (2002) conjugated cholera whole toxin and staphylococcal enterotoxin B (SEB) to tosyl-activated magnetic beads and used the corresponding specific aptamers in several assay formats. The limits of detection of the biotoxins were in the nanogram to the low picogram range. They also used the same strategy to select and amplify aptamers against non-pathogenic Sterne strain *Bacillus anthracis* spores.

As depicted in Figure 10.3, aptamers can also be used as both capture and signalling probes in the same sandwich assay, due to their ability to recognize different binding sites of the target analyte, and an example of this has been demonstrated by Drolet *et al.* (1996). In this work, the signalling probe was selected not simply against the target analyte but against the capture probe:target analyte complex.

Despite the immense promise of aptamers, their exploitation in ELISA-type assays has been quite limited to date. However, the results of the works cited here together with the ability of aptamers to discriminate between similar molecules only differing in small structural changes (Rehder and McGrown 2001), to discriminate between enantiomers (Geiger *et al.* 1996), and to detect small molecules (Frauendorf and Jäschke 2001) show the clear advantages and potential of aptamers over antibodies as analytical tools in diagnostic assays, and with increasing accessibility and commercial availability of aptamers, their use in affinity based assays should increase exponentially over the next few years.

Use of aptamers in affinity chromatography

Due to the high affinity offered by aptamers, these molecules have been used as tools in several analytical techniques, such as capillary electrophoresis (CE), capillary electrochromatography (CEC) and affinity chromatography (AC).

Capillary electrophoresis is a technique used to separate and detect analytes in a capillary format. This technique is capable of separating uncharged analytes, requires small volumes and short analysis times, can be automated, and presents high sensitivity. When aptamers are used in capillary electrophoresis (CE), the technique is called affinity probe capillary electrophoresis (APCE). The main advantage of using aptamers over other molecules is the ability to design the aptamer with specific characteristics, such as electrophoretic mobility. In this direction, German *et al.* (1998) used a fluorescent DNA aptamer to detect IgE in buffer and serum samples using APCE. This technique yielded well-defined peaks for both the aptamer and the aptamer-IgE complex. The detection limit for IgE was 46 pM with a linear dynamic range of 10^5, with the assay being highly specific, with IgG failing to be detected and IgE showing no interaction with alternate aptamers. Moreover, the presence of serum did not affect the analytical response. The same team used an equivalent strategy to detect thrombin, achieving a detection limit of 40 nM (this lower detection limit can be attributed to the lower binding constant of the anti-thrombin aptamers utilized).

Pavsky and Le (2001) also used APCE with fluorescent aptamer probes specifically to detect the reverse transcriptase (RT) of the type 1 human immunodeficiency virus (HIV-1). The aptamer did not exhibit cross-reactivity with other RTs and moreover, the calibration curve was linear to 50 nM.

Capillary electrochromatography (CEC) is a hybrid technique that combines chromatography with electro-osmotic flow to drive the solutes through the capillary. The advantage of this technique is the highly efficient separations achievable, the short times required for the assay, and the low volumes needed. Rehder and McGrown (2001) used an aptamer-derivatized stationary phase to separate variants LgA and LgB of β-lactoglobulin. These two variants differ by only two residues, but this small difference in sequence results in significant differences in thermal stability, structural flexibility and tertiary structure. The G-quartet aptamer structure plays an important role in the efficiency of the separation, as no separation was achieved with other structures. The results obtained using the aptamer in the stationary phase suggested that the technique could be useful for rapid and non-denaturing separation of proteins.

Aptamers have been also used in affinity chromatography (AC). In this technique, the aptamer immobilized on the stationary phase recognizes the target molecule in the mobile phase. Geiger *et al.* (1996) used a modified *in vitro* selection scheme based on affinity chromatography in the presence of L-arginine to allow enrichment of the specific RNA aptamers. The high enantioselectivity of the arginine by this aptamer was demonstrated by the 12 000-fold better recognition towards L-arginine than towards D-arginine. Romig *et al.* (1999) immobilized a DNA aptamer specific for human L-selectin to a chromatography support to create a chromatography column. The column was used to purify the recombinant human L-selectin-IgG fusion protein from Chinese hamster ovary cell-conditioned medium, since the aptamer can recognize the lectin domain of

L-selectin in a divalent cation-dependent manner. This matrix allowed a 1500-fold purification with a 83 per cent single step recovery. Deng *et al.* (2001) covalently attached a biotinylated DNA aptamer to porous chromatographic supports and packed the aptamer medium into fused-silica capillaries to perform affinity chromatography. After optimization of the mobile phase conditions, the column could selectively retain and separate cyclic-AMP, NAD^+, AMP, ADP, ADT and adenosine, even in tissue extracts.

Aptasensors

Due to the high molecular recognition capability of aptamers, their use in biosensors, coined aptasensors (O'Sullivan, 2002), has found increasing recent application. Most of the aptasensors use aptamers as recognition probes (unlike most of the ELISA/ELONA assays, which use them as signalling probes). The detector used to convert the biorecognition event into a measurable signal is usually optical, fluorescence being the most commonly used mode of transduction. Aptamers, as well as offering the previously described advantages of being small, simple, easy to manufacture, rationally designable and having high sensitivity, can also, when incorporated into a biosensor, be regenerated for repetitive use, a facet that has proved difficult to achieve with antibody-based biosensors.

Kleinjung *et al.* (1998) immobilized a biotin-labelled RNA aptamer specific against L-adenosine to a streptavidin-derivatized optical fibre core to use it as molecular recognition element in a biosensor. The binding of FITC-labelled L-adenosine was characterized using total internal reflection fluorescence and by the determination of the association and dissociation rates. The competitive binding assay allowed the selective detection of the chiral enantiomer L-adenosine with a discrimination of 1700-fold respect to the other enantiomer, with the analyte being quantified in the submicromolar range.

Potyrailo *et al.* (1998) covalently bound a fluorescently labelled anti-thrombin DNA aptamer to a glass support and detected the thrombin in solution measuring the changes in the evanescent wave-induced fluorescence anisotropy of the immobilized aptamer. They reported a 0.7 mol limit of detection in a 140 pL volume, with a dynamic range of three orders of magnitude, and in an assay requiring less than 10 min. Moreover, removing the bound thrombin with a denaturating agent, repetitive measurements were performed.

Lee and Walt (2000) developed a microarray using an anti-thrombin DNA aptamer. In this case, they immobilized the aptamer on the surface of silica microspheres and distributed them into microwells of optical imaging fibres, creating a microarray. The fluorescence emitted by the fluorescein-labelled thrombin was measured by a competitive binding assay. The beads allowed the regeneration of the biosensor and their re-use without any sensitivity change. The limit of detection achieved was 1 nM and the assay could be performed in 15 min (including regeneration). Although all the sites of the microarray were used to detect the same analyte, the results demonstrate the feasibility of using aptamer microarrays for simultaneous detection of several target analytes.

Non-labelled aptamers have been also used in biosensors, with the optical technique used to detect the binding event being Surface Plasmon Resonance (SPR). In one example, unlike the other biosensor formats, the target analytes were immobilized on the solid support with the aptamer in the mobile phase. Following identification of RNA aptamers specific for human $2'-5'$ oligoadenylate synthetase, Hartmann et al. (1998), used SPR and found that the binding dissociation constants were in the nanomolar range. By competition experiments, they discovered that the binding site for single-stranded RNAs at least partially overlaps that for the synthetic double-stranded RNA. Kraus et al. (1998) also used SPR to demonstrate the immobilization of aptamers against CD4 and the recognition of the binding site in the CDR2-like region in domain 1 of CD4 by these aptamers with high affinity and specificity.

Almost all the aptasensors reported to date have been based on optical transduction but aptamers have also been used as molecular probes with piezoelectric transduction. Liss et al. (2002) demonstrated a quartz crystal biosensor using a model system based on the immobilization of DNA aptamers and with human IgE as target analyte. In order to compare the aptasensors with an immunosensor, they also used anti-IgE antibodies as recognition molecules. Both sensors gave limits of detection of 0.5 nM. Moreover, the aptamer immobilization was dense and well-orientated, which extended the linear detection range to 10-fold higher concentrations of IgE, and tolerated regeneration with little loss of sensitivity. Finally, the aptasensor specifically detected and quantified the analyte in complex protein mixtures and the aptamers were stable to temperature and over several weeks.

Molecular aptamer beacons

Molecular beacons were initially described in 1996 as nucleic acids probes able to undergo spontaneous fluorogenic conformational change after hybridizing with specific nucleic acid targets (Tyagi and Kramer 1996). The discovery of aptamers allowed the application of such technology to protein/peptide detection by developing molecular aptamer beacons, which has immense potential both for affinity assays and aptasensors.

As reported by Yamamoto and Kumar (2002), any known aptamer can be engineered into a molecular aptamer beacon by adding a small nucleotide sequence to the 5'-end, complementary to a small sequence added to the 3'-end. A fluorophore can then be covalently attached to the 5'-end, while a fluorescence quencher is attached to the 3'-end. In the absence of ligand, the complementary extremes will form a duplex, forcing the beacon into a stem-loop structure. This places the fluorophore and quencher close to each other, resulting in no fluorescent signal as the fluorophore is quenched by energy transfer to the quencher (Figure 10.4). Ideally, when the ligand is present, the aptamer portion binds it, forming a probe-target hybrid that is stronger and more stable than the stem hairpin, and disrupting the stem. The resulting conformational change opens the structure, distancing the quencher and the fluorophore, and allowing the qualitative, or semi-quantitative measurement of the resulting fluorescent signal.

An alternative possibility reported was the use of a FRET (fluorescence resonance energy transfer) approach, labelling both ends of the aptamer with individual fluorophores that behave as a donor–acceptor pair (e.g. coumarin and carboxyfluorescein). In this case, the aptamer beacon shows an increased

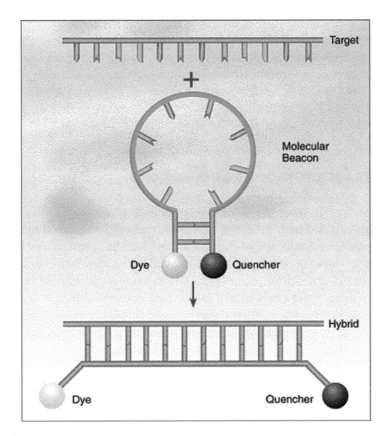

Figure 10.4 A molecular beacon begins as a stem-and-loop structure. The sequences at the ends of the probe match and bind, creating the stem, while the rest of the probe is unmatched and unbound, creating the loop. While folded this way, the fluorophore at one end of the probe is next to the quencher at the other end. When the probe binds to a single-stranded DNA template, the structure unfolds, separating the quencher from the dye and allowing fluorescence

fluorescence in the presence of target protein, resulting from an increase in acceptor intensity and simultaneous decrease in donor intensity (Li *et al.* 2002). Increased sensitivity and quantitation of the target was achieved with the FRET approach.

Molecular aptamer beacons have been successfully used to detect Tat HIV protein (Yamamoto and Kumar 2002) and thrombin (Hamaguchi *et al.* 2001; Li *et al.* 2002) in solution, and modified aptamer beacon models were used to detect cocaine (Stojanovic *et al.* 2001a) and theophylline (Frauendorf and Jäschke 2001). It has been suggested that aptamer beacons can be used for real-time protein detection, facilitating the monitoring of protein production and the study gene expression/regulation, perhaps even *in vivo* (Li *et al.* 2002). The use of molecular beacon structures with hammerhead ribozymes and catalytic RNA has also been reported (Stojanovic *et al.* 2001b; Frauendorf and Jäschke 2001).

For the realization of aptasensors, it is necessary to immobilize the aptamer probe (or molecular aptamer beacon) and to date this has not been reported. However, as DNA directed molecular beacons incorporating an immobilizable hairpin structure able directly to transduce molecular recognition into optical signal have been successfully developed (Fang *et al.* 1999; Henry *et al.* 1999; Liu and Tan, 1999; Li *et al.* 2001), it can only be anticipated that aptasensors exploiting molecular beacons will find application.

Aptamers for proteomic applications

Proteomics is derived from the word proteome that refers to the proteins encoded by an entire genome. The last decade of the 20th century was called the Decade of Genomics, and the first decade of the new millennium has been named the Decade of Proteomics. The sequencing of entire genomes, including the human genome, is resulting in the identification of a huge number of novel proteins whose functions are unknown. The major challenge of biomedical research during the next decade will include characterization of the properties and biological functions of about 100 000 different human proteins, and how these are involved in human diseases (Ryan and Patterson 2002). Proteomics can be divided into classical and functional proteomics. Classical proteomics is focused on studying complete proteomes, e.g. from two differentially treated cell lines, whereas functional proteomics studies more limited protein sets. Classical proteome analyses are usually carried out by using two-dimensional gel electrophoresis (2-DE) for protein separation followed by protein identification by mass spectrometry (MS) and database searches. The functional proteomics approach uses a subset of proteins isolated from the starting material, e.g. with an affinity-based method. This protein subset can then be separated by using normal SDS-PAGE or by 2-DE. Proteome analysis is complementary to DNA microarray technology: with the proteomics approach it is possible to study changes in protein expression levels and also protein–protein interactions and post-translational modifications (Sellers and Yates 2003; Jeffery and Bogyo 2003).

Consequently, there is currently a huge amount of interest in developing microarrays that facilitate delving into the challenging world of functional genomics and proteomics, the protein microarray technology being able to leverage years of improvements in genomic chip development. The use of protein arrays, or protein chips, allows automated, high-throughout analysis but are, however, limited by a variety of technical challenges that are difficult to overcome.

Antibody arrays, for example, can be expensive and time consuming to produce, because they require the synthesis and purification of large numbers of antibodies. Additionally protein arrays have the added complication of storage, as proteins are notoriously less stable than nucleic acids, and, moreover, the difficulties confronted when handling proteins without having an effect on their structure and binding ability presents further problems (Mitchell 2002).

To this end, alternatives to protein based chips are being explored and some advances have been made with cell microarrays, tissue culture microarrays, and with relevance to this chapter, aptamer microarrays. With the advances being made in automated aptamer selection, the cost and time required to produce vast numbers of aptamers will be reduced significantly, effectively eliminating the need for laborious and expensive antibody technology. Aptamers are also expected to be more specific and sensitive, as their chemical structure tolerates more aggressive washing, minimizing undesired background. Moreover, the relative ease of aptamer modification facilitates their facile immobilization via spotting technologies. Finally, the advances in the developments of photoaptamers and molecular aptamer beacons eliminate the requirement of adding a 'detecting' biocomponent, further simplifying microarrays for proteomic applications (Mitchell 2002; Green *et al.* 2001).

SomaLogic (Boulder, CO, USA) is already constructing arrays using the light-sensitive photoaptamers that were described earlier. SomaLogic's photoaptamers can be spotted on a glass surface and used to produce arrays for highly parallel protein profiling. Unlike antibody arrays, which require a secondary binding agent that can lead to problems of cross-reactivity, photoaptamer arrays require only one capture agent: proteins that do not cross-link to the surface of the arrays can be washed away. Preliminary results obtained have not indicated problems of cross-reactivity and the platform is predicted to expand exponentially. Such arrays could be used to detect, study, and even quantify, a vast number of proteins from complex mixtures at the same time (Petach and Gold 2002; Smith *et al.* 2003). Somalogic has recently forged a significant partnership with Celera Genomics Groups providing Celera early access to aptamers and aptamer arrays developed through SomaLogic's proprietary SELEX process. Celera is leveraging the genomic information it has produced to identify new candidate therapeutic targets and diagnostic markers and this partnership will contribute significantly to the realisation of proteomic applications.

Aptamers have been also applied in other formats for high throughout screening (HTS), an example being that reported by Green *et al.* (2001), where aptamers were used in competition binding HTS assays to identify and optimize small-molecule ligands to protein targets. To illustrate the application, the team used labelled aptamers to platelet-derived growth factor B-chain and wheat germ agglutinin to screen two sets of potential small-molecule ligands, and in both cases binding affinities of the ligands tested were strongly correlated with their inhibitory potencies in functional assays. This demonstration clearly indicates the potent possibilities of this type of HTS assay format for drug discovery, having particular application for protein targets that have no known binding partners, such as orphan receptors.

In vitro microarray systems can provide immense useful information in the area of quantitative proteomics but for true functional proteomics, *in vivo* systems for studying proteins in the complex cellular environment are required. Intramers are probably the most promising new tools in functional proteomics. As has been already discussed, intramers are expressed inside cells as part of expression vectors. They can then interact with targets in their natural environment and be used to study protein functions/interactions *in vivo* (Burke and Nickens 2002; Famulok *et al.* 2000; 2001; Mayer *et al.* 2001; Blind *et al.* 1999; Shi *et al.* 1999), thus showing immense promise for advances in *in vivo* functional proteomics.

Perspectives and future outlook

Since the inception of aptamers in 1990, there has been a plethora of exciting developments and discoveries reported and the area of aptamers technology is proving itself to be a dynamic machine whose cogs are continuously oiled by enabling advancements that are contributing to the true facilitation of aptamers as tools for diagnostics and bioanalysis, therapeutics and drug discovery, and quantitative and functional proteomics. The capacity of these powerful molecular tools is being further enhanced by recent improvements in their resistance to nucleases via chemical modification and the very promising area of spiegelmers, as well as the demonstration of the expression of aptamers within cells (intramers).

These advancements have guaranteed the widespread application of aptamers and coupled with the development of robotic platforms and microchips for the automated selection of aptamers, the application and use of aptamers can only be anticipated to increase exponentially. The recent establishment of several start-up and spin-off companies focused on aptamer technology further supports this anticipation. Aptamers were initially commercially exploited by NexStar (Colorado, USA) and in 1999 NexStar merged with Gilead Sciences (California, USA), with an agreement that Gilead would exploit the therapeutic use of aptamers and that NexStar's founder, Larry Gold would have the rights to exploit the diagnostic uses of aptamers. Gold went on to establish SomaLogic (Colorado, USA) in 1999, a company whose focus is on the use of aptamers for diagnostic and proteomic applications, and it is this company that has developed photo-SELEX and photoaptamers. Noxxon AG (Berlin, Germany), also established in 1999, is exploiting spiegelmers for therapeutics and drug discovery and Archemix, a company based in Massachusetts (USA) and established by Ron Breaker and Andrew Ellington is looking to exploit aptamers not only for therapeutic applications but also as components for single-molecule detection biosensors. Finally, Vitex (also based in Massachusetts, USA), a biotechnology company dedicated to developing products that inactivate and remove pathogens in blood, is working together with William James to exploit aptamers for the diagnosis of prion diseases.

Thus, the perspectives of aptamer technology are clear. Before 2005, we can anticipate the launch of these companies' technologies onto the market and their incorporation into our daily lives. When one considers that both liposomes and monoclonal antibodies were reported approximately 20 years before the first therapeutic applications were approved for clinical use, it appears that the medical applications of aptamers are on a very fast historical track (Brody and Gold 2000), and the potential impact on human life is immense with the personalized medicine and theranostics aptamers support being on the imminent horizon (Jain 2002).

Acknowledgements

This review has been carried out with financial support from the Commission of the European Communities, specific RTD programme 'Quality of Life and Management of Living Resources', QLKI-2002-02077 'Quantification of Coeliac Disease toxic gluten in foodstuffs using a Chip system with integrated Extraction, Fluidics and biosensoric detection'. It does not necessarily reflect its views and in no way anticipates the Commission's future policy in this area.

Abbreviations

2-DE	two-dimensional gel electrophoresis
3D	three dimensional
AC	affinity chromatography
APCE	affinity probe capillary electrophoresis

CE	capillary electrophoresis
CEC	capillary electrochromatography
DNA	deoxyribonucleic acid
ELISA	enzyme linked immunosorbent assay
ELONA	enzyme linked oligonucleotide assay
FITC	fluorescein isothiocyanate
HIV	human immunodeficiency virus
HNE	human neutrophil elastase
HTS	high throughput screening
MS	mass spectrometry
PCR	polymerase chain reaction
RNA	ribonucleic acid
RT	reverse transcriptase
SELEX	systematic evolution of ligands by exponential enrichment
SPR	surface plasmon resonance
UV	ultra-violet

References

Blind, M., Kolanus, W. and Famulok, M. (1999) Cytoplasmic RNA modulators of an inside-out signal-transduction cascade. *Proc Natl Acad Sci USA* **96**: 3606–3610.

Brody, E.N. and Gold, L. (2000) Aptamers as therapeutic and diagnostic agents. *J Mol Biotechnol* **74**: 5–13.

Bruno, J.G. and Kiel, J.L. (2002) Use of magnetic beads in selection and detection of biotoxin aptamers by electrochemiluminescence and enzymatic methods. *BioTechniques* **32**: 178–183.

Burke, D.H. and Gold, L. (1997) RNA aptamers to the adenosine moiety of S-adenosyl methionine: structural interferences from variations on a theme and the reproducibility of SELEX. *Nucleic Acids Res* **25**: 2020–2024.

Burke, D.H. and Nickens, D.G. (2002) Expressing RNA aptamers inside cells to reveal proteome and ribonome function. *Brief Funct Genom Prot* **1(2)**: 169–188.

Butz, K., Denk, C., Fischer, B., Crnkovic-Mertens, I., *et al.* (2001) Peptide aptamers targeting the hepatitis B virus core protein: a new class of molecules with antiviral activity. *Oncogene* **20**: 6579–6586.

Celis, J.E. and Gromov, P. (2003) Proteomics in translational cancer research: toward an integrated approach. *Cancer Cell* **3(1)**: 9–15.

Chambers, J.J. and Valdes, J.J. (2002) In vitro selection of lysozyme specific aptamers on silicon chips. Presented as a poster at the 23rd Army Science Conference, 2–5, Dec 2002, Florida, USA.

Cohen, B.A., Colas, P. and Brent, R. (1998) An artificial cell-cycle inhibitor isolated from a combinatorial library. *Proc Natl Acad Sci USA* **95**: 14272–14277.

Colas, P., Cohen, B., Jessen, T., Grishina, I., *et al.* (1996) Genetic selection of peptide aptamers that recognize and inhibit cyclin-dependent kinase 2. *Nature* **380**: 548–550.

Colas, P., Cohen, B., Ko Ferrigno, P., Silver, P.A. and Brent, R. (2000) Targeted modification and transportation of cellular proteins. *Proc Natl Acad Sci USA* **97**: 13720–13725.

Cook, G.P. and Greenberg, M.M. (1996) A novel mechanism for the formation of direct strand breaks upon anaerobic photolysis of duplex DNA containing 5-bromo-deoxyuridine. *J Am Chem Soc* **118**: 10025–10030.

Cox, J.C. and Ellington, A.D. (2001) Automated selection of anti-protein aptamers. *Bioorg Med Chem* **9**: 2525–2531.

Cox, J.C., Rudolph, P. and Ellington, A.D. (1998) Automated RNA selection. *Biotechnol Prog* **14(6)**: 845–850.

Davis, K.A., Abrams, B., Lin, Y. and Jayasena, S.D. (1996) Use of a high affinity DNA ligand in flow cytometry. *Nucleic Acids Res* **24**: 702–706.

Davis, K.A., Abrams, B., Lin, Y. and Jayasena, S.D. (1998) Staining of cell surface human CD4 with 2′-F-pyrimidine-containing RNA aptamers for flow cytometry. *Nucleic Acids Res* **26**: 3915–3924.

Deng, Q., German, I., Buchanan, D. and Kennedy, R.T. (2001) Retention and separation of adenosine and analogues by affinity chromatography with an aptamer stationary phase. *Anal Chem* **73**: 5415–5421.

Dietz, T.M. and Koch, T.H. (1987) Photochemical coupling of 5-bromouracil to tryptophan, tyrosine and histidine, peptide-like derivatives in aqueous fluid solution. *Photochem Photobiol* **46**: 971–978.

Dietz, T.M. and Koch, T.H. (1989) Photochemical reduction of 5-bromouracil by cystine derivatives and coupling of 5-bromouracil to cystine derivatives. *Photochem Photobiol* **49**: 121–129.

Drolet, D.W., Moon-McDermott, L. and Romig, T.S. (1996) An enzyme-linked oligonucleotide assay. *Nat Biotechnol* **14**: 1021–1025.

Ellington, A.D. and Szostak, J.W. (1990) In vitro selection of RNA molecules that bind specific ligands. *Nature* **346**: 818–822.

Famulok, M. (1994) Molecular recognition of amino acids by RNA-aptamers: an L-citrulline binding RNA motif and its evolution into an L-arginine binder. *J Am Chem Soc* **116**: 1698–1706.

Famulok, M., Mayer, G. and Blind, M. (2000) Nucleic acid aptamers – from selection in vitro to applications in vivo. *Accounts Chem Res* **33**: 591–599.

Famulok, M., Blind, M. and Mayer, G. (2001) Intramers as new promising new tools in functional proteomics. *Chem Biol* **8**: 931–939.

Fang, X., Liu, X., Schuster, S. and Tan, W. (1999) Designing a novel molecular beacon for surface-immobilised DNA hybridisation studies. *J Am Chem Soc* **121**: 2921–2922.

Frauendorf, C. and Jäschke, A. (2001) Detection of small organic analytes by fluorescing molecular switches. *Bioorg Med Chem* **9**: 2521–2524.

Gebhardt, K., Shokraei, A., Babaie, E. and Lindquist, B.H. (2000) RNA aptamers to S-adenosylhomocysteine: kinetic properties, divalent cation dependency, and comparison with anti-S-adenosylhomocysteine antibody. *Biochemistry* **39(24)**: 7255–7265.

Geiger, A., Burgstaller, P., von der Eltz, H., Roeder, A. and Famulok, M. (1996) RNA aptamers that bind L-arginine with sub-micromolar dissociation constants and high enantioselectively. *Nucleic Acids Res* **24**: 1029–1036.

German, I., Buchanan, D.D. and Kennedy, R.T. (1998) Aptamers as ligands in affinity probe capillary electrophoresis. *Anal Chem* **70**: 4540–4545.

Geyer, C.R., Colman-Lerner, A. and Brent, R. (1999) 'Mutagenesis' by peptide aptamers identifies genetic network members and pathway connections. *Proc Natl Acad Sci USA* **96(15)**: 8567–8572.

Golden, M.C., Collins, B.D., Willis, M.C. and Koch, T.H. (2000) Diagnostic of photo-SELEX-evolved ssDNA aptamers. *J Biotechnol* **81**: 167–178.

Good, P.D., Krikos, A.J., Li, S.X., Bertrand, E., *et al.* (1997) Expression of small, therapeutic RNAs in human cell nuclei. *Gene Ther* **4(1)**: 45–54.

Green, L.S., Bell, C. and Janjic, N. (2001) Aptamers as reagents for high-throughput screening. *BioTechniques* **30**: 1094–1110.

Haller, A.A. and Sarnow, P. (1997) In vitro selection of a 7-methyl-guanosine binding RNA that inhibits translation of capped mRNA molecules. *Proc Natl Acad Sci USA* **94(16)**: 8521–8526.

Hamaguchi, N., Ellington, A. and Stanton, M. (2001) Aptamer beacons for the direct detection of proteins. *Anal Biochem* **294**: 126–131.

Hartmann, R., Norby, P.L., Martensen, P.M., Jorgensen, P., *et al.* (1998) Activation of $2'-5'$ oligoadenylate synthetase by single-stranded and double stranded RNA aptamers. *J Biol Chem* **273**: 3236–3246.

Henry, M.R., Wilkins Stevens, P., Sun, J. and Kelso, D.M. (1999) Real-time measurements of DNA hybridisation on microparticles with fluorescence resonance energy transfer. *Anal Biochem* **276**: 204–214.

Hesselberth, J., Robertson, M.P., Jhaveri, S. and Ellington, A.D. (2000) *In vitro* selection of nucleic acids for diagnostic applications. *Rev Mol Biotechnol* **74**: 15–25.

Holeman, L.A., Robinson, S.L., Szostak, J.W. and Wilson, C. (1998) Isolation and characterization of fluorophore-binding RNA aptamers. *Fold Des* **3(6)**: 423–431.

Ito, S., Saito, I. and Matsuura, T. (1980) Acetone.sensitized photocoupling of 5-bromouridine to tryptophan derivatives via electrontransfer process. *J Am Chem Soc* **102**: 7535–7541.

Jain, K.K. (2002) Personalized medicine. *Curr Opin Mol Ther* **4(6)**: 548–558.

James, W. (2000) Aptamers. In *Encyclopedia of Analytical Chemistry*. R.A. Meyers (Ed), pp. 4848–4871. John Wiley and Sons Ltd, Chichester.

James, W. (2001) Nucleic acid and polypeptide aptamers: a powerful approach to ligand discovery. *Curr Opin Pharmacol* **1**: 540–546.

Jayasena, S.D. (1999) Aptamers, an emerging class of molecules that rival antibodies in diagnostics. *Clin Chem* **45(9)**: 1628–1650.

Jeffery, D.A and Bogyo, M. (2003) Chemical proteomics and its application to drug discovery. *Curr Opin Biotechnol* **14(1)**: 87–95.

Jeong, S., Eom, T., Kim, S., Lee, S. and Yu, J. (2001) In vitro selection of the RNA aptamer against the Sialyl Lewis X and its inhibition of the cell adhesion. *Biochem Biophys Res Commun* **281(1)**: 237–243.

Jhaveri, S., Olwin, B. and Ellington, A.D. (1998) In vitro selection of phosphorothiolated aptamers. *Bioorg Med Chem Lett* **8(17)**: 2285–2290.

Kau, T.R. and Silver, P.A. (2003) Nuclear transport as a target for cell growth. *Drug Disc Today* **8(2)**: 78–85.

Kawakami, J., Imanaka, H., Yokota, Y. and Sugimoto, N. (2000) In vitro selection of aptamers that act with Zn^{2+}. *J Inorg Biochem* **82(1–4)**: 197–206.

Kawazoe, N., Ito, Y. and Imanishi, Y. (1997) Bioassay using a labelled oligonucleotide obtained by *in vitro* selection. *Biotechnol Prog* **13**: 873–874.

King, D.J., Ventura, D.A., Brasier, A.R. and Gorenstein, D.G. (1998) Novel combinatorial selection of phosphothioate oligonucleotide aptamers. *Biochemistry* **37(47)**: 16489–16493.

Kleinjung, F., Klussmann, S., Erdmann, V.A., Scheller, F.W., *et al.* (1998) High-affinity RNA as recognition element in a biosensor. *Anal Chem* **70**: 328–331.

Klevenz, B., Butz, K. and Hoppe-Seyler, F. (2002) Peptide aptamers: exchange of the thioredoxin-A scaffold by alternative platform proteins and its influence on target protein binding. *Cell Mol Life Sci* **59(11)**: 1993–1998.

Klussman, S., Nolte, A., Bald, R., Erdmann, V.A. and Fürste, J.P. (1996) Mirror-image RNA that binds D-adenosine. *Nat Biotechnol* **14(9)**: 1112–1116.

Kolonin, M.G. and Finley, R.H. (2000) A role for cyclin J in the rapid nuclear division cycles of early *Drosophila* embryogenesis. *Develop Biol* **227(2)**: 661–672.

Konopka, K., Duzgunes, N., Rossi, J. and Lee, N.S. (1998) Receptor ligand-facilitated cationic liposome delivery of anti-HIV-1 Rev-binding aptamer and ribozyme DNAs. *J Drug Target* **5(4)**: 247–259.

Kraus, E., James, W. and Barclay, A.N. (1998) Novel RNA ligands able to bind CD4 antigen and inhibit CD4(+) T lymphocyte function. *J Immunol* **160**: 5209–5212.

Kubik, M.F., Bell, C., Fitzwater, T., Watson, S.R. and Tasset, D.M. (1997) Isolation and characterization of 2′-fluoro-, 2′-amino-, and 2′-fluoro-/amino-modified RNA ligands to human IFN-gamma that inhibit receptor binding. *J Immunol* **159(1)**: 259–67.

Kujau, M.J. and Wölfl, S. (1998) Intramolecular derivatization of 2′-amino-pyrimidine modified RNA with functional groups that is compatible with re-amplification. *Nucleic Acids Res* **26(7)**: 1851–1853.

Kusser, W. (2000) Chemical modified nucleic acid aptamers for in vitro selections: evolving evolution. *J Biotechnol* **74**: 27–38.

Lee, M. and Walt, D.R. (2000) A fiber-optic microarray biosensor using aptamers as receptors. *Anal Biochem* **282**: 142–146.

Leva, S., Burmeister, J., Muhn, P., Jahnke, B., *et al.* (2002) GnRH binding RNA and DNA spiegelmers: a novel approach toward GnRH antagonism. *Chem Biol* **9(3)**: 351–359.

Li, J., Tan, W., Wang, K., Xiao, D., *et al.* (2001) Ultrasensitive optical DNA biosensor based on surface immobilization of molecular beacon by a bridge structure. *Anal Sci* **17**: 1149–1153.

Li, J.J., Fang, X. and Tan, W. (2002) Molecular aptamer beacons for real-time protein recognition. *Biochem Biophys Res Commun* **292**: 31–40.

Liss, M., Petersen, B., Wolf, H. and Prohaska, E. (2002) An aptamer-based quartz crystal protein biosensor. *Anal Chem* **74**: 4488–4495.

Liu, X. and Tan, W. (1999) A fiber-optic evanescent wave DN biosensor based on novel molecular beacons. *Anal Chem* **71**: 5054–5059.

Mayer, G., Blind, M., Nagel W., Böhm, T., *et al.* (2001) Controlling small guanine-nucleotide-exchange factor through cytoplasmic RNA intramers. *Proc Natl Acad Sci USA* **98(9)**: 4961–4965.

Meisenheimer, K.M. and Koch, T.H. (1997) Photocross-linking of nucleic acids to associated proteins. *Crit Rev Biochem Mol Biol* **32**: 101–140.

Mitchell, P. (2002) A perspective on protein microarrays. *Nat Biotechnol* **20**: 225–229.

Nolte, A., Klussman, S., Bald, R., Erdmann, V.A. and Fürste, J.P. (1996) Mirror-design of L-oligonucleotide ligands binding to L-arginine. *Nat Biotechnol* **14(9)**: 116–119.

Osborne, S.E., Matsumura, I. and Ellington, A.D. (1997) Aptamers as therapeutic and diagnostic reagents: problems and prospects. *Curr Opin Chem Biol* **1**: 5–9.

O'Sullivan, C.K. (2002) Aptasensors – the future of biosensing? *Anal Bioanal Chem* **372**: 44–48.

Padilla, R. and Sousa, R. (1999) Efficient synthesis of nucleic acids heavily modified with non-canonical ribose 2′-groups using a mutant T7 RNA polymerase (RNAP). *Nucleic Acids Res* **27(6)**: 1561–1563.

Pagratis, N.C., Fitzwater, T., Jellinek, D. and Dang, C. (1997) Potent 2′-amino- and 2′-fluoro-2′-deoxyribonucleotide RNA inhibitors of Keratinocyte Growth Factor. *Nat Biotechnol* **15**: 68–73.

Paul, C.P., Good, P.D., Winer, I. and Engelke, D.R. (2002) Effective expression of small interfering RNA in human cells. *Nat Biotechnol* **20(5)**: 505–508.

Paul, C.P., Good, P.D., Li, S.X., Kleiauer, A., *et al.* (2003) Localized expression of small RNA inhibitors in human cells. *Mol Ther* **7(2)**: 237–247.

Pavsky, V. and Le, X.C. (2001) Detection of human immunodeficiency virus type 1 reverse transcriptase using aptamers as probes in affinity capillary electrophoresis. *Anal Chem* **73**: 6070–6078.

Petach, H. and Gold, L. (2002) Dimensionality is the issue: use of photoaptamers in protein microarrays. *Curr Opin Biotechnol* **13**: 309–314.

Pieken, W.A., Olsen, D.B., Benseler, F., Aurup, H. and Eckstein, F. (1991) Kinetic characterization of ribonuclease-resistant 2′-modified hammerhead ribozymes. *Science* **253**: 314–317.

Potyrailo, R.A., Conrad, R.C., Ellington, A.D. and Hieftje, G.M. (1998) Adapting selected nucleic acid ligands (aptamers) to biosensors. *Anal Chem* **70**: 3419–3425.

Rehder, M.A. and McGrown, L.B. (2001) Open-tubular capillary electrochromatography of bovine ß-lactoglobulin variants A and B using and aptamer stationary phase. *Electrophoresis* **22**: 3759–3764.

Robertson, D.L. and Joyce G.F. (1990) Selection in vitro of an RNA enzyme that specifically cleaves single-stranded DNA. *Nature* **344**: 467–468.

Romig, T.S., Bell, C. and Drolet, D.W. (1999) Aptamer affinity chromatography: combinatorial chemistry applied to protein purification. *J Chromatogr* **731**: 275–284.

Ruckman, J., Green, L.S., Waugh, S., Gillette, W.L., *et al.* (1998) 2′-fluoropyrimidine RNA-based aptamers to the 165-amino acid form of Vascular Endothelial Growth Factor (VEGF165). Inhibition of receptor binding and VEGF-induced vascular permeability through interactions requiring the exon 7-encoded domain. *J Biol Chem* **273**: 20556–20567.

Ryan T.E. and Patterson, S.D. (2002) Proteomics: drug target discovery on an industrial scale. *Trends Biotech* **20(12 Suppl)**: S45–51.

Rye, P.D. and Nustad, K. (2001) Immunomagnetic DNA aptamer assay. *BioTechniques* **30**: 290–295.

Schmidt, S., Diriong, S., Mery, J., Fabbrizio, E. and Debant, A. (2002) Identification of the first Rho-GEF inhibitor, TRIP alpha, which targets the RhoA-specific GEF domain of Trio. *FEBS-Letters* **523**(1–3): 35–42.

Schürer, H., Buchynskyy, A., Korn, K., Famulok, M., *et al.* (2001) Fluorescence correlation spectroscopy as a new method for the investigation of aptamer/target interactions. *Biol Chem* **382**: 479–481.

Sellers T.A. and Yates J.R. (2003) Review of proteomics with applications to genetic epidemiology. *Genet Epidemiol* **24**(2): 83–98.

Shi, H., Hoffman, B.E. and Lis, J.T. (1999) RNA aptamers as effective protein antagonists in a multicellular organism. *Proc Natl Acad Sci USA* **96**: 10033–10038.

Smith, D., Collins, B.D., Heil, J. and Koch, T.H. (2003) Sensitivity and specificity of photoaptamer probes. *Mol Cell Proteomics* **2**(1): 11–18.

Sooter, L.J. and Ellington, A.D. (2002) Reflections on a novel therapeutic candidate. *Chem Biol* **9**: 857–858.

Stojanovic, M.N., de Prada, P. and Landry, D.W. (2001a) Aptamer-based folding fluorescent sensor for cocaine. *J Am Chem Soc* **123**: 4928–4931.

Stojanovic, M.N., de Prada, P. and Landry, D.W. (2001b) Catalytic molecular beacons. *Chem Biochem* **2**: 411–415.

Swanson, B.J., Kutzer, J.C. and Koch, T.H. (1981) Photoreduction of 5-bromouracil. Ionic and free-radical pathways. *J Am Chem Soc* **103**: 1274–1276.

Tuerk, C. and Gold, L. (1990) Systematic evolution of ligands by exponential enrichment: RNA ligands to bacteriophage T4 DNA polymerase. *Science* **249**: 505–510.

Tyagi, S. and Kramer, F.R. (1996) Molecular beacons: probes that fluoresce upon hybridisation. *Nat Biotechnol* **14**: 303–308.

Wiegand, T.W., Williams, P.B., Dreskin, S.C., Jouvin, M.H., *et al.* (1996) High-affinity oligonucleotide ligands to human IgE inhibit binding to Fc epsilon receptor I. *J Immunol* **157**(1): 221–230.

Williams, D.M., Benseler, F. and Eckstein, F. (1991) Properties of 2'-fluorothymidine-containing oligonucleotides: interaction with restriction endonuclease EcoRV. *Biochemistry* **30**(16): 4001–4009.

Willis, M.C., LeCuyer, K.A., Meisenheimer, K.M., Uhlenbeck, O.C. and Koch, T.H. (1994) An RNA-protein contact determined by 5-bromouridine substitution, photo-crosslinking and sequencing. *Nucleic Acids Res* **22**: 4947–4952.

Wilson, D.S., Keefe, A.D. and Szostak, J.W. (2001) The use of mRNA display to select high-affinity protein-binding peptides. *Proc Natl Acad Sci USA* **98**(7): 3750–3755.

Wlotzka, B., Leva, S., Eschgfäller, B., Burmeister, J., *et al.* (2002) *Proc Natl Acad Sci USA* **99**(13): 8898–8902.

Xu, C.W. and Luo, Z.J. (2002) Inactivation of Ras function by allele-specific peptide aptamers. *Oncogene* **21**(37): 5753–5757.

Yamamoto, R. and Kumar, P.K.R. (2002) Molecular beacon aptamer fluoresces in the presence of Tat protein of HIV. *Genes to Cells* **5**: 389–396.

11

Chip Based Proteomics Technology

Mikhail Soloviev, Richard Barry and Jon Terrett

Introduction

The 20th century has witnessed a number of revolutions, of which the most peaceful and useful was the birth and development of molecular biology, which radically changed our understanding of life and of its molecular machinery. Major milestones of the molecular biology revolution include the discovery of the molecular structure of DNA by J.D. Watson, F. H. C. Crick and M. H. F. Wilkins (in 1953, Cambridge, UK), chain termination sequencing of DNA by F. Sanger (MRC LMB, Cambridge, UK), and sequencing of the human genome achieved in 2001 through a massive parallel effort of academic and commercial establishments world-wide (Lander *et al.* 2001; McPherson *et al.* 2001). The most recent breakthrough in this area has been the development of chips and technologies for whole genome re-sequencing (in a 'one genome at a time' fashion) by Solexa Ltd (Little Chesterford, UK; www.solexa.co.uk). Such rapid progress was possible due to a number of reasons, with the development and use of chip-based technologies being crucial in allowing the unprecedented increase in the throughput of molecular biology research.

Proteins (from the Greek proteios, meaning 'of first importance') were first purified in the middle of the 19th century, by the Dutch chemist Gerardus Mulder, who defined them to be 'without doubt the most important of all substances of the organic kingdom, and without it life on our planet would

Molecular Analysis and Genome Discovery edited by Ralph Rapley and Stuart Harbron
© 2004 John Wiley & Sons, Ltd   ISBN 0 471 49847 5 (cased)   ISBN 0 471 49919 6 (pbk)

probably not exist'. Having been discovered almost a century earlier than nucleic acids, proteins were routinely been studied using traditional technologies, which have long since reached their limits of throughput. These techniques do not allow a highly parallel approach due to their physical limitations, and because of their cost, poor reproducibility and large sample consumption. That is why proteomics required the development of new truly multiplex approaches for protein research.

Proteomics

Characterization of the complement of expressed proteins from a single genome is a central focus of the evolving field of proteomics. Monitoring the expression and properties of a large number of proteins provides important information about the physiological state of a cell and an organism. A cell can express a large number of different proteins and the expression profile (the number of proteins expressed and the expression levels) vary in different cell types, explaining why different cells perform different functions.

High throughput DNA microarray technology is well established and allows the simultaneous measurement of the expression of tens of thousands of genes. These can be measured in both normal and diseased states and the differential gene expression profile can provide fundamental insights into specific diseases and help to identify novel therapeutic targets. While genes contain the nucleic acids, which are the codes for expression of proteins, it is generally accepted that proteins rather than nucleic acids mediate most of the functions of the cell and protein expression levels in the cell are more indicative of disease critical changes and drug response. Moreover, there is sometimes little direct correlation between mRNA expression levels and amounts of the respective proteins in the cell (Gygi *et al.* 1999). Since many diseases associate with or even result from changes in the protein expression pattern, a comparison between normal and disease conditions may reveal proteins whose changes are critical in the disease, thus making it possible to identify appropriate diagnostic and therapeutic targets. Methods of detecting protein expression profiles also have important applications in tissue typing, forensic identification, clinical diagnosis, pharmacoproteomics and drug target discovery.

Beginnings of modern proteomics

Proteomics can be defined as the systematic analysis of proteins to determine their identity, quantity and function. Until recently proteins were studied individually using either a single or combination of established techniques which could be generally divided into *identification* techniques (i.e. indirect identification by the size on a gel, through a functional assay, or a ligand binding assay or by affinity reagents) and *separation* techniques (such as chromatography, isoelectric focusing, electrophoresis and their variations). Purified proteins could also be identified directly, i.e. through protein sequencing (from the N-termimus by Edman degradation or from the C-terminus using carboxypeptidases or thiocyanate degradation, etc.). In the traditional approach a study of an individual protein would go through a number of consecutive purification stages each followed by identification and quantitation steps.

The central concept of modern proteomics is 'multiplexing' or a simultaneous analysis of all proteins in a defined protein population. However, since one genome produces many proteomes and the number of expressed genes in a single cell may exceed 10 000, the characterization of thousands of proteins in order to evaluate proteomes ideally requires a high-throughput, automated process. This is why new improved methods for high-throughput protein identification and quantitation were needed. Many such technologies have been developed recently, led by advances in the electronics industry and more recently by the successes of the DNA chip-based technologies.

Protein separation and purification approaches

Most of the modern developments in the area of proteomics, including chip-based proteomics, are however based on traditional and established protein separation and purification techniques. We will briefly summarize these here to enable a better understanding of the capabilities and limitations of the modern chip based proteomics. All major 'old technologies' listed below are used to various degrees in modern 'protein chips'.

Gel-electrophoresis

Electrophoresis is the most widely applied method for quantitative separation of complex protein mixtures and has been known and used for decades. It is based on the concept of separating proteins in one dimension by their molecular weight. An important advantage of electrophoresis is the ability to separate biological polymers (proteins as well as nucleic acids) without the need for their sequence information.

Isoelectric focusing (IEF)

Isoelectric focusing utilizes the zwitterionic nature of proteins for their separation in an electric field, depending on the protein's pI. Unlike traditional electrophoresis, the IEF is an end point technique, employing a pH gradient to focus individual proteins into sharp bands, where they remain stable and stationary in the gel until removal of the electric field. An ability to focus protein samples is a unique capability of IEF and is not achievable using other separation techniques.

Chromatography

Chromatography is another very common process, which in general can be described as a combination of adsorption and desorption steps. There are a number of distinct chromatographic techniques including ion exchange, reverse phase (hydrophobic), affinity, size-exclusion (gel filtration), lectin affinity (Concanavalin A) and metal chelate (IMAC) in addition to a few other less common chromatographic techniques. Unlike traditional technologies, where chromatography was used as both an analytical and preparative tool, chip applications utilize only an analytical scale approach.

Two-dimensional electrophoresis

Although a combination of IEF followed by conventional electrophoresis, two-dimensional (2D) gels have nevertheless earned a rightful place as a separate and unique technique in protein analysis. Despite being known and used for decades, the two-dimensional electrophoretic separation technology can be considered a culmination of the traditional approaches for studying proteins. The two-dimensional character of separation dramatically increases the number of potentially resolvable proteins, which may reach several thousand, and this is not achievable using any of the techniques to date. Because of their relative simplicity, 2D gels have been used for many decades and most of differential proteomics so far has been done using 2D gels (Page *et al.* 1999; Gygi *et al.* 2000). This approach has resulted in massive databases of protein sequences and their expression levels. These databases allow selection of proteins or protein isoforms, which can be used in disease diagnostics or other related applications. More recently, a number of 'hybrid' approaches have appeared on the proteomics horizon, each being a combination of two or more of the independent separation dimensions like IEF, electrophoresis, or chromatographic separation. These may look novel but in fact all are variations on the theme of the good old 2D gels.

Chip based proteomics

Protein chips, more so than DNA chips, represent a huge opportunity for technological development in many areas. DNA arrays were relatively easy to commoditize for expression analyses and genotyping following the production of huge cDNA/EST libraries and DNA sequence databases. Even though Affymetrix (Affymetrix, Inc., CA USA; www.affymetrix.com) have developed a very advanced and complex DNA array system using photolithography, the methodologies and physics for complementary DNA/DNA and DNA/RNA binding had been well known for 20 years and PCR provided a cheap and easy mechanism for producing purified 'affinity' ligands. These DNA fragments are also stable and perform similarly on a variety of substrates. Outside of the biological processes, spotting, scanning and analysis methods developed very soon afterwards. For protein arrays some of the downstream technologies are available but there is a huge shortage of affinity reagents, and even for proteins where antibodies are known to work for IHC, ELISA and western blotting, the conversion of these into a chip format for homogeneous assays of protein expression is not straightforward. Despite the considerable technical and financial hurdles, functional protein chips will undoubtedly have a dramatic impact on our ability to find disease related protein changes, ascribe functionality to many proteins, and elucidate pathways and interactions at an unprecedented pace. Described here are the opportunities and issues facing the development of protein chips for quantitative proteomics

Miniaturization

There are a few obvious and easily acceptable reasons for miniaturization. Miniaturization allows one to pack more material into the same volume. For example, hundreds of different protein affinity reagents may be spotted on a chip instead of using Western blots or Sepharose affinity columns. High-throughput examples of this kind have been developed for molecular biology applications, where thousands of DNA hybridization reactions are performed on a single glass slide instead of Southern and Northern blots. Other, less obvious but equally important advantages of miniaturization include a potential to increase the reaction kinetics due to much smaller reaction volumes (and therefore faster reagent diffusion times) and significantly increased surface to volume ratios. This feature becomes especially useful for studying proteins since one of the two major approaches to protein separation, chromatography, relies on surface-mediated interactions. Truly miniature chromatographic applications do not require the use of porous resins and a careful selection of pore sizes, since a high surface to volume ratio of a small capillary channel, which is also easier to control, may be sufficient. In addition to these, most of the microfluidic applications allow capillary forces to manipulate the reagent solution. This often simplifies the design of the experiment (i.e. of the chip) and eliminates the need for additional equipment (i.e. for high precision pumping and mixing of the reagents). When capillary force is not sufficient for manipulating the reagents, spinning of a whole CD based laboratory (e.g. www.gyros.com) could be employed to finish the job. In addition, the lower power consumption of miniaturized devices and the possibility of portable applications, i.e. small chip-based devices with disposable chips, is also more attractive for routine diagnostic applications when compared with a desk-top sized HPLC system.

Miniaturization is attractive financially. First, and most importantly, one requires less reagents for smaller sized reactions. Secondly, it becomes possible with small chips to produce fully integrated devices, comprising not only sample manipulation and separation (i.e. capillaries, micro-pumps, micro-valves, reaction chambers, micro-columns), but also sample detection (i.e. integrated optics, integrated electrodes) and potentially data analysis and storage (for example, an integrated laboratory could be 'built' inside an ordinary looking CD-ROM like disk). Also, miniature chips are generally cheaper to produce in large quantities. This leads to better results (since more chips could be run in parallel to improve statistics) as well as higher reproducibility in general (since chips produced in bulk would match each other better than individually made large-scale gels or chromatography columns). Currently, price tags attached to the new chip-based proteomic applications may seem prohibitive, but the prices are likely to decrease with a widening use and increased general acceptance of these technologies.

Miniature chips for proteomics applications may be produced in a number of ways. The chip design and production technique depends mostly on the application sought. Some applications, like microarrays, often utilize widely available and cheap glass microscopic slides as the base of the chip. For other applications, usually involving truly miniaturized 'labs' (i.e. microfluidic applications) chips are often made in a totally different way. Technical advances in the electronics industry have made new materials (such as silicon) and new etching techniques available in life sciences. Silicon is a material of choice for truly miniaturized Laboratory-on-Chip applications, because of the expertise and technology currently available for microfabrication of silicon devices. Silicon allows precise etching to be achieved (which could be chemical, photochemical or mechanical). Silicon is also a very convenient material to provide an interface to the 'on board' electronics for further integration. Unfortunately, silicon is not a very chemically stable material and together with relatively high manufacturing costs the use of silicon is confined to a limited number of special applications or for development purposes. These days, plastics have become more popular due to their ever increasing toughness, chemical inertness, stability and decreasing production costs. Unlike silicon, polymers can be micromachined and processed in a variety of ways, such as laser ablation, hot embossing, microinjection moulding, casting and bonding. The possibility of moulding a whole chip with all the microfluidic channels in one go, instead of using long and expensive silicon etching processes, has resulted in moulding currently becoming the most attractive option due to the very high degree of accuracy achievable (i.e. low micrometre range) and the low costs involved. Glass, however, remains the first choice of material if chemical stability is required.

Limitations of continuous miniaturization

Miniaturization is a continuous trend not only in electronics, but also in life sciences. Reduction in size often results in significant changes, with most being extremely useful and desirable in proteomic research. However, there are also some important limitations. Ever decreasing sample and reaction volumes result in proportionally decreased signals and protein detection and quantitation techniques have their limitations. For example, UV absorption, which is the oldest and simplest of protein measurement methods, is limited to microgram quantities of protein (Goldfarb *et al.* 1951). Protein measurements based on biochemical reactions developed by Lowry (Lowry *et al.* 1951), Bradford (1976), or using bicinchoninic acid (BCA) assay (Smith *et al.* 1985) are all limited to low microgram amounts of protein. Using fluorescent dyes brings the sensitivity detection threshold down to the high nanogram range, whereas direct fluorescent labelling of proteins allows low nanogram to high picogram levels of detection. Affinity reagent-based detection and quantitation, especially if it relies on additional amplification of signal (i.e. ELISA principle), has lowered the detection sensitivity by almost an order of magnitude and is the principle used in some affinity microarrays (described further below). However, any additional signal amplification stage introduced results in the proportional increase of noise and background signals. Therefore, the sensitivity and selectivity of any such system (e.g. ELISA, rolling cycle amplification technology (RCAT)) will primarily depend on the specificity and affinity of the primary antibody used. Another recently developed detection technique utilizes a phenomenon called surface plasmon resonance (SPR), and allows the measurement of ca. 10 picogram of the surface immobilized protein. Importantly, proteins do not require special labelling. However, the sensitivity of this detection method is also compromised by its reliance on the protein separation technique used to immunoprecipitate proteins on the surface of the chip (i.e. affinity separation). In molecular biology research, fluorescent dyes capable of 'lighting up' a single DNA molecule, such that it becomes detectable by optical imaging, are now available (i.e. TOTO, YOYO, BOBO and POPO dyes from the Molecular Probes Inc, OR; www.probes.com). These techniques are not yet developed for direct labelling of proteins; therefore an unlimited reduction of reaction and sample volumes is not yet possible for protein applications.

Another important implication of working at micro- or nano-scales is that at these dimensions liquid flow becomes laminar and even a simple task of mixing two samples becomes a problem. This is applicable to both nucleic acid and protein chips. Especially vulnerable are microfluidic applications, but even simple spotted arrays (whether DNA or proteins) require sample mixing, for example using surface acoustic waves (Advalytix AG, Germany; www.advalytix.de) or sophisticated sample re-circulation (i.e. Memorec Stoffel, GmbH, Germany; www.memorec.com) in order to improve efficiency of hybridization.

In the microworld, capillary forces, viscosity and surface tension become major factors determining liquid behaviour. These properties could be taken advantage of, but could also take over the whole microfluidic chip, e.g. by preventing liquid penetration through narrow capillaries with hydrophobic surfaces. Miniaturization therefore requires a careful control of chemical and physical properties of the material-to-liquid interface surface (i.e. capillary effects).

Capillary channels the size of a single molecule (such as DNAs or large proteins) can now be manufactured. This may allow for single molecule applications to be carried out. As yet, there is little data available on how DNA and large proteins would behave under such conditions. However, because of physical limitations (i.e. prevalence of surface tension and viscosity over gravitational forces and absence of turbulence) miniaturization of liquid handling seems unreasonable below the low micrometer range.

Problems of data analysis

Unlike traditional 'one at a time' approaches, when most of the researcher's time was spent doing experimental work where results were relatively straightforward and simple to interpret, highly parallel multiplexed experiments require less time to run and much longer to analyse. Being directly imported or copied from the DNA chip designs, protein chips (especially antibody arrays) could generate vast amounts of information, which requires careful interpretation and sensible handling. A great deal of effort has been spent on improving and automating DNA array data analysis and a number of commercial products are available (e.g. from Affymetrix, Inc., BD Biosciences Clontech (Palo Alto, CA), Memorec Stoffel Gmbh (Köln, Germany), etc.) (www.affymetrix.com, www.bdbiosciences.com, www.memorec.com). Listed below are four major problems one will face when working with protein chips. First, most of the protein chip results come as images (e.g. of a protein array on a slide), which require a reproducible procedure of separating the information (i.e. spots) from the rest of the image (i.e. background or other unrelated parts of the image). Although spotting has been improved since the early days of array technology, a great deal of image recognition is sought (similar to Optical Character Recognition, used to extract text from scanned images). Because this procedure will generate the basis of the raw data, image analysis is probably the most critical step in chip informatics. Following on, the digitized data will be used to interpret differential expression, to cluster the differential signal patterns into families and categories or to compare different sets of protein array data. Clustering (Bozinov *et al.* 2002; Wang *et al.* 2002) is mostly a mathematical problem and, providing that biologically meaningful criteria are applied, it is a fairly straightforward task. However, interpreting the differential expression is a more challenging task, even more complicated than that for DNA array data. We will detail the peculiarities of protein array data quantitation later in this chapter. Despite the odds, protein chips offer a unique opportunity to achieve statistically significant results, which could not be physically achieved before (one would agree that running 100 Western blots in parallel is more time and resource consuming than processing a single protein array). However, new chip-based technologies need to be complemented by bioinformatics that support the compilation and meaningful analysis of the vast amounts of data output generated using chip technologies.

Protein arrays

A protein array is an ordered arrangement of protein samples immobilized on a solid substrate. In general, protein arrays are a direct import of the DNA array technology, applied to proteins. These could also be divided into macroarrays (hundreds of ~300 μm spots) which can be imaged with standard gel scanning scanners, or microarrays (thousands of <200 μm spots) which require specifically designed robotics and imaging equipment which, in general, are not commercially available as a complete system. Protein microarrays are generated by high-speed robotics on a wide range of substrates (discussed below) for which probes (e.g. antibodies) with known identity are used to bind experimental samples (i.e. proteins being assayed) enabling parallel protein profile studies to be undertaken. An experiment with a single protein chip can therefore supply information on thousands of proteins simultaneously and this provides a considerable increase in throughput.

There are two main types of protein chip assays, direct binding and sandwich-type (ELISA) assays. In a direct binding assay the unknowns, i.e. proteins being assayed, are labelled directly with a detection reagent such as a fluorophore. This enables the labelled proteins to be bound by antibodies immobilized on the array and their relative abundance to be measured directly by a fluorescent scanner, following a brief wash step, without further processing of the chip. The advantages of this method lie with its simplicity, speed and capacity for quantitation. However, its disadvantage relates to the actual labelling process in that it may interfere with recognition epitopes on the protein thus preventing its antibody binding. Direct binding assays are suitable for both antibody and antigen screenings as well as for studying protein–protein interactions.

A sandwich assay negates the necessity for labelling the proteins of interest with a detection reagent. In a sandwich-based assay the array is also comprised of capture proteins (e.g. antibodies) which bind proteins in their unmodified native form. The detection step requires a second antibody possessing specificity to a second epitope on the captured protein, which is distinct from the capture antibody. The second antibody is conjugated to a detection molecule, i.e. either a fluorophore for primary detection, or an enzyme (e.g. horseradish peroxidase) or ligand (e.g. biotin) for secondary detection through a chemiluminescent molecule or a fluorophore. An advantage of this technique is in the degree of sensitivity achieved through amplification of the signal thus allowing lower abundance proteins to be identified. However, this is paralleled by a loss in the quantitative nature of the assay and requires double the amount of antibodies.

There is no doubt that the potential of protein array based proteomics could have many advantages over 2D gels and chromatography. Protein arrays enable the scale up of analysis by running multiple affinity recognition steps in a single chip (i.e. higher throughput analysis). Protein-affinity arrays should also allow better reproducibility and more quantitative protein expression analysis. Described below are a few important points to consider prior to the development and use of protein arrays for quantitative proteomics.

Materials and surfaces

Unlike nucleic acids, which can be either spotted or synthesised on the surface of the chip, protein arrays in the majority of cases involve the actual spotting of proteins (i.e. antibodies) onto a surface of the chip. The choice of a substrate plays a critical role in determining the ultimate performance of the chip. A good substrate provides plenty of binding sites for the protein spotted (e.g. antibodies), provides access for labelled antigens to these antibodies, and secures that all of the proteins are kept in their native state and are not denatured at any stage of the assay process (i.e. binding). To date good results have been achieved using activated polyacrylamide pads immobilized on glass or oxidized silicon. These are commercially available from PerkinElmer Life Sciences (PerkinElmer, Inc., Boston, MA formerly Packard Biosciences, Meriden, CT; www.perkinelmer. com) under the trade name Hydrogel™ (Figure 11.1). Proteins are typically dispensed into a discrete area within the polyacrylamide pad to create a two-dimensional array in a three-dimensional structure. Although such Hydrogels will bind proteins without any pre-treatment, we found that pre-treating the Hydrogels with glutaraldehyde greatly increases the amount of retained spotted antibodies. (Schriver et al., 2003). Once the protein immobilization reaction is complete (typically 30 min after dispensing) the remaining aldehydes are reduced with sodium borohydride or blocked using reagents containing amino groups (e.g. TRIS buffer) and/or free amino acids. Hydrogel immobilization has given better results in terms of signal to noise ratio and fluorescence specifically when compared with glass or silicon substrates (Scrivener et al. 2003; Barry et al. 2002). Another possibility is to cast a thin agarose gel on a glass slide, using suitable gaskets or other spacers, and a siliconized cover glass (or cover slips or similar solid surface for producing a uniform thickness of gel). Following casting, the gel could be activated with CNBr (Axen et al. 1971; Yunginger et al. 1972). An advantage of any three-dimensional gel-based immobilization techniques over flat glass or plastic surface is that target-capture agent binding would most closely

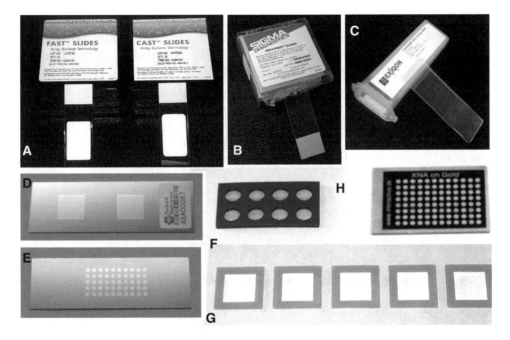

Figure 11.1 Commercially available microarray substrates, suitable for protein immobilization. Fast™ and Cast™ slides from Schleicher&Schuel have nylon and nitrocellulose (respectively) synthesized on glass surface (A). High protein binding capacity is their major advantage, which is offset by strong non-specific sorption of fluorescent dyes to these substrates (so that even a small amount of free label will outstrip specific signals). SIGMA Poly-Prep™ poly-Lysine coated glass slides (B). 'Immobilizer' all-plastic slides from EXIQON (C). Hydrogel™ on silicon slides from Packard Biosciences (currently Perkin-Elmer). 1 cm square (D) and 50 × 2 mm diameter spots (E) polyacrylamide-based Hydrogels are suitable for both fluorescence based assays and mass-spectrometry detection. CoverWell™ perfusion chambers (from GRACE Bio-Labs www.gracebio.com) useful to forming miniature hybridization chambers on glass slides (F). Sticky gaskets for making sub-miniature reaction chambers (G) on flat slides (for example as on panel D). The sides of chambers (66 µl total volume) are formed by the slide (one side) and a cover slip (another side). 'XNA on Gold' microarray slide (until recently available from ThermoHybaid) is made by depositing PTFE spacers (black) onto the gold plated glass slide (H). Gold surface allows for easy derivatization, whilst PTFE prevents cross-contamination between neighbouring chambers (half a slide shown)

approximate the binding likely to occur in free solution i.e. less impeded by solid surfaces.

Other surfaces include thin membranes (such as nitrocellulose, nylon (charged or uncharged), PVDF and/or their derivatives) attached to or immobilized on glass slides or other solid surfaces. Solid supports include silicon with modified surface chemistries to enable directional orientation of capture reagents (e.g. www.zyomyx.com), and aluminium with chemically or biochemically treated surfaces (www.ciphergen.com). Nitrocellulose coated glass slides (FAST™ slides) and nylon coated (CAST™ slides) substrates are available commercially from Schleicher & Schuell BioScience GmbH (Germany; www.schleicher-schuell.com). Nylon and nitrocellulose have proven their utility for many years for protein immobilization. But even today these remain one of the best choices for protein arrays as well. A major advantage of using nitrocellulose or nylon based supports is that they have a high protein binding capacity or flat 2D solid supports (glass, silicon, plastic, etc.). A higher binding capacity results in a stronger fluorescent signal, and therefore allows for the use of binders having lower affinity. Below we list and compare some of the best-known protein array supports:

Binding capacity (highest to lowest):
(a) immobilized nitrocellulose ~ immobilized nylon (charged > uncharged)
(b) FAST™ ~ CAST™
(c) Hydrogel
(d) derivatized solid surfaces (glass, silicon, plastics)

Spot geometry (best to worst):
(a) derivatized solid surfaces (glass, silicon, plastics)
(b) FAST™ ~ CAST™
(c) immobilized nitrocellulose ~ immobilized nylons

Variability in spotting (least variable to most variable):
(a) FAST™
(b) CAST™
(c) immobilized nitrocellulose
(d) immobilized nylons
(e) Hydrogels

Moisture retaining (best to worst):
(a) agarose pads
(b) acrylamide pads (Hydrogels)
(c) immobilized membranes or derivatized solid surfaces (glass, silicon, plastics)

Tolerance to glycerol in the spotting mixtures (highest to lowest):
(a) FAST™ ~ immobilized nitrocellulose
(b) CAST™ ~ immobilized nylons

Content

In order to make an array of biological molecules, the molecules have to be available and purified (prior to spotting) or able to be synthesized *in situ*. This task is relatively easy with nucleic acids, for which most of the sequencing information is available, and there are only 4 different but chemically similar bases. Nucleic acids are not only information carriers, but also perfect affinity reagents via base-pairing.

Faster analysis time is a major factor influencing the decision to use protein arrays. However, they do have one important disadvantage, and that is the 'closed' nature of the system. This restricts measurements to those targets which have their interacting partners present within the array. Therefore, the usefulness and applicability of a protein chip (and a DNA chip) is determined by its content, i.e. which, and how many, proteins or antibodies are present.

Antibodies

Antibodies are the most commonly used affinity reagents for making protein arrays. Currently ca. 50 000 to 100 000 various antibodies are available commercially world-wide representing ca. 5 000 to 10 000 different genes/proteins. Generating traditional polyclonal or monoclonal antibodies suitable for use on protein arrays (to assay native proteins) is time consuming and requires that sufficient amounts of appropriate antigens are available in their native non-denatured state. For the last decade other methods of generating affinity reagents were actively sought. These include phage display systems (Figure 11.2). Phage display libraries are being exploited commercially to generate recombinant affinity reagents for use in diagnostics as well as for protein array applications by Cambridge Antibody Technology (CAT) Ltd (Cambridge, UK) (www.cambridgeantibody.com), (Dyax – www.dyax.com), (Morphosys – www.morphosys.de). Another way to generate quickly a large number of recombinant antibodies with desired properties is by utilization of the Antibody-Ribosome-mRNA (ARM) complexes, the technique used in the Babraham Institute (Cambridge, UK). Oxford Glyco-Sciences (Oxford, UK) has recently offered another way to generate recombinant proteins (e.g. antibodies or protein fragments), with the desired binding properties, using the principle of self-assembly for in vitro synthesized affinity reagents (Fletcher *et al.* 2003). Phylos (www.phylos.com) have also developed their PROfusion™ self-assembly protein array technology which covalently links proteins to their own mRNA and so enables attachment via a 'capture oligonucleotide' to a solid surface. Self-assembly allows direct linking of proteins to their DNAs to bypass the limitation of cellular systems for cloning and affinity maturation of functional proteins (e.g. antibodies). Self-assembly technology offers an opportunity quickly to generate cheap protein affinity reagents, which can also be efficiently labelled, for use in traditional affinity assays or for protein arrays instead of conventional antibodies. Because proteins can be directly linked to their nucleic acids, such self-assembled complexes can be used for cloning proteins or protein affinity reagents (antibody, their fragments or antibody mimics, etc.).

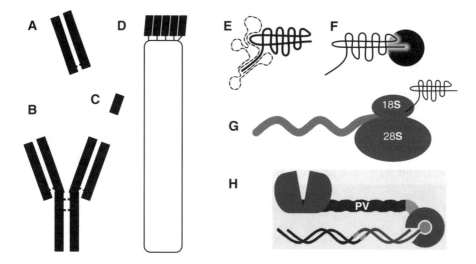

Figure 11.2 Diversity of affinity reagents used in proteomics research. The cartoons depict typical affinity reagents and include antibody Fab fragments (A), full size immuno-globulins (B), single chain variable antibody domains scFv's (C), phage display particles expressing scFv's (D), aptamers, formed by single chain nucleic acid fragments (dotted line) which fold into a definitive 3D structure which may bind proteins (E), artificial affinity reagents produced using molecular imprinting technology (F) and *in vitro* generated antibody-ribosome-mRNA complexes (G) or self-assembled protein–nucleic acid complexes (H)

The ability quickly to generate thousands of affinity reagents may be a crucial factor in the development of protein affinity arrays (Scrivener *et al.* 2003; Barry *et al.* 2002; Soloviev *et al.* 2001). Also, the ability quickly to determine nucleic acid sequence and therefore to identify the associated proteins could be extremely helpful for protein affinity selection or directed protein evolution. Alternatively, proteins can be assembled with labelled nucleic acids, which can be done either co-translationally or post-translationally; the proteins thus become labelled to a high specific activity through their association with their nucleic acids, without being chemically modified. This could result in not only a higher specific activity of labelling but would also avoid the chemical modification that takes place when proteins are labelled directly. Nucleic acid molecules associated with proteins could also improve sensitivity of detection of such proteins down to a single molecule level, by enabling detection using PCR.

Competition and intellectual property rights have caused some to look into protein scaffolds alternative to native immunoglobulins. For example Phylos Inc. (Lexington, MA) has chosen the tenth fibronectin type III domain (10Fn3) (Xu *et al.* 2002) as a protein scaffold to generate antibody-mimic molecules with an ability to bind their 'antigens' with high affinity. Affibody AB (Bromma, Sweden; www.affibody.com) have chosen a more radical approach by utilizing a Protein-A scaffold for their Affibodies™. These are small (only 6 kDa) and stable molecules, containing a single Fc binding domain of the Protein A. The specificity of protein–protein interactions mediated by such an Affibody™ (i.e. binding) can be manipulated through mutagenesis of the 13 amino acids, which are spread throughout this protein sequence. Additional variability could be achieved by using two or more different Affibodies designed to bind the same protein target. A great advantage of Affibodies for chip based proteomics over traditional antibodies is that because of the smaller size (6 kDa Affibodies vs. 150 kDa immunoglobulins), the small Affibodies allow for a much larger number of molecules to be immobilized inside the same space, potentially leading to stronger binding signals if affinities similar to full antibodies can be achieved. Despite having such a great advantage, Affibodies, just like any other protein affinity reagents (including recombinant immunoglobulins) require cloning and affinity maturation to be performed, followed by a larger scale *in vitro* or *in vivo* expression to generate sufficient amounts of the reagents.

A radically different ProteinPrint™ approach to affinity reagents generation is used by Aspira Biosystems, Inc (San Francisco, CA, USA; www.aspirabio.com). This is a molecular imprinting technology, and relies on vinyl or acrylic monomer-containing compounds, which upon binding to a target protein and polymerization will keep the shape complementary to the target molecule. A limitation of molecular imprinting, however, is the need actually to have a template to imprint on. This means that sufficient amounts of purified proteins have to be available and must be immobilized in a fashion suitable for molecular imprinting. In principle, molecularly imprinted compounds could be generated against any protein, but in practice the technology is limited to short peptides or proteins with the exposed C-termini. This greatly limits the usefulness of the technology. However, if taken together with a Peptidomics approach (Schrivener *et al.* 2003), the molecular imprinting technology may in the future become the method of choice when a large number of different affinity reagents are required.

It is also possible to use nucleic acids as protein affinity reagents. An obvious example would be an array of double-stranded DNAs containing transcription factor binding sites. Such arrays can be obtained from BioCat GmbH (Heidelberg, Germany; www.biocat.de) and could be used in an analysis of the transcription protein machinery. Unlike double-stranded nucleic acids, single stranded DNAs or RNAs have a unique secondary structure which depends on the nucleic acid sequence (good examples of such single-stranded nucleic acids include tRNA, ribozymes or ribosomal RNAs). This property of single-stranded nucleic acids could be used to generate artificial protein affinity reagents from short fragments of single stranded DNA, called aptamers. Although affinities often are not as good as with traditional antibodies, they could be increased by using more than one individual aptamer molecule per protein. Such photocross-linked individual aptamers, i.e. photoaptamers are being developed at SomaLogic, Inc (Boulder, Colorado; www.somalogic.com). A great disadvantage of any aptamer-based system is that nucleic acids carry a strong negative charge, which will bias their binding to mostly positively charged proteins.

Above, we have mentioned several technologies used to generate affinity reagents for use with protein arrays. But there is one problem common to all anti-protein affinity reagents regardless of their nature. This lies in the heterogeneity of their binding affinities. Therefore, running 'DNA-style' antibody chips is not practical.

Proteins

Protein arrays can display antigens or other proteins. Such arrays may be used for high-throughput analysis of an antibody repertoire (i.e. from patients' serum) or for functional protein analysis, ligand binding or to investigate other protein–protein interactions. Such protein arrays, just like antibody arrays, need a content consisting of reasonably pure and characterized proteins. These could either be purified from natural sources or produced individually using traditional recombinant techniques, and spotted on glass slides as done at Sense Proteomics Ltd (Cambridge, UK). Throughput limitations of traditional techniques can be lifted if proteins are produced *in vitro* (Fletcher *et al.* 2003). Another example of successful protein *in situ* arrays (PISA) comes from the Babraham Institute (Cambridge, UK; www.babraham.ac.uk). PISA technology utilizes affinity tags to bind proteins to modified surfaces. PISA is simple to perform but is limited to one protein (or protein mixture) per well of a multiwell plate (He *et al.* 2001).

Spotting

Except for the few applications which allow protein synthesis *in situ* the majority of protein arrays are produced by spotting. This process is critical in determining the quality of a protein array, protein and array stability and ultimately signal to noise ratio. Parameters such as temperature and humidity require careful optimization, because unlike nucleic acids proteins are less tolerant to repeated cycles of drying and re-hydration or temperature variations and may denature irreversibly. There are two major ways of 'printing' the protein arrays, these are contact and non-contact printing.

Contact spotting

Compared with DNA, which is relatively robust in nature, proteins are much more susceptible to irreversible functional damage through dehydration events or upon contact or attachment to a solid surface. Many proteins exhibit an inherent 'stickiness' to a range of surfaces (chip and dispenser surface) and this has the potential to induce contact-based conformational changes and impair protein function, e.g. loss of epitope recognition in an antibody. Thus the manipulation and treatment of proteins for building arrays cannot be reproduced directly from DNA array protocols. Therefore, in addition to making a careful selection of the surface on which to print a protein array the method of delivering the proteins onto the chosen surface must also be considered.

In both contact and non-contact printing the protein sample (e.g. antibody in solution) being arrayed is initially aspirated from a source plate. In contact printing this is generally achieved using a specific metal pin device which simply dips into the sample and draws it up, into a channel cut in the pin, by capillary action.

The traditional 'tweezer' style pins, which are also used in DNA printing, are comprised of a metal pin with a relatively imprecisely defined channel, i.e. groove cut up the middle of the pin, which holds the sample. In order to dispense the protein onto the array surface the pin has to contact the substrate and requires an actual tapping force plus surface tension to expel the sample. Alternative 'flat tips' are also available (e.g. Telechem International, Inc.) where the sample forms at the end of the tip and is dispensed by a more gentle 'ink stamping' mechanism. These tips have a range of tip, channel and reservoir sizes allowing both spot diameter and loading volume to be manipulated. A strength of contact pin spotters is that by combining a number of pins in one 'printing head' very high numbers of proteins can be arrayed within a relatively short time-frame, i.e. high throughput.

A potential problem encountered with contact spotting pins can be a build up of protein material within the pin channel, therefore the washing steps (generally a combination of ethanol and water washes plus sonication) have to be extremely meticulous. It is also important to avoid carry-over contamination between protein samples compromising the array. In general, volumes dispensed are higher using contact spotters than with non-contact spotters, nanolitres versus sub-nanolitres respectively. Additionally, there is also less scope for programming the volume of protein sample being spotted with a contact spotter since the mechanism of dispensing does not readily lend itself to regulation.

A manual spotting device for the production of low-scale microarrays has been developed by Schleicher & Schuell BioScience GmbH (Germany; www.schleicher-schuell.com). The 'MicroCASTer' comprises an 8-pin manual spotter and enables arrays to be generated without a requirement for expensive robotics. The system is particularly suited to novices or for initial use in evaluation of the array technology. Both 96 and 384 well source plates can be accommodated and up to 768 spots can be arrayed on one slide. An index system also allows exact spot positioning for duplicating arrays.

Non-contact spotting

When printing proteins it is preferable to keep any contact between the dispenser and the array to a minimum to avoid possible denaturation, cross-contamination or surface damage which is likely with hydrogel and nylon surfaces. Therefore, devices such as the BioChip Arrayer (PerkinElmer) which use piezo-tip glass capillary dispensers to both aspirate and dispense the protein sample are preferable.

Defined 'large' volumes can be aspirated and the aliquots are dispensed. To dispense, an electric current is applied and this propagates a pressure wave within the capillary dispenser causing a defined, measured volume (droplet) of sample to be emitted from a predetermined height above the array surface. The arrayer can be programmed to dispense any number of droplets (1000 s per second) of a particular sample and so the exact volume is known. This provides excellent reproducibility between spots of approximately 100 μm diameter. Volumes dispensed are in the picolitre range, compared with nanolitres for pin-spotters, and this reduces sample consumption. Following deposition the remainder of the sample held in the capillaries can be recovered to the source plate reducing sample use further: this cannot be achieved using standard pin spotters.

Since a pressure difference can be measured with each dispense event any error in sample deposition can be logged for reference, again this is not possible with contact dispensing. A further advantage of capillary dispensing relates to the flow-through washing system available, this flushes the system between sample pick-ups and so eliminates cross-over contamination. In common with pin-spotters a series of piezo-capillaries can aspirate and dispense protein samples in unison which maximises the throughput of generating protein arrays.

PerkinElmer Life Sciences supply their own sub-microlitre non-contact, drop-on-demand piezoelectric dispensing technology in addition to their array surfaces (the Hydrogel™). The PiezoTip™ is capable of dispensing 180 μm spots at 250 μm spacing with 10 μm spatial resolution. Because the tips never come in direct contact with surface, the risk of carry-over contamination is greatly reduced and use of non-traditional or delicate substrates becomes possible, in addition to more traditional slides, membranes and microplates.

Another non-contact system is TopSpot, developed by HSG-Institute of Micromachining and Information Technology (HSG-IMIT, Germany; www.imit.de) and The Institute for Microsystem Technology (IMTEK) at The Albert-Ludwig University of Freiburg (www.imtek.de), in cooperation with GeneScan AG (www.genescan.com). The system incorporates microfabricated printheads that hold 24 or 96 reservoirs (each containing several microlitres) and print nozzles. All of the nozzles are placed within an area of 18 mm² with a pitch of a few hundred micrometres. It is capable of depositing up to 96 distinct analysing substances on a chip, although by positioning consecutive prints alongside one another more complex biochips can be generated. Substances are applied with high precision as nanolitre droplets and spot sizes are approximately 200 μm. The company offers a choice of products to address microarraying requirements in terms of throughput. For lower throughput requirements the TopSpot Entry arrayer is a manually operated machine and is able to print up to 96 different substances on the substrate of choice. If a larger number of microarrays are required the TopSpot Modula arrayer is capable of spotting up to 480 substances on up to 300 slides per hour.

Arrayjet Ltd (www.arrayjet.co.uk) offers a multiple (over 100) nozzle inkjet printing technology capable of highly parallel printing of ~100 pl volumes into ~100 μm spots. The Arrayjet printheads are capable of a firing frequency of ~1000 spots per second but the technology is somewhat limited because of the loading problems, i.e. hundreds of nozzles have to be preloaded with samples prior to spotting.

Detection

Detecting proteins is often performed using techniques applicable for both protein and nucleic acid detection. For example, an efficient and well established method for measuring the presence and abundance of a protein is through attaching a fluorescent molecule, i.e. a fluorophore. In addition to such a generic approach, protein detection is often carried out using techniques such as mass spectrometry. There are also techniques, which combine protein detection (typically through antibodies) and signal amplification steps. The latter employ either enzyme-mediated signal amplification (e.g. ELISA) or rely on DNA amplification (e.g. rolling circle amplification techniques (Schweitzer *et al.* 2002), see also below)

Fluorescence

There are three stages involved in measuring fluorescence. Initially, a photon of energy is supplied by an external source (laser, incandescent lamp) and is absorbed by a fluorophore molecule (attached to a protein of interest), i.e. excitation. During the lifetime of the excited state (a period of nanoseconds), the fluorophore undergoes conformational changes and interacts with its molecular environment which dissipates energy. As the fluorophore returns to its ground state a photon of energy is emitted.

However, due to energy dissipation, not all the molecules initially excited by absorption return to the ground state, i.e. the energy of these photons is lower and of longer wavelength than the excitation photons. This difference in energy or wavelength (Stokes Shift), allows emission photons to be detected against excitation photons. Importantly, since the same fluorophore (conjugated to a protein or incorporated into DNA) can be excited and detected repeatedly a high degree of sensitivity can be achieved with fluorescence detection techniques.

Whereas the fluorescent labelling of mRNA samples for application to DNA arrays can be incorporated during the cDNA synthesis step, i.e. using a fluorescent-tagged nucleotide, there are no practical equivalents available for labelling proteins. Therefore, proteins must be fluorescently labelled by attaching fluorophore molecules to either pre-existing reactive side-chains of the proteins own amino acids or to introduced or modified groups. Currently, there is an extensive range of reactive fluorophore products available (e.g. Molecular Probes) and these span a wide range of excitation – emission spectra wavelengths. This allows attachment of fluorophores virtually to any protein or peptide through amines (e.g. succinimide dyes), thiol groups (e.g. iodoacetamide, maleimide dyes), carbohydrate groups (e.g. hydrazide dyes) or via an intermediate molecule, e.g. by introducing biotin which subsequently can be complexed to a streptavidin–fluorophore conjugate.

A number of scanning-based instruments for detecting fluorescence from DNA/protein arrays are available. The most effective are confocal laser microarray readers (e.g. ScanArray from PerkinElmer Life Sciences, www. perkinelmer.com), which acquire a high relevant signal to low background noise, i.e. good signal-to-noise ratios. CCD based detection is used in scanners produced by Berthold Technologies GmbH (Germany; www.bertholdtech.com). Having much faster scanning rates (compared with confocal laser systems), CCD based detectors lack spatial resolution and confocal capabilities of laser scanners.

The latest laser instruments offer up to five excitation wavelengths and each can be configured to detect several emission wavelengths simultaneously in one experiment. This allows dual (or greater) -labelled experiments to be carried out where individual samples being profiled and compared are labelled with fluoro-phores of independent emission wavelengths and are simultaneously hybridized to a single array. This enables differential protein expression data from two samples to be collated from one protein array and avoids any putative expression changes measured to be a result of marginal variations in the arrays themselves. Current scanners also possess precise 5 μm excitation beams making it possible to generate fluorescence from discrete 5 μm pixels. This produces precise detail from standard 200 μm diameter spotted proteins and therefore enables accurate measurements to be made of protein abundance.

One problem concerning the use of antibodies to capture fluorescent labelled proteins regards an inability to determine any cross-reactivity between antibodies immobilized on the array. Based purely on a fluorescent image there is no method of determining if a specific antibody has captured its cognate ligand. This can be of particular concern since the vast majority of antibodies have been raised against a native protein or peptide. Unlike DNA, where incorporation of a fluorophore does not directly alter the base-pairing (hybridization) of the comple-mentary cDNA strands, it is possible that direct attachment of a fluorescent label to a protein may mask or alter a recognized antibody binding site (epitope). Therefore, an antibody exhibiting good kinetics of binding for the native protein may show lower kinetics of binding for the fluorescently labelled protein, or may fail to recognize it at all. Conversely, an unrelated antibody may bind a labelled protein it would not normally recognize in its native form. This can be overcome by using a sandwich type assay to capture unlabelled proteins although this requires a second detection antibody for each protein being assayed. Also, the second antibody either needs to be labelled itself which may modify its binding capabilities, or a third fluorophore-labelled antibody with ubiquitous recognition for the detection antibodies will be required. Obviously, this increases both the cost and complexity of the assay and ultimately may adversely affect its quantitative nature.

A further concern regarding the use of fluorescence relates to the removal of unconjugated dye molecules following the labelling procedure. Since the dyes used are reactive and conjugate to specific groups on proteins in the sample being assayed, the antibodies assembled in the arrays also have the potential to become labelled through their equivalent groups. The removal of unconjugated 'free' dye molecules following protein labelling is therefore very important to avoid antibodies 'lighting-up' with fluorescence and producing false-positives. This situation does not arise with DNA arrays since the fluorescent nucleotides used for labelling cannot incorporate into cDNA immobilized on the array.

Although fluorescence can be very effective for use with protein arrays alternative techniques negating the requirement for labelling proteins or peptides have been developed (e.g. peptidomic approach (Scrivener *et al.* 2003)) and are discussed below.

Planar waveguides

Zeptosens (www.zeptosens.com) presents an alternative approach to both the structure of the array and the excitation light pathway. The technology is based on thin film planar waveguides (PWG).

In this system a thin film (Ta_2O_5) with a high refractive index is deposited on a transparent support with a lower refractive index. The excitation light is a parallel laser light beam integrated into the film by diffraction, via a diffractive grating imprinted in the substrate, and this creates a strong evanescent field with highest intensity close to the surface. This effect can be exploited to selectively excite only those fluorophores situated in close proximity to the surface of the waveguide, i.e. at the chip surface where capture reagents are immobilized. The metal oxide layer can be modified to allow for the stable immobilization of DNAs and proteins for a wide range of bioassay applications.

Therefore, when a fluorescently labelled analyte becomes bound to an immobilized capture reagent, excitation of the fluorophore by the evanescent field and detection of the analyte is limited only to the array surface. Since signals from unbound fluorophore molecules in solution are not detected there is no require-ment for washing steps allowing the analysis to be carried out in real time. This also produces a significant improvement (x10–x100fold) in the signal to noise ratio and greatly increases the sensitivity of the assay in comparison to traditional fluorescence scanners.

Mass spectrometry

Proteins can be detected and identified using mass spectrometry (MS) techniques. To increase accuracy of detection, sample proteins are digested (typically with trypsin) prior to MS. The peptides obtained are typically identified using matrix-assisted laser desorption ionization-time of flight (MALDI-TOF) MS technique followed by database mass matching. Further confirmation can be obtained using MS/MS techniques with collision-induced dissociation (CID) to fragment the peptide enabling an amino acid sequence to be generated. MS-based detection has been used by Gygi *et al.* (1999a) in their ICAT (isotope-coded affinity tags) approach, although this is not really a chip-based technology. MALDI-MS is used in addition to fluorescence detection of peptides in the peptidomics approach developed by Scrivener *et al.* (2003). Whereas the Ciphergen ProteinChip® system (www.ciphergen.com) uses an array surface to capture whole proteins from biological samples, which are then analysed by surface enhanced laser desorp-tion/ionization (SELDI) mass spectrometry

Signal amplification

To increase sensitivity of detection signal amplification techniques such as ELISA (enzyme-linked immunosorbent assay) or ECL (enhanced chemiluminescence) can be applied. Rolling Circle Amplification (RCAT) is another sensitive tech-nique, which can be incorporated into a chip format and used directly to detect the presence of a relatively small number of RNA and DNA molecules (Schweitzer *et al.* 2002; Lizardi *et al.* 1998). RCAT can also be used to measure protein molecules present in low abundance on an array, i.e. an on-chip signal amplification model. This is of particular value since molecules may be present at very low levels and direct fluorescence detection may not be of adequate sensitiv-ity. To detect proteins a circularized DNA oligonucleotide probe is attached to a recognition antibody, and is used to generate an extended single-stranded DNA product consisting of thousands of copies of the oligonucleotide probe. Along the length of the newly synthesized DNA several thousand binding sites for fluores-cent dyes are sited. An amplification of up to four orders of magnitude can be achieved with RCAT, although there is still a dependence on affinities of antibodies for those analytes being assayed. The signal to noise ratios will also still be dependent on antibody affinity and crossreactivity.

Efficiency of antibody labelling could be greatly enhanced if protein–DNA fusions are used (Fletcher *et al.* 2003; Weng *et al.* 2002). In such a case an antibody could be labelled effectively through the associated DNA to a much higher level compared with traditional protein labelling techniques.

Other detection techniques

Other detection methods applicable to microarrays include Surface Plasmon Resonance (SPR) and ELISA. SPR biosensors monitor interactions by measuring the mass concentration of biomolecules close to a surface of the sensor. This effect arises when light is reflected under certain conditions from a conducting film at the interface between two media of different refractive indices. The measured SPR response (which is typically the angle of minimum reflected light intensity) is directly proportional to the mass of molecules that bind to the surface. Therefore SPR enables one to discriminate between a sample in solution and molecules associated with the surface of the sensor and so measurements can be made in real-time (Cullen *et al.* 1987; Fagerstam *et al.* 1990). In addition, SPR allows detection of unlabelled molecules, which makes this technique very attractive for use in biomedical research and diagnostics. SPR-based detection is exploited commercially by BIAcore AB (Sweden; www.biacore.com), XanTec bioanalytics GmbH (Germany; www.xantech.com) and Genoptics (France; www.genoptics.fr).

A microarray platform based on high-throughput micro-ELISA arrays has been developed at Genometrix, Inc (currently High Throughput Genomics Inc. www.htgenomics.com). This type of array has been used to profile a series of cytokine proteins. This system enables the simultaneous analysis of up to 25 analytes based on a starting sample of 25 µl volume. A good dynamic range of approximately 400 000-fold (2 mg/ml to 4 pg/ml) can be achieved with this technique.

Instrumentation

Some manufacturers now provide 'complete solutions' which utilize or are built around chip-based proteomics. One such complete solution incorporating custom-built microarrays and instrumentation is provided by Zeptosens Sg., Switzerland. The ZeptoMARK™ high performance protein microarrays are based on planar waveguide (PWR) technology for evanescent field detection. Array surfaces are modified to produce an accessible substrate, which is also stable and minimizes non-specific adsorption. This allows arrays to be applied in the analysis of antigen–antibody, enzyme–substrate and membrane receptor–ligand interactions. Arrays can be imaged using the ZeptoREADER™, a high-throughput fluorescence imaging microarray readout system tailored to planar waveguide (PWG) technology and analysed using specialised ZeptoVIEW™ software.

PerkinElmer Life Sciences (PerkinElmer, Inc., Boston, MA formerly Packard Biosciences, Meriden, CT; www.perkinelmer.com) offers a complete package, which includes their own gel-based substrates (Hydrogels™), BioChip Arrayer robot and fluorescence laser scanner –ScanArray® for array imaging and analysis (all as discussed above in Non-contact spotting and Fluorescence).

Randox Laboratories Ltd (www.randox.com) have developed a multianalyte biochip array technology referred to as Evidence® platform. This incorporates low-density (ca. 8–20 spots) diagnostic arrays on 1 cm square chips. Evidence® comprises a series of panel-profiling diagnostic tests, e.g. drugs of abuse (panel of nine), tumour markers (eight markers), cardiac markers, fertility hormones, thyroid hormones, antibiotic drug residues and allergens.

Ciphergen Biosystems Ltd (UK; www.ciphergen.com) offer their ProteinChip® system, which consists of the instrumentation, protein arrays and software to compare the presence and abundance of individual proteins in crude samples. The ProteinChip Array® comprises a series of non-selective affinity matrices, e.g. hydrophobic, normal phase, ion exchange, metal ion exchange, which are incubated with crude protein preparations. The bound samples are subjected to a series of wash steps of differing properties, e.g. water, organic, salt, pH, detergents, denaturants, etc. This leads to the sample becoming split into sub-fractions. The captured proteins are then analysed by mass spectrometry (Chapman 2002). Such crude fractionation is suitable for taking 'snapshots' of the protein mixtures (e.g. quality control of cell culture media), but may not be very useful for studying differential protein expression due to the poor mass resolution of this technique, which is largely limited to lower MW proteins (i.e. < 25 kDa). Such, or similarly obtained, crude subfractions could also be analysed by 1D-PAGE.

Other applications and products

BD Biosciences Clontech (www.bdiosciences.com) have produced an antibody array on glass. As is true of any antibody array, the BD protein array represents a 'closed' system, and is limited in its capacity to those antibodies present on the array.

Holt et al. (2000) have developed a Matrix Screening™ platform for high-throughput screening of protein–protein interactions. In this system grids of intersecting lines are used to bring different $V_H V_L$ antibody fragments together on the surface of a 'combinatorial' array. This enables thousands of protein–

protein interactions or antibody heavy and light chain pairings to be screened. This technique may be used in affinity maturation studies of antibodies, proteomics assays and combinatorial library screening. Bussow *et al.* (1998) and Holz *et al.* (2001) have reported a technique useful for establishing catalogues of protein products of arrayed cDNA clones and for antibody repertoire screenings by gridding a bacterially expressed cDNA libraries onto high-density filters. Simone and co-workers (2000) have coupled a laser capture microdissection (LCM) with sensitive quantitative chemiluminescent immunoassays in an attempt to quantitate the number of prostate-specific antigen molecules. Individual cells (ca 30 μm in diameter) were captured under direct microscopic visualization from the heterogeneous tissue section onto a polymer transfer surface. The cellular macromolecules from the captured cells were solubilized in a microvolume of extraction buffer and directly assayed using an automated (1.5 h) sandwich chemiluminescent immunoassay. This technique is capable of measuring actual numbers of prostate-specific antigen molecules in microdissected tissue cells and has a broad applicability in the field of proteomics. Some other formats for functional proteomics on chips have also been published. Zhu *et al.* (2000) have expressed and immobilized almost the entire set of possible protein kinases (119 of 122) from *Saccharomyces cerevisiae*. These kinases were incubated with 17 different substrates to characterize their phosphorylation activity and this identified both known and previously uncharacterized activities of established and speculative kinases (homology predictions). Large arrays of immobilized proteins can be examined for protein–protein interactions in a system in some ways analogous to yeast two-hybrid experiments (Walter *et al.* 2000), and in a somewhat ironic assay, genome wide protein–DNA interactions have been assessed using antibodies to immunoprecipitate DNA–protein complexes which are then characterized on a DNA microarray (Kodadek 2001).

Problems of protein microarrays

So why have protein arrays not progressed to the extent of DNA arrays? The reason is there are no complete sets of affinity reagents with the required specificities and affinities. This is due firstly, to an absence of information on exactly which protein and which post-translational modification pattern is expressed differentially under particular conditions. Having access to such data enables the choice of relevant protein targets. Secondly, a protein array for quantitative proteomics is effectively an array of high affinity reagents, e.g. antibodies. The development of each antibody, unlike the synthesis of an oligonucleotide or purification of a PCR product, requires significantly more time and resources with a variable success rate. Therefore it is unlikely that a generic set of affinity reagents against all proteins (even from one species) will be available in the near future. But this is not the only obstacle to making practical protein chips. DNA can be dried, re-hydrated, frozen, boiled, or even partially degraded without the loss of its unique information content (sequence identifier) or high affinity binding to its complementary strand. Proteins, except in very rare cases, will not withstand such harsh treatment. Unlike nucleic acids, which can attach to the surface of a solid support in any direction, protein affinity reagents must be correctly orientated and the correct epitopes must be accessible. In addition, the protein binding steps may not be carried out under denaturing conditions, unlike DNA hybridization, and this creates additional difficulties with background staining. So the development of protein arrays requires additional effort.

Current array-based proteomics techniques are in their infancy and suffer from a variety of problems including the isolation of sufficient numbers of high affinity ligands cognate for the proteins of interest (Scrivener *et al.* 2003; Soloviev 2001; Kodadek 2001). In addition, there are non-specific binding and background issues where the process of labelling proteins can obscure specific binding. There are also sample labelling issues where the inherent heterogeneity of proteins, polypeptides and peptides means that they all take up label to different extents, making absolute quantitation of expression levels complex (Abbott 2002). Inherent in labelling a sample is the need to purify the sample from unbound reactive label. Because of sample losses during purification larger amounts of starting sample are required, although there have been developments in this area in cDNA microarray applications (Causton 2002). Another principal difficulty highlighting the difference between protein–antibody interactions and nucleic acid complementary chain interactions is that protein or antibody labelling methods use reactive amino acid side chains to attach the label. Labelling of such reactive side chains can interfere with binding to affinity reagents.

Another major difference between nucleic acid arrays and protein arrays is that for any arrayed nucleic acid, the affinity of its interaction with a corresponding complementary strand is a product of length up to a plateau value. The affinity of interaction is thus similar for all nucleic acid fragments of similar length. However, antibody–antigen pairs show a wide range of affinities irrespective of size. In the context of an antibody array, these differences in affinities can result in a wide range of signal intensities even if equimolar amounts of a labelled sample are added to an array of cognate antibodies (Figure 11.3). Moreover, highly abundant proteins may fully saturate the immobilized antibody and mask any differential binding signal between two samples. The latter problems taken together mean that any apparent differences in the direct binding signals obtained between two different samples, for example diseased versus normal tissue, should not be used to indicate the absolute abundance of the proteins in the two samples; the relationship between the binding signal strength and the protein concentration requires calibration. The problem is exacerbated if the aim is to compare differential expression between different cell or tissue types, where the generation of calibration curves for each protein would be a daunting task. Arraying different amounts of antibodies in inverse proportions to the affinity of a given antibody is also fairly impractical.

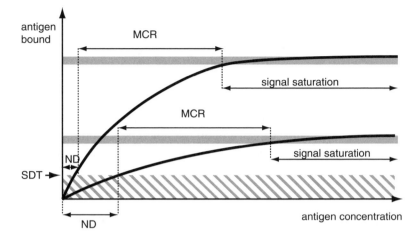

Figure 11.3 A schematic diagram illustrates two different antibody binding isotherms. Hatched area and the SDT indicate signal detection threshold, which depends on antibody affinities and protein detection technique (e.g. fluorescence, radioactivity, surface plasmon resonance). At concentrations below their respective SDTs, bound proteins will not be detected (ND). They greyed out areas depict signal saturation, where quantitation of binding is not possible due to a binding saturation. Sensitivity of the detector and signal saturation define a narrow MCR window (measurable concentration range), where the amount of bound proteins could be calculated based on the signal readouts

Microfluidics

Agilent Technologies (Palo Alto, CA; www.agilent.com) have developed a Lab-Chip, used in conjunction with their 2100 Bioanalyser system, for the electrophoretic separation of proteins, DNA and RNA. The 2100 Bioanalyser is designed to streamline the processes of protein expression, gene expression analysis and RNA isolation. The technology can therefore be applied to either electrophoretic chip applications or cell based assays. Essentially the system is an automated capillary gel runner and still requires gels to be loaded, i.e. not a definitive dry-run technique. However, the system is suitable for routine sizing and quantitation of proteins (as well as DNAs).

Gyros AB (Uppsala, Sweden; www.gyros.com) have produced a CD-based laboratory (CD-Lab), which integrates multiple steps into a streamlined process. A flexibility of design enables CD products to be built to suit specific processes. Each CD-Lab can be comprised of hundreds of application-specific microstructures. As the CD spins, samples are processed in parallel in nanolitre amounts within the microstructures. Currently, the range of products includes a 'MALDI-CD', which can be used to prepare samples for MALDI-TOF mass spectrometry analysis.

An eTag™ Assay System from Aclara Biosciences (Mountain View, CA, USA; www.aclara.com) has been developed for simultaneously measuring the abundance of multiple proteins in a sample. Each eTag reporter used is a fluorescent label, coupled to a probe via a cleavable linkage, and has a distinct electrophoretic mobility. When an eTag reporter-labelled probe binds an analyte, the coupling linkage is cleaved and the eTag is released. Multiple binding reactions can be combined followed by capillary electrophoresis, and each eTag reporter is used to identify the probe to which it was originally attached. Since binding events occur in solution this also avoids problems associated surface immobilization, denaturation and non-specific binding of proteins.

Other chip-based technologies in proteomics

SPR based BIAcore instruments (BIAcore AB, Sweden; www.biacore.com) combines an SPR detection system with a microfluidics system (see also above) and data handling software. BIAcore also supplies gold plated glass chips for substrate immobilization for use with BIAcore machines. Users are offered a number of protein immobilization techniques such as through amine groups, carboxyl groups, thiol groups, cis-diols on carbohydrate residues, histidine tags, or biotin-mediated. Underivatized gold surfaces are also available for custom applications. The sensitivity of detection using BIAcore reaches ca. 10 picogram of the surface immobilized unlabelled protein. However, the exact conversion factor between the reflective angle and surface concentration is dependent on properties of the sensor surface and buffer system as well as the nature of the molecules bound, thus limiting the method's capability of absolute quantitation of the bound protein. BIAcore instruments are capable of measuring protein binding in real-time (kinetic binding) as well as performing traditional end point binding analysis. BIAcore chips are compatible for use with MALDI mass spectrometry thus allowing additional integration. BIAcore analysis can be applied to membrane proteins and such studies have been attempted (through immobilizing cell membrane preparations or proteoliposomes or by means of a lipid deposition and surface reconstitution approach) but such experiments are less reliable and result in somewhat less quantitative data. Unfortunately SPR-based detection is not suitable for measuring small molecule binding, since an SPR signal is proportional to a total mass associated with the sensor surface, not the number of molecules.

Genoptics (Orsay, France; www.genoptics.fr) has an SPR-based system based on their spotted Bio-Micro array. This allows the parallel detection of up to several hundreds of protein (antibody) – ligand interactions with no requirement for labelling. This permits a real-time analysis of differential protein profiles. Biotrove Inc. (Cambridge, MA; www.biotrove.com) provides nanofluidics platforms for storage and ultra-high density screening. The technology is based on the

LivingChip™ – an array of 10 000 micro-channels (containing up to 50 nanolitres), which relies on surface tension and hydrophobic barriers to hold fluids in place and prevent mixing. The LivingChip™ is primarily used for storage: samples are recovered into microtitre plates using air-jets. The system could potentially be used for protein expression, such as cells, phages and analysis, i.e. ELISA. Sample loading and recovery seem to be the limiting steps offsetting all or most of the potential advantages of the platform. Graffinity Pharmaceuticals AG (Heidelberg, Germany; www.graffinity.com) have produced a chemical microarray platform for screening proteins in order to identify small molecule ligands. The immobilized compounds (small drug fragments) are incubated with a purified and solubilised target protein to generate an affinity fingerprint. This delivers information on the potential for small molecules as drug leads.

Peptidomics (Scrivener *et al.* 2003] is a further chip based proteomics method (Figure 11.4). The chip consists of an array of capture reagents (antibodies) that are immobilized on a solid support. Each antibody specifically recognizes a region of a target peptide fragment, generated by enzymatic or chemical cleavage of the protein of interest. Protein cleavage results in a homogeneous mixture of peptides and these can be analysed in a consistent environment. Thus all proteins from any class can be assayed in parallel from any given starting material *(e.g.* body fluid,

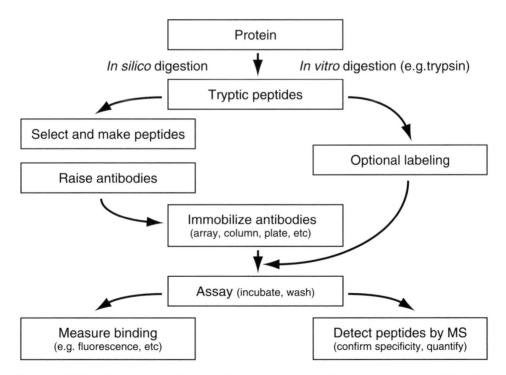

Figure 11.4 A schematic diagram illustrating the peptidomics approach, which relies on protein digestion and analysis of peptides

whole cells, mixed cell types from organs, microbial). Detection is by mass spectrometry (MALDI-TOF-MS) which allows the mass to charge ratio of the peptide to be identified and hence provides verification that the capture is specific. Following identification of the peptide, the proteins responsible for differential signals are easily established. The use of mass spectrometry avoids protein/peptide labelling, which is an advantage with any protein assay system, since labelling may alter recognition epitopes. Ideally, capture agents are generated against peptides or are specific for a particular region of a protein so that high affinities can be obtained. Such peptide sequences can be determined theoretically, using (software) tools to predict enzymatic digestion. For antibody generation, peptides can either be produced synthetically or generated from a protein of interest in a reproducible manner, e.g. by tryptic digestion. However, chemical synthesis allows for higher throughput and reproducibility. If a peptide contains an amino acid that may undergo a post-translational modification the epitope chosen should avoid inclusion of that particular amino acid to allow the fragment to be bound irrespective of any modification. The technique allows relative quantitation by measuring the amount of a captured peptide fragment relative to the amount of a peptide fragment of a constitutively expressed protein, or relative to exogenous controls, etc. Quantitation could also be achieved using isotopically labelled peptides, similarly to the ICAT principle described earlier by Gygi *et al.* (1999).

Peptidomics enables high-throughput screening of proteins in a microarray format and has several advantages over more widely used methods involving affinity capture of intact proteins. First, the homogeneity of digested proteins (typically in the form of tryptic peptides) results in a more uniform pool of target species, allowing for a higher degree of quantitation to be possible. Also, as peptides are much more stable and robust than proteins they are less prone to degradation and so handling is easier. Peptide species selected for antibody capture can be readily generated and so their specificity and affinity can be easily selected. As each protein is broken down into many smaller components, ranges of peptides become available from which to choose the most antigenic species. Peptides are also particularly suited to detection by mass spectrometric techniques, such as MALDI-TOF MS. Their masses can be accurately determined, allowing for mass matching database searches to be performed with high confidence. Peptides are also amenable to tandem mass spectrometry techniques, which allow the peptide sequence to be obtained

Conclusion

The development of protein expression profiling (proteomics) techniques based on a chip format has expanded rapidly in recent years. A wide range of companies have entered the market and an increasingly large range of products is becoming available to the proteomics researchers. Although an analysis of a whole proteome, as is potentially attainable for genomics using DNA chips, still remains in the future for proteins, a number of methodologies and technologies to realise this objective are now in place. This type of analysis for proteins will undoubtedly emerge within the next few years as the number of capture agents (antibodies, aptamers, etc.), which are suitable for application to an array format continues to increase. In the interim, both specialized chips for a focused analysis of groups of proteins, or peptides, and diagnostic chips for patient sample-screening are available, being used, and their complexity is expanding rapidly.

A combination of almost 15 years of experience of DNA microarray technology and the realization that proteins need to be analysed for expression and function to speed up advances in medicine has driven the automation, miniaturization, investment and technological advances necessary to begin to make functioning protein chips. Will the development of protein chips go along the familiar way paved by DNA chip developers, through continuing improvements in the surface immobilization techniques, further increases in affinity and density of the immobilized antibody and steady growth in the public acceptance of the new technology or will it take other routes? Early evidence from conferences and published reports suggests much greater variability in the style of protein arrays in all areas including surfaces, arrayers, affinity reagents, assays, analyses, and readers.

References

Abbott, A. (2002) Betting on tomorrow's chips. *Nature* **415**:112–114.

Axen, R. and Ernback, S. (1971) Chemical fixation of enzymes to cyanogen halide activated polysaccharide carriers. *Eur J Biochem* **18**(3):351–360.

Barry, R., Scrivener, E., Soloviev, M. and Terrett, J. (2002) Chip-based proteomics technologies. *Int Genomic Proteomic Technology* 14–22.

Bozinov, D. and Rahnenfuhrer, J. (2002) Unsupervised technique for robust target separation and analysis of DNA microarray spots through adaptive pixel clustering. *Bioinformatics* **18**(5):747–756.

Bradford, M.M. (1976) A rapid and sensitive method for the quantitation of microgram quantities of protein utilizing the principle of protein-dye binding. *Anal Biochem* **72**: 248–254.

Bussow, K., Cahill, D., Nietfeld, W., Bancroft, D., *et al.* (1998) A method for global protein expression and antibody screening on high-density filters of an arrayed cDNA library. *Nucleic Acids Res* **26**(21):5007–5008.

Causton, H. (2002) Applications of microarrays. *Trends Mol Med* **8**(6):308.

Chapman, K. (2002) The ProteinChip Biomarker System from Ciphergen Biosystems: a novel proteomics platform for rapid biomarker discovery and validation. *Biochem Soc Trans.* **30**(2):82–87.

Cullen, D.C., Brown, R.G. and Lowe, C.R. (1987/88) Detection of immuno-complex formation via surface plasmon resonance on gold-coated diffraction gratings. *Biosensors* **3**(4):211–225.

Fagerstam, L.G., Frostell, A., Karlsson, R., Kullman, M., *et al.* (1990) Detection of antigen–antibody interactions by surface plasmon resonance. Application to epitope mapping. *J Mol Recognit* **3**(5–6):208–214.

Fletcher, G., Mason, S., Terrett, J. and Soloviev, M. (2003) Self-assembly of proteins and their nucleic acids. *J Nanobiotechnol* **1**: 1.

Goldfarb, A.R., Saidel, L.J. and Mosovich, E. (1951) The ultraviolet absorption spectra of proteins. *J Biol Chem* **193**, 397–404.

Gygi, S.P., Rist, B., Gerber, S.A., Turecek, F., *et al.* (1999a) Quantitative analysis of complex protein mixtures using isotope-coded affinity tags. *Nat Biotechnol.* **17**(10):994–999.

Gygi, S.P., Rochon, Y., Franza, B.R. and Aebersold, R. (1999b) Correlation between protein and mRNA abundance in yeast. *Mol Cell Biol* **19**(3):1720–1730.

Gygi, S.P., Corthals, G.L., Zhang, Y., Rochon, Y., and Aebersold, R. (2000) Evaluation of two-dimensional gel electrophoresis-based proteome analysis technology. *Proc Natl Acad Sci USA* **97**:9390–9395.

He, M. and Taussig, M.J. (1997) Antibody-ribosome-mRNA (ARM) complexes as efficient selection particles for in vitro display and evolution of antibody combining sites. *Nucleic Acids Res.* **25**(24):5132–5134.

He, M. and Taussig, M.J. (2001) Single step generation of protein arrays from DNA by cell-free expression and in situ immobilisation (PISA method). *Nucleic Acids Res* **29**(15): E73.

Holt, L.J., Bussow, K., Walter, G. and Tomlinson, I.M. (2000) By-passing selection: direct screening for antibody-antigen interactions using protein arrays. *Nucleic Acids Res* **28**(15): E72.

Holz, C., Lueking, A., Bovekamp, L., Gutjahr, C., *et al.* (2001) A human cDNA expression library in yeast enriched for open reading frames. *Genome Res* **11**(10): 1730–1735.

Iyer, V.R., Horak, C.E., Scafe, C.S., Botstein, D., *et al.* (2001). Genomic binding sites of the yeast cell-cycle transcription factors SBF and MBF. *Nature* **409**:533–538.

Kodadek T. (2001) Protein microarrays: prospects and problems. *Chem Biol* **8**(2):105–115.

Lander, E.S., Linton, L.M., Birren, B., Nusbaum, C., *et al.* (2001) Initial sequencing and analysis of the human genome. *Nature* **409**:860–921.

Liu, W.M., Mei, R., Di, X., Ryder, T.B., *et al.* (2002) Analysis of high density expression microarrays with signed-rank call algorithms. *Bioinformatics* **18**(12): 1593–1599.

Lizardi, P.M., Huang, X., Zhu, Z., Bray-Ward, P., *et al.* (1998). Mutation detection and single-molecule counting using isothermal rolling-circle amplification. *Nat Genet* **19**(3):225–232.

Lowry, O.H., Rosebrough, N.J., Farr, A.L. and Randall, R.J. (1951) Protein measurement with the Folin phenol reagent. *J Biol Chem* **193**:265–275.

McPherson, J.D., Marra, M., Hillier, L., Waterston, R.H., *et al.* (2001) A physical map of the human genome. *Nature* **409**:934–941.

Page, M.J., Amess, B., Townsend, R.R., Parekh, R., *et al.* (1999) Proteomic definition of normal human luminal and myoepithelial breast cells purified from reduction mammoplasties. *Proc Natl Acad Sci USA* **96**:12589–125940.

Schweitzer, B., Roberts, S., Grimwade, B., Shao, W., *et al.* (2002) Multiplexed protein profiling on microarrays by rolling-circle amplification. *Nat Biotechnol* **20**(4):359–365.

Scrivener, E., Barry, R., Platt, A., Calvert, R., *et al.* (2003) Peptidomics: a new approach to affinity protein microarrays. *Proteomics* **3**(2):122–128.

Simone, N.L., Remaley, A.T., Charboneau, L., Petricoin, E.F. 3rd, *et al.* (2000) Sensitive immunoassay of tissue cell proteins procured by laser capture microdissection. *Am J Pathol* **156**(2):445–452.

Smith, P.K., Krohn, R.I., Hermanson, G.T., Mallia, A.K., *et al.* (1985) Measurement of protein using bicinchoninic acid. *Anal Biochem* **150**: 76–85.

Soloviev, M. (2001) EuroBiochips: spot the difference. *Drug Discov Today* **6**:775–777.

Walter, G., Bussow, K., Cahill, D., Lueking, A. and Lehrach, H. (2000) Protein arrays for gene expression and molecular interaction screening. *Curr Opin Microbiol* **3**(3):298–302.

Wang, J., Delabie, J., Aasheim, H., Smeland, E. and Myklebost, O. (2002) Clustering of the SOM easily reveals distinct gene expression patterns: results of a reanalysis of lymphoma study. *BMC Bioinformatics* **3**(1):36.

Weng, S., Gu, K., Hammond, P.W., Lohse, P., *et al.* (2002) Generating addressable protein microarrays with PROfusion™ covalent mRNA-protein fusion technology. *Proteomics* **2**(1):48–57.

Xu, L., Aha, P., Gu, K., Kuimelis, R.G., *et al.* (2002) Directed evolution of high-affinity antibody mimics using mRNA display. *Chem Biol* **9**(8):933–942.

Yunginger, J.W. and Gleich, G.J. (1972) Comparison of the protein-binding capacities of cyanogen bromide-activated polysaccharides. *J Allergy Clin Immunol* **50**(2):109–116.

Zhu, H., Klemic, J.F., Chang, S., Bertone, P., *et al.* (2000) Analysis of yeast protein kinases using protein chips. *Nat Genet* **26**(3):283–289.

12

Infectomics Overview: Holistic and Integrative Studies of Infectious Diseases

Sheng-He Huang and **Ambrose Jong**

Introduction

Despite the advances of modern science and the availability of effective anti-microbial agents over the last 50 years, the threat of infectious diseases or death that can result from microbial infections is still a most important subject of medical research and clinical practice (Fauci 2001; Huang *et al.* 2002). The continual emergence of previously undescribed new infectious diseases and re-emergence of old organisms will certainly heighten the global impact of infectious diseases in the 21st century. Another significant problem in medicine is the development of microbial resistance to antimicrobial drugs, due to the widespread and often inappropriate use of these antimicrobials. As clinical practice tends towards greater use of invasive interventions and with patients living longer, there is a continually growing proportion of older and immunocompromised patients. These groups are predisposed to opportunistic infections caused by non-pathogenic microbes such as yeast. In addition, how to deal with and prevent bioterrorism is becoming a very serious issue in the 21st century. The development of new anti-infective agents against various microbial organisms and in favour of the host defence has emerged as an urgent issue in modern medicine. The availability of genome sequences, and the development of microarrays and

Molecular Analysis and Genome Discovery edited by Ralph Rapley and Stuart Harbron
© 2004 John Wiley & Sons, Ltd   ISBN 0 471 49847 5 (cased)   ISBN 0 471 49919 6 (pbk)

computational tools offer holistic and integrative strategies for resolving the above major medical problems.

The traditional approach to human infectious diseases has limited research to individual virulence genes, the important organisms, and a single or a limited number of the host defence components. Currently there is an urgent need for holistic approaches that can efficiently, precisely and integratively study microbial infections. The term 'infectomics' was recently coined and defined as an integrative omic approach globally to study microbial infections (Huang *et al.* 2002). The most important omic approaches include genomics and proteomics. Recently, glycomics has become a valuable addition to the general omic approaches (Wang *et al.* 2002). Each branch of omics embraces two essential aspects: structural and functional studies. The global genotypic and phenotypic changes (infectomes) in microbes and their hosts contributing to infections are encoded by the genomes of microbial organisms and their hosts. These changes, including the levels of replication (gene), transcription (mRNA), translation (protein) and post-translational modifications (e.g. glycosylation and phosphorylation), in both pathogen and infected host are believed to be patterned and stereotyped. Recently, DNA microarrays have been used as the major high-throughput approaches globally to monitor genotypic and phenotypic changes in both microbial organisms and their infected hosts contributing to infection (Cummings and Relman 2000; Huang *et al.* 2002; Lucchini *et al.* 2001). Protein and carbohydrate microarrays are becoming very promising and complementary tools globally to dissect infectious diseases (MacBeath 2002; Mitchell 2002; Schweitzer and Kingsmore 2002; Walter *et al.* 2002; Wang *et al.* 2002). The fundamental issue of infectious diseases is how globally and integratively to understand the interactions between microbial organisms and their hosts by using infectomics (Figure 12.1).

Investigating microbial infection with 'omic' approaches

Genomic studies of microbial infections

Since the first reported complete sequence of the genome of a living organism *Haemophilus influenzae* RD was published, genomic projects have been moving in all aspects. The complete genomes of human (*http://www.ncbi.nlm.nih.gov/genome/seq/*), over 1000 viruses and more than 100 non-viral microbes, including many human pathogens, have been sequenced (*http://www.tigr.org/tdb/mdb/mdbcomplete.html*) (Venter *et al.* 2001; Hood and Galas 2003). An online database (http://www.integratedgenomics.com) continues monitoring both completed and ongoing genome projects worldwide. In parallel to the tremendous amount of data accumulated, the pace of analysis and gene screening methods rapidly accelerates. The challenge in this post-genomic era is how to 'mine' these data

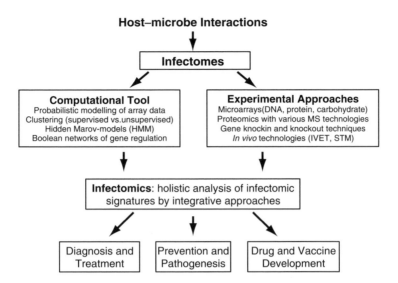

Figure 12.1   An outline diagram of infectomics. Infectomes are induced in microbial organisms and their hosts as the outcomes of host–microbe interactions. These infectomic signatures can be holistically and integratively dissected by a melding of computational tools and experimental approaches. IVET: *in vivo* expression technology; STM: signature-tagged mutagenesis

not only for understanding the bacterial physiology and pathogenesis, but also for drug and vaccine development (Emilien *et al.* 2000; Loferer 2000; Ohlstein *et al.* 2000). Among new high-throughput methods, DNA microarrays are prominent for their simplicity, comprehensiveness, and data consistency. Instead of detecting and studying one gene at a time, microarrays allow thousands or tens of thousands of specific DNA or RNA sequences to be detected simultaneously on a square of glass or silica only 1–2 cm (see Chapters 7, 8 and 11).

Two common approaches, cDNA and oligonucleotide microarrays, have been used for high-density array studies. In one, cDNA microarrays are constructed by physically attaching DNA fragments (PCR products) to a solid phase. By using a robotic arrayer and capillary printing tips, one can print >30 000 DNA fragments on a microscopic slide. Techniques that describe microarray construction methods and hybridization protocol are available (*http://cmgm.stanford.edu/ pbrown/index.html*) and in Chapter 7. Another approach is to construct the arrays by use of single-stranded oligonucleotides *in silico* using photolithographic techniques, primarily used by the commercial company, Affymetrix, Inc. In order to measure relative gene expression by using cDNA microarrays, RNA is prepared from the two samples to be compared, and labelled cDNA is made by reverse transcription, incorporating either Cy3 (green) or Cy5 (red) fluorescent dye. The two labelled cDNA mixtures are mixed and hybridized to the microarray, and the slide is scanned. In the case of green Cy3 and red Cy5 signals are overlaid, yellow spots indicate equal intensity for the dyes. With the use of image analysis software, signal intensities are determined for each dye at each element of the array, and the logarithm of the ratio of Cy5 intensity to Cy3 intensity is calculated. Positive log (Cy5/Cy3) ratios indicate relative excess of the transcript in the Cy5-labelled sample, and negative log (Cy5/Cy3) ratios indicate relative excess of the transcript in the Cy3-labelled sample. In oligonucleotide microarrays, a series of cRNA biotinylated samples were prepared to measure the gene expression profiles individually under various conditions, such as disease stages. After several such experiments have been performed, the data set can be analysed by various clustering or other computational analyses to identify the overall gene expression profile or lists of up- or down-regulated genes.

The DNA microarray has made it possible to tackle qualitatively different questions in biology and medicine. The ability to measure the expression level of thousands of genes in any tissue sample allows exploration of gene function on a scale previously impossible. On the one hand, DNA microarray has been used to monitor microbial infectomes during infection (Huang 2003; Huang *et al.* 2002). A microarray containing 1534 ORF of *H. pylori* was used to explore its acid-induced gene expression (Ang *et al.* 2001). Eighty genes increase their expression levels significantly during acid stress. For example, Omp11 encodes a member of the proton-translocating ATPase family. This is in agreement with its functions that Omp11 plays a role in pH regulation by extruding protons from cytoplasm. Monitoring the response of *H. pylori* genes during acid stress may be helpful in understanding the pathogenesis under acidic stomach conditions. Similar microarray approaches have been performed on infectious viruses, such as HIV (Geiss *et al.* 2000; Vahey *et al.* 1999), cytomegalovirus (CMV) (Zhu *et al.* 1998), and *Pseudomonas aeruginosa* (Ichikawa *et al.* 2000). On the other hand, DNA microarray is an extremely effective way to monitor the host infectomes during pathogenic invasion. Healthy individuals protect themselves against microbial pathogens by two types of defence mechanism: innate and adaptive immunity. Mucosal epithelial surfaces of the intestinal, respiratory, and genitourinary tracts are important tissue barriers that are the most common entry routes for microbial invasion. These epithelial cells can function as sensors for microbial infections (Kagnoff and Eckmann 1997). The host infectomes are believed to be patterned and stereotyped. These phenotypic signatures or expression profiles of the hosts can be used to distinguish among different infectious agents or different pathogenic mechanisms. Recent human cDNA microarrays have been used to monitor globally the host response to viruses (e.g. human immunodeficiency virus, cytomegalovirus, and coxsackievirus), bacteria (e.g. *Salmonella typhimurium, Pseudomonas aeruginosa,* and *Listeria monocytogenes*), and parasites (e.g. *T. gondii*) (Blader *et al.* 2001; Cleary *et al.* 2002). Gene expression profiles of human foreskin fibroblasts in response to *T. gondii* infection have been monitored by using human cDNA microarrays consisting of ~22,000 known genes and uncharacterized expressed sequence tags. These studies suggest that the early response (1–2 h) does not require parasite invasion and expression of the late phase genes is mainly dependent on the direct presence of the parasite. The difference in the early and late phase gene expression may not only facilitate dissecting pathogenesis of this parasitic disease but also provide important information on the progress of the disease. The host responses to various infectious agents were analysed by using microarrays containing cDNAs derived from more than 15 000 human genes (Relman 1999). mRNA levels in multiple samples are compared with differential fluorescence labelling. We have recently performed a time-course study of an opportunistic *Cryptococcus neoformans* invasion in human endothelium cells, using oligonucleotide microarray to monitor the alteration of 12 599 gene profiles. Each time-point can be compared by scattering analyses (Huang *et al.* 2002). An ontological analysis reveals gene expression patterns of different functional classes. For example, an immune-related class (64 genes) is compared at 0 (uninfected) and 4 h infection status (Huang *et al.* 2002). Profiling gene expression patterns in host cells during infection provides global and accurate information for building a comprehensive framework to interpret pathogenic processes.

Global approaches based on genomes have also facilitated a fundamental shift from the classical methods of vaccine development and direct antimicrobial screening programmes toward rational and genomewide target-based strategies (Huang *et al.* 2002). The use of high-throughput cloning and expression of candidate genes permits a comprehensive evaluation of all predicted gene products. The candidate proteins can then be directly tested in animals for protection against challenge or time to resolution of infection. The advantages of using the genomic approach are that the method does not rely on knowledge of any protein function and it conducts a global search. An excellent example of using DNA microarrays for vaccine development is to follow *N. meningitidis* serogroup B (MenB) gene regulation during interaction with human epithelial cells (Grifantini *et al.* 2002) (Figure 12.2). Five new vaccine candidates with the capability of inducing bactericide antibodies in mice were identified with this procedure. Whole-genome DNA microarray analysis has been used to compare the genomes of variants of the tuberculosis vaccine strain *Mycobacterium bovis bacillus* Calmette-Guérin (BCG), methicillin-resistant *Staphylococcus aureus* (MRSA) and other pathogens. These data provide new information about the evolution of these human pathogens and, more importantly, suggest rational approaches to the design of improved diagnostics and antimicrobial agents. Similar genomic algorithms were used for the prediction of surface-localized proteins from the genomic sequence of *S. pneumoniae* (Wizemann *et al.* 2001).The vaccine candidates were directly applied in an animal model to screen the proteins for vaccine efficacy. Two vaccine candidates were identified that conferred protection in mice. These proteins represent a novel class of proteins that are only expressed during *in vivo* infection. Use of a whole genome approach to search for new vaccine candidates has also been applied to *Mycobacterium tuberculosis* (Gomez

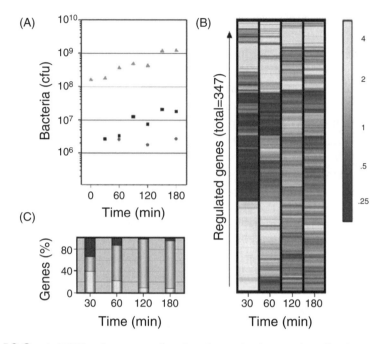

Figure 12.2 A DNA microarray showing dynamic changes in adhesion, growth and infectomes of *N. meningitidis* serogroup B (MenB) during bacteria adhering to epithelial cells. (A) Bacteria counts. Adhering bacteria were counted after washing and lysis of the host cells. (■) Adhering bacteria; (▲) bacteria freely growing in the medium; (●) Growth of cell-associated bacteria. (B,C) RNA was isolated at different times from both adhering and freely growing bacteria, and used to probe DNA chips carrying the entire MenB genome. (B) Clustered expression profiles of genes whose regulation differs from freely growing bacteria by at least two folds. (C) Regulated genes in (B) were further analysed according to activation state (light grey, upregulated; dark grey, downregulated) to give a visual indication of the persistence of gene regulation. Light grey to dark grey scale represents fold difference. Reproduced by courtesy of G. Grandi from Grifantini *et al.* (2002)

et al. 2000), *Helicobacter pylori* (Chakravarti *et al.* 2001), etc. These few examples highlight current approaches whereby genomic analysis and screening can be done *in silico* before any wet laboratory research is implemented. Although we have become faster and better at predicting vaccine candidates, the need and the time required for experimental laboratory and clinical studies still remains unchanged in the development of safe and efficacious vaccines. The mining of microarray data for microbial organisms will certainly help in the identification of many potential vaccine candidates and antimicrobial targets. In the future, a combination of the strengths and advantages of both the genome-wide microarray technologies and the conventional screening strategies will greatly revolutionize vaccine development and drug discovery (Rosamond and Allsop 2000).

Proteomic approaches to microbial infections

Almost simultaneous with the advancement of genomic studies, there has been an increased interest in high-throughput protein analyses in the post-genomic era. The ultimate goal of genomics is the global elucidation of the functional partners of genes and genomes (proteins and proteome) (Huang *et al.* 2002). The proteome represents the functional status of a cell in response to environmental stimuli and thus provides more direct information on functional changes. Therefore, as an alternative and complementary approach to genomic-based technologies, proteomics is essential for the identification and validation of proteins and for the global monitoring of infectomic changes in protein expression and modification during infections. There are two major approaches to proteomics, two-dimensional gel electrophoresis (2-DE)-based technology and protein microarrays (see Chapter 11).

The 2-DE approach to proteomics begins with identification of protein spots on 2-DE gels. Subsequently, the identified spots are linked to the images of the gels. Proteins can be further identified by matrix-assisted laser desorption ionization (MALDI) and mass spectrometry (MS). This is able to separate and to detect several thousand protein spots in a good gel with combination of MS and computer technologies. Commercial robots are now available for staining gels, spot excision and subsequent proteolysis before MS. MALDI, MS and special software such as Melanie 3 (*http://www.expasy.ch/melanie/*) can be used further to characterize proteins of interest. Proteomics has been successfully used for comparative analysis of protein expression profiles of organisms and the infected host cells. For example, the protein expression patterns of virulent *Mycobacterium tuberculosis* strains and attenuated vaccine strains were compared using a combination of 2-DE and MS. Among 1800 protein spots isolated, six new gene products, not previously predicted by the genomic study of *M. tuberculosis*, were identified with proteomics. A combination of proteomics and genomics has been used to identify unknown regulons and proteins in *Bacillus subtilis* and gram-positive organisms (Huang *et al.* 2002). Recently, 2-DE-MS-based proteomics has been a powerful tool for infectomic studies of infectious diseases (Huang *et al.* 2003). One example is the proteomic investigation to compare the protein compositions of virulent *M. tuberculosis* strains with attenuated vaccine strains using a combination of two-dimensional electrophoresis (2-DE) (Figure 12.3A) and mass spectrometry (MS) approach (Figure 12.3B,C) (Jungblut *et al.* 2001). Six new genes, not previously detected in the genome of *M. tuberculosis*, were identified in this study. The studies demonstrate the value of proteomics in identifying gene products undetected by the genomic approach. Proteomics is also a powerful tool in identifying proteins of importance for diagnosis in proteome analysis of pathogenic microorganisms, such as *Borrelia burgdorferi* (Lyme disease) and *Toxoplasma gondii* (toxoplasmosis) (Jungblut *et al.* 1999). Sera from patients with early and late symptoms of Lyme borreliosis contained antibodies of various

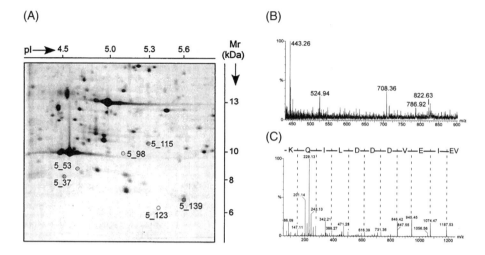

Figure 12.3 A proteomic analysis showing open reading frames (ORF) in *M. tuberculosis* H37Rv that are not predicted by genomics. (A) Identification of previously unpredicted ORFs from *M. tuberculosis* H37Rv by two-dimensional electrophoresis (2-DE). Proteins were stained with silver nitrate. The M_r range between 6 and 15 kDa and the pI range between 4 and 6 are indicated. The six *de novo* spots indicated by numbers were sequenced by nanospray MS/MS. Results revealed that ORFs were not predicted previously. B and C, MS analysis of spot 5_98. (B) Spectrum of the trypsinized protein. Peptides for sequencing were labelled and fragmented. (C) fragmentation pattern of the peptide with an *m/z* of 708.36 identified as VEIEVDDDLIQK [Reprinted with permission from Infect Immun (Jungblut *et al.* 2001)]

classes against about 80 antigens each in the 2-DE, including the known antigens OspA, B, and C, flagellin, p83/100 and p39. Similarly, antibody reactivity to seven different marker antigens of *T. gondii* allowed differentiation between acute and latent toxoplasmosis, an important diagnostic tool in both pregnancy and immunosuppressed patients. A combination of 2-DE with MALDI-TOF-MS has been successfully used for infectomic analyses of the *Chlamydia pneumonia* elementary body (EB) in Hep-2 cells and *Chlamydia trachomatis* reticulate body (RB) in HeLa 229 cells (Vandahl *et al.* 2001; Molestina *et al.* 2002; Shaw *et al.* 2002). Several novel features of *C. pneumonia*, such as energy-producing enzymes and type III secretion proteins, have been implicated in its proteomics maps (Vandahl *et al.* 2001). A novel 7-kDa RB protein was identified in *C. trachomatis* but was not found in *C. pneumonia* (Shaw *et al.* 2002). However, there are a number of limitations with the 2-DE-MS approach (Huang *et al.* 2003). The resolution of the 2-DE-MS approach is limited when a total cell lysate is analysed (Graves and Haystead 2002). In the crude extract, it is difficult to detect the low-copy proteins as the most abundant proteins may dominate the gel. It was demonstrated in yeast crude extracts that no low-copy proteins were detected by 2-DE (Graves and Haystead 2002). In order to overcome this limitation, it is better to use 2-DE for dissecting subproteomes in partially purified protein complexes instead of whole proteomes in a total cell lysate. For example, the analysis of subproteomes for organisms (such as stimulons, regulons and genetic or pathogenicity islands) and hosts (e.g. caveolae, endosomes, phagosomes and mitochondria) not only makes the work more manageable but frequently provides more detailed information that is lost if a total cell lysate is used. Protein phosphorylation is an important regulatory mechanism contributing to microbial pathogenesis. MS is one of the most powerful tools for detecting phosphoproteomes (Graves and Haystead 2002). The use of *in vivo* labelling of proteins with inorganic ^{32}P is a common approach to studying protein phosphorylation. The infectomes of cells that differ in protein phosphorylation can be analysed by culturing normal and infected cells in inorganic ^{32}P and preparing cell lysates. Alterations in the phosphorylation state of proteins can then be detected by 2-DE and autoradiography. Proteins of interest are isolated from the gel and then microsequenced by MS (Graves and Haystead 2002). Another limitation of 2-DE is that the procedure for protein analysis is labour-intensive and time-consuming. A number of approaches have been developed to bypass gel electrophoresis. Among them, protein microarrays have emerged as the most interesting technique alternatives to 2-DE (Graves and Haystead 2002). See Chapter 11.

The protein microarray technologies are undergoing rapid development for various applications (Delehanty and Ligler 2002; Walter *et al.* 2002). Proteins are immobilized by being attached either on flat solid phases or in capillary systems (microfluidic arrays). Preferred solid phase materials are modified glass or filter membranes with low-fluorescence background. Protein microarrays are commonly printed and imaged using the same commercially available arrayers and scanners as for DNA microarrays (Walter *et al.* 2002). The most common arrayers are pin-based systems. Fluorescence-based imaging systems are commonly used for protein microarrays by either direct labelling of ligands or indirect detection via labelled antibodies. Extremely low levels of analyte can be detected by surface-confined fluorescence, such as the fluorescence planar waveguide (FPWG) technique. Using MS, proteins and their post-translational modifications can be identified in complex protein mixtures (Walter *et al.* 2002). Protein microarrays appear to be promising tools for infectomic studies of microbial infections. CD8 T lymphocytes from certain immunologically stable HIV-1 patients secrete a soluble factor, termed CAF, that suppresses HIV-1 replication. However, the identity of CAF remained elusive despite an extensive search since 1986. By means of a protein-chip technology, a cluster of secreted proteins (alpha-defensin 1, 2 and 3) were identified as CAF (Zhang *et al.* 2002). The identification was based on the specific antibody recognition and amino acid sequencing. CAF activity was abolished or neutralized by an antibody specific for human alpha-defensins. Replication of HIV-1 *in vitro* was also blocked by synthetic and purified preparations of alpha-defensins. Protein microarrays containing malaria parasite surface protein are currently being developed for studies of parasitic diseases (Bacarese-Hamilton *et al.* 2002). Two types of protein chips have been designed. The first one contains most of the *P. falciparum* antigens currently being studied in vaccine trials for their ability to elicit protective immunity in animal and human. The second type of protein microarrays contain recombinant proteins that are shown or predicted to be either secreted or exposed on the surface of the malaria parasites. These proteins are the ones most likely to have the capability directly to react with serum antibodies, functioning as targets of natural acquired immunity. The preliminary results show that the immune response detected by protein chips against a set of arrayed malaria antigens is consistent with the prior data obtained by testing the same group of sera in ELISA and immunoblot (Bacarese-Hamilton *et al.* 2002). Most of the protein microarrays are still largely experimental but their development is progressing rapidly. The recent progress in the arrayed library will allow gene expression analysis and protein interaction screening on a whole-genome scale (Walter *et al.* 2002). In contrast to DNA microarrays, dynamic gene action can be studied by protein arrays directly, provided the natural structure and activity of proteins are maintained. However, a number of challenging issues remain. These include novel systems required for high-throughput protein expression and functionally active proteins arrayed at high density (Walter *et al.* 2002).

Glycomic studies of infectious diseases

Analogous to the terms that characterize the genome and proteome, the glycome is defined as the total carbohydrate complement. Glycomics is to decipher the information content and expression patterns of cellular glycomers under various conditions. Carbohydrates are the key components of glycoproteins, glycolipids, glycosaminoglycans and proteoglycans, which are important glycoconjugates responsible for recognition, adhesion, and signalling between cells (Fukui *et al.* 2002; Drickamer and Taylor 2002; Love and Seeberger 2002; Wang *et al.* 2002). Expression of these molecules with certain carbohydrate structures is frequently specific to cell or tissue type. Cellular glycoconjugates are essential for normal tissue growth and repair, bacterial and viral invasion of host organisms, as well as tumour-cell motility and progression. Carbohydrate structures are required for host–microorganism interactions. Many microorganisms use host cell glycoconjugates as receptors or co-receptors. Their structural diversity and selectivity of host tissue expression play a significant role in the tropism of microbial infections. On the other hand, microbial polysaccharides are important components of many major antigenic structures that are recognized by their host cells. For example, polysaccharides are the main structures of O and K antigens, and endotoxin in *E. coli* pathogens. It has been shown that the microbial antigens mimic the structures of host components, assisting a microbe to escape from the host's immune defence (Wang *et al.* 2002). Glycomic identification and characterization of glycoconjugates from microbial organisms and their hosts should be important for dissecting microbial pathogenesis, and improving diagnostic and therapeutic tools. Recently, a carbohydrate-based microarray has been developed for glycomic analysis of microbial infections and anti-infection responses (Wang *et al.* 2002). Microbial polysaccharides were immobilized on a surface-modified glass slide without chemical conjugation. Using this procedure, a large repertoire of microbial antigens (~20 000 spots) can be spotted on a single micro-glass slide, reaching the capacity to include most common organisms. This system is highly sensitive, allowing simultaneous detection of a broad spectrum of antibody specificities with as little as a few microlitres of serum specimen (Figure 12.4). Carbohydrate or oligosaccharide microarrays are becoming very promising and complementary tools holistically to dissect infectious diseases (Huang 2003).

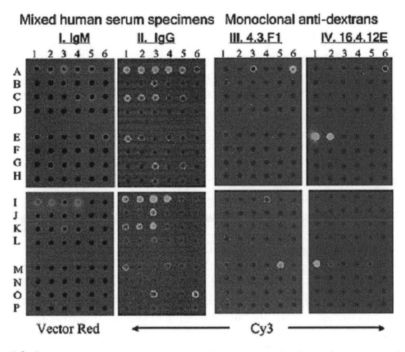

Figure 12.4 A carbohydrate microarray for characterization of human and murine antibodies. A total of 48 distinct antigen preparations were spotted on slides at antigen concentrations of 0.5 mg/ml and 0.02 mg/ml. Each preparation has a unique antigen ID number. They were incubated with combined human serum specimens at a concentration equivalent to 1 : 100 dilution of each specimen or with mouse mAbs at 1 μg/ml. The human IgM captured by microarrays was visualized using an anti-human IgM–AP conjugate and the colour developed using Vector Red. The human IgG anti-carbohydrates were detected using a biotinylated anti-human IgG and visualized by a Cy3–streptavidin. The isotype profile of human anti-carbohydrate antibodies identified by microarray assay was confirmed by a carbohydrate-based ELISA binding assay. Reproduced by courtesy of D. Wang from Nature Biotechnology (Wang *et al.* 2002)

Dissecting the microbiome: ecological solutions to infectious diseases

Microbiome is the collective genome of our normal microbiota. From birth to death, we share a benign coexistence with a vast, complex, and dynamic consortium of microbes. Most of our microbial commensals reside in our gastrointestinal (GI) tract, which harbours a rich microbiota of >500 different bacterial species (Hooper and Gordon 2001). Some of them have important health functions. These include stimulating the immune system, protecting the host from invading bacteria and viruses, and aiding digestion. The gut microbiota is established rapidly after birth and remains relatively stable throughout life (Alvarez-Olmos and Oberhelman 2001). The GI mucosa provides a protective interface between the internal environment and the constant external challenge from food-derived antigens and microbes. The composition and effect of the normal microbiota may be altered by a number of environmental factors. These include use of antibiotics, immunosuppressive therapy, irradiation, hygiene and imbalance of nutrition. All the mentioned factors can contribute to a decline in the incidence of microbial stimulation that may dampen host defence and predispose us to infectious diseases (Alvarez-Olmos and Oberhelman 2001). Therefore, the introduction of beneficial live bacteria into the GI tract (probiotics) may be a very attractive rationale for providing a microbial stimulus to the host immune system against organisms (Isolauri 2001). The microbes frequently used as probiotic agents include *Lactobacillum* and *Bifidobacterium*. It has been shown that enterally-administered *Lactobacillus casei* GG decreases the frequency of *E. coli* translocation in human intestinal epithelial monolayers (Caco-2) and a neonatal rabbit model (Lee *et al.* 2000; Mattar *et al.* 2001). Multiple mechanisms of probiotic therapy have been postulated, including production of antimicrobial agents, competition for space or nutrients, and immunomodulation. For example, *Lactobacillum* spp. is able to attenuate colitis in IL-10-deficient mice; probiotic agents containing *Lactobacillum*, *Bifidobacterium* and *Streptococcus* spp. are effective in treatment of chronic 'pouchitis', a complication subsequent to surgical therapy for ulcerative colitis (Hooper and Gordon 2001). Various studies suggest that the use of lactic acid bacteria (LAB) as live vectors is a promising approach for delivering drugs, antimicrobial agents and vaccines to defined host niches, due to their safety, ability to persist within the indigenous microbiota, adjuvant properties, and low intrinsic antigenicity (Alvarez-Olmos and Oberhelman 2001).

The development of microbial infections is determined by the nature of host–microbe relationships. These include host–pathogen, host–commensal and pathogen–commensal interactions. A holistic balance of these relationships is essential to our health. However, this balance remains poorly defined. Little attention has been paid to commensal microbes that may be beneficial to the host defence systems. Dissecting the role of normal microbiota in microbial infection and exploring probiotics as ecological approaches to infectious diseases will be challenging and may require decoding the microbiome, and simultaneously monitoring of host and microbial gene expression profiles during the course of colonization. Whole-genome genotyping of *H. pylori* isolates using DNA microarrays reveal that the *cag* pathogenicity island (PAI) and gastric microbiota play a significant role in colonization. The *cag* PAI-negative strain is unable to colonize conventionally raised Lewis[b] transgenic mice harbouring a normal gastric microbiota, whereas the *cag* PAI-positive isolate colonizes 74 per cent of the animals (Björkholm *et al.* 2001). DNA microarray analysis of intestinal transcriptional responses to colonization of *Bacteroides thetaiotaomicron* suggest that this commensal bacterium modulates expression of genes involved in several important intestinal functions. These include nutrient absorption, mucosal barrier fortification, xenobiotic metabolism, angiogenesis and postnatal intestinal maturation. Therefore, holistic studies on microbiome are important for our understanding of microbial infection (Huang *et al.* 2002).

Animal models and gene knockin/knockout technologies

Animal models of human infectious diseases are crucial for studies on infectomics due to the limited opportunities for controlled experimental manipulation in human. Among the model organisms amenable to genetic analysis, the mouse has been by far the most well-developed and physiologically relevant system for study of human host defence (Bedell *et al.* 1997; Glusman *et al.* 2001; Harris and Foord 2000; Paigen 1995). Not only are the mouse and human genomes very similar, but also gene knockin and knockout technologies, developed during the past two decades, allow gene function studies in the mouse that are not possible in most other organisms. For virtually every gene in the human genome, a counterpart can readily be identified in the mouse. Genetic manipulation within the living mouse is routine and can be performed with extraordinary precision (Bradley 2002). Null mutations, as well as subtle missense or gain-of-function mutations, can be introduced into virtually any gene in the mouse germline using homologous-recombination-based, gene-targeting technology (Bradley 2002). The conditional inactivation of gene expression *in vivo* can be used to inactivate a gene in only a subset of cells or at well-defined stages of development. Data sets of the mouse–human synteny are deposited in the major human and mouse databases and are becoming even more comprehensive as the completion of the sequencing of the mouse genome is achieved (*www.ncbi.nlm.nih.gov/Homology*) (Waterston *et al.* 2002). The comparative map of the mouse and human genomes is the most well developed of all species. A comparison of mouse and human genomes followed by experimental verification yields an estimated 1019 additional genes (Guigo *et al.* 2003). A comprehensive summary of mouse/human homology has placed 1416 loci on both maps by using human physical mapping data and mouse genetic maps (Qureshi *et al.* 1999). One hundred and eighty-one conserved linkage groups have been established. Comparative analysis of whole genome sequences of both mouse and human can be used for determining their fine chromosomal and genetic relationships.

Mouse models have been successfully used for studies on the pathogenesis of various microbial infections. Since mice have short generation times and high breeding efficiency this animal model is extremely useful for dissecting and identifying the most important individual genetic elements that govern the host response to important organisms (Qureshi *et al.* 1999). The first requirement for functional genomic study of host defence systems is to identify commercially available inbred strains of mice that can show differential responses to a well-defined infectious challenge. This provides short-cuts to gene identification, unequivocal proof that a mutation in that gene makes the host resistant or susceptible to infection, and rapid dissection of the molecular pathway for which the mutant gene codes. Once different phenotypes are identified, controlled breeding will be performed to determine the mode of inheritance of the phenotype (simple or complex). Linkage analysis can be carried out to correlate the inheritance of resistance or susceptibility to a specific infectious challenge with one or more chromosomal locations. Knowledge of the genomic structure of human and mouse can be used to facilitate localization and identification of the human orthologues of resistance or susceptibility genes identified through animal experiments. These candidate genes can further be tested for human infection susceptibility through genetic analysis. For example, the identification and characterization of genetic factors influencing natural susceptibility or resistance to infectious diseases in humans and mouse can provide new insight into the basic mechanisms of host defence against infections. Studies on inbred strains of mice have demonstrated unambiguously that Nramp1 (natural resistance-associated macrophage protein 1) plays an important role in host resistance to infections. Based on this information, the human homologue of Nramp1 (NRAMP1) has been cloned and characterized. Mouse Nramp1 and human NRAMP1 share significant sequence homology (85 per cent identity and 92 per cent similarity). The high degree of sequence homology, the presence of homologous regulatory elements within the promoter regions, and similar tissue expression profiles suggest that the NRAMP1 has similar functions in both mouse and human (Qureshi *et al.* 1999). Nramp1 has been reported to confer resistance or susceptibility to *M. bovis*, *S. typhimurium* and *L. donovani* in the mouse (Ables *et al.* 2002).

Computational and mathematic tools for infectomics

Genomics- and microarray-based technologies are rapidly creating mountains of data that are overwhelming conventional techniques of biological research.

Mathematical or computational analysis of omic and microarray data is becoming an essential part of modern biomedical sciences including infectious diseases. The application of mathematical approaches to infectious diseases dates back at least as far as Daniel Bernoulli's mathematical theory on smallpox control in 1760 (Levin *et al.* 1997). By far the most important result to come out of mathematical studies of infectious diseases is the mathematical epidemic theory (Levin *et al.* 1997; Levin 2002). Mathematical models have been essential to the analysis of transmission dynamics of infectious diseases and informed decision-making (Levin *et al.* 1997; Levin 2002). However, like most biomedical sciences, heretofore mathematical and computational methods have not played a significant role in infectious diseases. The melding of biomedical sciences, including infectious diseases, with mathematical and computational approaches is desperately needed in the postgenomic era. It is impossible holistically to monitor and dissect infectomes without the aid of mathematical and computational methods. The central issues in infectious diseases involve the interplay between host resistance and microbial infectivity. The most promising hope for the future for dissecting both these issues is network information embedded in microbial and host genomes. Information for thousands of gene products can be produced in a single experiment due to the availability of genomic and proteomic approaches. Therefore, mathematical or computational analysis of biological data (bioinformatics) is becoming an essential part of biomedical sciences. Bioinformatics is an integral part of infectomics (Table 12.1). Various mathematical approaches have been developed to model gene regulatory networks, such as linear models, Bayesian networks, differential equations and probabilistic Boolean networks (Shmulevich *et al.* 2002). Most microarray data analyses so far have focused, more or less, on *ad hoc* application of unsupervised or supervised clustering algorithms. In unsupervised clustering, no predefined set of classes is required. If a coregulated class of genes or proteins is known, supervised clustering algorithms are used to assign uncharacterized molecules to that class. However, none of these approaches capture all the dimensions of gene regulation networks (Baldi and Brunak 2001). The use of infectomic approaches for the genotypic and phenotypic study of microbial infection is still at an early stage of development. The future challenges for the study of microbial infection include global dissection of microbial virulence determinants and host defence systems, development of mathematical approaches to model the interplay between host resistance and microbial infectivity, and construction of databases to manage and analyse vast amount of data derived from genotypic and phenotypic infectomics.

Prospects of infectomics

The availability of complete genome sequences of human and a growing list of microbial pathogens, and the rapid development of computational tools have

Table 12.1   Selected Web sources for infectomics

Host and model organism genomes
Human Genome Project: *http://www.ncbi.nlm.nih.gov/genome/seq/*
Mouse Genome Project: *http://www.nih.gov/science/models/mouse*
Microbial pathogen genomes
Microbial genomes: *http://www-fp.mcs.anl.gov/~gaasterland/genomes.html*
Pathogen genomics (NIAID): *http://www.niaid.nih.gov/dmid/genomes/*
Parasite genomes: *http://www.ebi.ac.uk/parasites/parasite-genome.html*
WHO-TDR Parasite Genome Committee: *http:www.who.ch/tdr/workplan/genome.htm*
HIV database: *http://hiv-web.lanl.gov/*
Virus database: *http://life.anu.edu.au*
DNA microarrays and genomics
AMAD software package: *http://www.microarrays.org/software.html*
Microarray market place: *http://www.lab-on-a-chip.com/files/corepage.htm*
DeRisis lab Protocol and software: *http://www.microarrays.org*
Brown lab guide to microarray: *http://cmgm.standford.edu/pbrown*
Cyber-T software: *http://www.genomics.uci.edu/software.html*
Stanford Microarray Database: *http://genome-www4.stanford.edu/MicroArray/SMD/*
EBI software: *http://www.ebi.ac.uk/arrayexpress/*
Protein microarrays and proteomics
A Human Pathogen Microarray: *http://www.noabdiagnostics.com/ANMicroarrays.htm*
Nagayama Protein Array: http://*www.jst.go.jp/erato/project/nts_P/nts_P.html*
Biochain's protein Array: *http://www.biochain.com/proteinarray.htm*
Protein Microarray Instruments: *http://www.biocompare.com/spotlight.asp*
Proteomics software: *http://www.expasy.ch/ch2d/*
Carbohydrate microarrays and glycomics
Lectin Based Biosensor Array: *http://www.science.uwaterloo.ca/~mikkels/lectin.html*
Glycome: *http://glycome.unh.edu/*
Glycome: *http://www.glycoforum.gr.jp/science/now/now3E.html*
Special issues of infectious diseases
Antimicrobial resistance: *http://www.who.int/emc/amr.html; www.earss.rivm.nl;*
http://www.cdc.gov/ncidod/dbmd/antibioticresistance
Emerging infectious diseases: *http://www.cdc.gov/ncidod/emergplan*

made a tremendous impact on pathogenesis, diagnosis, prevention and treatment of infectious diseases. The most fundamental problem of infectious diseases is how holistically and integratively to dissect the interplay between microbial infectivity and the host resistance by using infectomics. The melding of computational science, omic approaches and the conventional disciplines of infectious diseases is the most challenging issue in infectomics. The progress of infectomics will allow us greatly to revolutionize the ways to study microbial pathogenesis, to manage infectious diseases, to identify unrecognized organisms, to dissolve the present crisis of antibiotic resistance and to expand the range of vaccine candidates and potential antimicrobial targets, facilitating a fundamental shift from reductionist approaches toward holistic strategies. The prevention and treatment of diseases, including microbial infections, will eventually enter an era where holistic solutions to health problem can be efficiently individualized.

References

Ables, G.P., Nishibori, M., Kanemaki, M. and Watanabe, T. (2002) Sequence analysis of the NRAMP1 genes from different bovine and buffalo breeds. *J Vet Med Sci* **64**:1081–1083.

Alvarez-Olmos, M.I. and Oberhelman, R.A. (2001) Probiotic agents and infectious diseases: a modern perspective on a traditional therapy. *Clin Infect Dis* **32**:1567–1576.

Ang, S., Lee, C.Z., Peck, K., Sindici, M., *et al.* (2001) Acid-induced gene expression in *Helicobacter pylori*: study in genomic scale by microarray. *Infect Immun* **69**:1679–1686.

Bacarese-Hamilton, T., Bistoni, F. and Crisanti, A. (2002) Protein microarrays: from serodiagnosis to whole proteome scale analysis of the immune response against pathogenic microorganisms. *Biotechniques* (Supp) December:24–29.

Baldi, P. and Brunak, S. (eds) (2001) *Bioinformatics: The Machine Learning Approach* (2nd edn). pp.299–321. Cambridge. The MIT Press.

Bedell, M.A., Jenkins, N.A. and Copeland, N.G. (1997) Mouse models of human disease. Part I: techniques and resources for genetic analysis in mice. *Genes Dev* **11**:1–10.

Bjorkholm, B., Lundin, A. Sillen, A., Guillemin, K., *et al.* (2001) Comparison of genetic divergence and fitness between two subclones of *Helicobacter Pylori*. *Infect Immun* **69**:7832–7838.

Blader, I.J., Manger, I.D. and Boothroyd, J.C. (2001) Microarray analysis reveals previously unknown changes in Toxoplasma gondii-infected human cells. *J Biol Chem* **276**:24223–24231.

Bradley, A. (2002) Mining the mouse genome. *Nature* **420**:512–514.

Chakravarti, D.N., Fiske, M.J., Fletcher, L.D. and Zagursky, R.J. (2000) Application of genomics and proteomics for identification of bacterial gene products as potential vaccine candidates. *Vaccine* **19**:601–612.

Cleary, M.D., Singh, U., Blader, I.J., Brewer, J.L. and Boothroyd, J.C. (2002) *Toxoplasma gondii* asexual development: identification of developmentally regulated genes and distinct patterns of gene expression. *Eukaryot Cell* **1**:329–340.

Cummings, C.A. and Relman, D.A. (2000) Using DNA microarrays to study host–microbe interactions. Emerging Infectious Diseases **6**:513–525.

Delehanty, J.B. and Ligler, F.S. (2002) A microarray immunoassay for simultaneous detection of proteins and bacteria. *Anal Chem* **74**:5681–5687.

Drickamer, K. and Taylor, M.E. (2002) Glycan arrays for functional glycomics. *Genome Biol* **3**:1034.

Emilien, G., Ponchon, M., Caldas, C., Isacson, O. and Maloteaux, J.-M. (2000) Impact of genomics on drug discovery and clinical medicine. *Q J Med* **93**:391–423.

Fauci, A.S. (2001) Infectious diseases: considerations for the 21st century. *Clin Infect Dis* **32**:675–685.

Fukui, S., Feizi, T., Galustian, C., Lawson, A.M. and Chai, W. (2002) Oligosaccharide microarrays for high-throughput detection and specificity assignments of carbohydrate-protein interactions. *Nat Biotechnol* **20**:1–7.

Geiss, G.K., Bumgarner, R.E., An, M.C., Agy, M.B., *et al.* (2000) Large-scale monitoring of host cell gene expression during HIV-1 infection using cDNA microarrays. *Virology* **266**:8–16.

Glusman, G., Rowen, L., Lee, I., Boysen, C., *et al.* (2001) Review comparative genomics of the human and mouse T cell receptor loci. *Immunity* **15**:337–349.

Gomez, M., Johnson, S. and Gennaro, M.L. (2000) Identification of secreted proteins of *Mycobacterium tuberculosis* by a bioinformatic approach. *Infect Immun* **68**:2323–2327.

Graves, P.R., Haystead, T.A. (2002) Molecular biologist's guide to proteomics. *Microbiol Mol Biol Rev.* **66**:39–63.

Grifantini, R., Bartolini, E., Muzzi, A., Draghi, M., *et al.* (2002) Previously unrecognized vaccine candidates against group B meningococcus identified by DNA microarrays. *Nat Biotechnol* **20**:914–921.

Guigo, R., Dermitzakis, E.T., Agarwal, P., Ponting, C.P., *et al.* (2003) Comparison of mouse and human genomes followed by experimental verification yields an estimated 1019 additional genes. *Proc Natl Acad Sci USA.* **4**: 1140–1145.

Harris, S. and Foord, S.M. (2000) Transgenic gene knock-outs: functional genomics and therapeutic target selection. *Pharmacogenomics* **1**:433–443.

Hood, L. and Galas, D. (2003) The digital code of DNA. *Nature* **421**:444–448.

Hooper, L.V. and Gordon, J.I. (2001) Commensal host-bacterial relationships in the gut. *Science* **292**:1115–1118.

Huang, S.H. (2003). Infectomics: the study of response to infection using microarrays, In *Encyclopedia of the Human Genome* (Editor-in-Chief: D.N. Cooper), Nature Publishing Group, pp. 453–458.

Huang, S.H., Triche, T. and Jong, A.Y. (2002) Infectomics: genomics and proteomics of microbial infections. *Funct Integr Genomics* **1**:331–344.

Huang, S.H., Jong, A.Y. and Summersgill, J.T. (2003) Infectomic analysis of microbial infection using proteomics. *Handbook of Proteomics.* (Edited by P.M. Conn), Humana Press Inc., Totowa, New Jersey, pp. 347–356.

Ichikawa, J.K., Norris, A., Bangera, M.G., Geiss, G.K., *et al.* (2000) Interaction of pseudomonas aeruginosa with epithelial cells: identification of differentially regulated genes by expression microarray analysis of human cDNAs. *Proc Natl Acad Sci USA* **97**:9659–9664.

Isolauri, E. (2001) Probiotics in human disease. *Am J Clin Nutr* **73**:1142S–1146S.

Jungblut, P.R., Zimny-Arndt, U., Zeindl-Eberhart, E., Stulik, J., *et al.* (1999) Proteomics in human disease: cancer, heart and infectious diseases. *Electrophoresis* **20**:2100–2110

Jungblut, P.R., Muller, E.C., Mattow, J. and Kaufmann, S.H. (2001). Proteomics reveals open reading frames in *Mycobacterium tuberculosis* H37Rv not predicted by genomics. *Infect Immun* **69**:5905–5907.

Kagnoff, M.F. and Eckmann, L. (1997) Epithelial cells as sensors for microbial infection. *J. Clin Invest* **100**:6–10.

Lee, D.J., Drongowski, R.A., Coran, A.G. and Harmon, C.M. (2000) Evaluation of probiotic treatment in a neonatal animal model. *Pediatr Surg Int* **16**:237–242.

Levin SA (2002) New directions in the mathematics of infectious disease, in *Mathematical Approaches for Emerging and Reemerging Infectious Diseases* (Castillo-Chavez, C., Blower, Sally and van den Driessche, Paulie eds), Springer, New York, Berlin, pp.1–5.

Levin, S.A., Grenfell, B., Hastings, A. and Perelson, A.S. (1997) Mathematical and computational challenges in population biology and ecosystems science. *Science* **275**:334–343.

Loferer, H. (2000) Mining bacterial genomes for antimicrobial targets. *Mol Med Today* **6**:470–474.

Love, K.R. and Seeberger, P.H. (2002) Carbohydrate arrays as tools for glycomics. *Angew Chem Int d Engl* **41**:3583–3586.

Lucchini, S., Thompson, A. and Hinton, J.C. (2001) Microarrays for microbiologists. *Microbiology* **147**(6):1403–1414.

MacBeath, G. (2002) Protein microarrays and proteomics. *Nat Genet* **32**:S526–S532.

Mattar, A.F., Drongowski, R.A., Coran, A.G. and Harmon, C.M. (2001) Effect of probiotics on enterocyte bacterial translocation in vitro. *Pediatr Surg Int* **17**:265–268.

Mitchell, P. (2002) A perspective on protein microarrays. *Nat Biotechnol* **20**(3):225–229.

Ohlstein, E.H., Ruffolo, R.R., Jr. and Elliott, J.D. (2000) Drug discovery in the next millennium. *Annu Rev Pharmacol Toxicol* **40**:177–191.

Paigen, K. (1995) A miracle enough: the power of mice. *Nat Med* **1**:215–220.

Olden, K. and Guthrie, J. (2001) Genomics: implications for toxicology. *Mutation Res* **473**:3–10.

Qureshi, S.T., Skamene, E. and Malo, D. (1999) Comparative genomics and host resistance against infectious diseases. *Emerg Infect Dis* **5**:36–47.

Relman, D.A. (1999) The search for unrecognized pathogens. *Science* **292**:1308–1310.

Rosamond, J. and Allsop, A. (2000) Harnessing the power of the genome in the search for new antibiotics. *Science* **287**:1973–1976.

Schweitzer, B. and Kingsmore, S.F. (2002) Measuring proteins on microarrays. *Curr Opin Biotechnol* **13**(1):14–19.

Shaw, A.C., Larsen, M.R., Roepstorff, P., Christiansen, G. and Birkelund, S. (2002) Identification and characterization of a novel *Chlamydia trachonatis* reticulate body protein. *FEMS Microbiol Lett* **212**:193–202.

Shmulevich, I., Dougherty, E.R., Kim, S. and Zhang, W. (2002) Probabilistic Boolean Networks: a rule-based uncertainty model for gene regulatory networks. *Bioinformatics* **18**:261–274.

Vahey, M., Nau, M.E., Barrick, S., Cooley, J.D., *et al.* (1999) Performance of the Affymetrix GeneChip HIV PRT 440 platform for antiretroviral drug resistance genotyping of human immunodeficiency virus type 1 clades and viral isolates with length polymorphisms. *J Clin Microbiol* **37**:2533–2537.

Vandahl, B.B., Birkelund, S., Demol, H., Hoorelbeke, B., *et al.* (2001) Proteome analysis of the *Chlamydia pneumoniae* elementary body. *Electrophoresis* **22**:1204–1223.

Venter, J.C., Adams, M.D., Myers, E.W. *et al.* (2001) The sequence of the human genome. *Science* **291**:1304–1351.

Walter, G., Bussow, K., Lueking, A. and Glokler, J. (2002) High-throughput protein arrays: prospects for molecular diagnostics. *Trends Mol Med* **8**(6):250–253.

Wang, D., Liu, S., Trummer, B.J., Deng, C. and Wang, A. (2002) Carbohydrate microarrays for the recognition of cross-reactive molecular markers of microbes and host cells. *Nat Biotechnol* **20**:275–281.

Waterston, R.H., Lindblad-Toh, K., Birney, E., Rogers, J., *et al.* (2002). Initial sequencing and comparative analysis of the mouse genome. *Nature* **420**:520–562.

Wizemann, T.M., Heinrichs, J.H., Adamou, J.E., *et al.* (2001) Use of a whole genome approach to identify vaccine molecules affording protection against *Streptococcus pneumoniae* infection. *Infect Immun* **69**:1593–1598.

Zhang, L., Yu, W., He, T., Yu, J., *et al.* (2002) Contribution of human alpha-defensin 1, 2, and 3 to the anti-HIV-1 activity of CD8 antiviral factor. *Science* **298**:995–1000.

Zhu, H., Cong, J.P., Mamtora, G., Gingeras, T. and Shenk, T. (1998) Cellular gene expression altered by human cytomegalovirus: global monitoring with oligonucleotide arrays. *Proc Natl Acad Sci USA* **95**:14470–14475.

13
The Drug Discovery Process

Roberto Solari

Introduction

Pharmaceuticals play a major role in the practice of modern medicine and the companies that produce and market them represent some of the world's largest industrial corporations. The pharmaceutical industry as we know it is still relatively young and only started to emerge about 100 years ago as advances in chemistry and biology made possible the discovery and synthesis of drugs. In the post-war years up to the late 1980s the pharmaceutical industry grew at a spectacular rate and launched 44 new drugs in one year at its peak in 1994. The success was based on a deep understanding of physiology and pharmacology coupled to brilliant medicinal chemistry.

In the 1980s, the ability to rapidly clone and manipulate genes heralded the start of the molecular biology era that culminated in 2001 with the publication of the first draft of the human genome sequence. This had an impact on the pharmaceutical industry that underwent a dramatic revolution with the advent of the molecular pharmacology era. The speed of the change is highlighted by the fact that the cDNA for the $\beta2$-adrenergic receptor was only cloned as recently as 1986 (Dixon *et al.* 1986), even though the first β-adrenergic drug was launched in 1965. These advances in molecular biology coincided with breakthroughs in compound screening technology and chemical synthesis. Engineering and robotics solutions allowed hundreds of thousands of chemicals to be tested in biological assays every day, so called high-throughput screening (HTS). Chemistry also changed to meet the growing demands for more compounds and more diversity

Molecular Analysis and Genome Discovery edited by Ralph Rapley and Stuart Harbron
© 2004 John Wiley & Sons, Ltd   ISBN 0 471 49847 5 (cased)   ISBN 0 471 49919 6 (pbk)

to fuel these screening engines and strategies for combinatorial chemical synthesis evolved in this period. Thus, due to the introduction of high-throughput technologies in the 1980s the drug discovery process was changed from a methodical, empirical science to a highly automated industry.

These dramatic changes to the drug discovery process were, and still are, widely expected to improve productivity in the industry. However this has not yet materialized and there are some uncertainties about the future direction of the industry. In 2001 only 25 new drugs were launched, the lowest number for almost three decades. The industry is spending ever-increasing amounts of money on Research and Development (R & D), close to $50 billion per year, and according to the Tufts University Centre for the Study of Drug Development (http://csdd.tufts.edu/) the cost of developing a new drug has also risen from $231 in 1987 to $802 million by 2000. Investors expect major pharmaceutical companies to achieve 10 per cent annual growth and in order to achieve this each of the major companies must launch on average four new chemical entities (NCEs) per year each with average sales of $350 million. However from 1996 to 2001 the industry launched on average less than one NCE per year per company and of all the drugs launched in 1996 only 25 per cent had sales in excess of $350 million. Added to these pressures are the looming patent expirations for many of the world's top selling drugs and a large number of company mergers that inevitably generate both problems and solutions.

Another significant trend in the late 20th century is the emergence of a biotechnology industry. Traditionally the pharmaceutical industry has made chemical drugs but molecular biology opened the way for a new generation of biological drugs such as erythropoietin, G-CSF and interferon-beta, a number of which have now achieved blockbuster status in terms of sales. The FDA (Food and Drugs Administration) approved the first recombinant protein in 1982 and by 2001 Biotechnology products contributed 35 per cent of new product launches. Many new vaccines and monoclonal antibodies are currently in late stages of clinical trials and represent a significant growth area for new biological medicines.

The aim of this chapter is to give an outline of the drug discovery process as it is currently configured in the pharmaceutical industry. It is not an exact blueprint since practices may vary from company to company, between different therapeutic areas and between chemical and biological drugs. It also represents a snapshot in time as this is a highly dynamic and rapidly evolving industry.

The process

Going from a research idea to a marketed drug takes from 10 to 15 years and involves a vast range of technologies and specialities along the way. Between 1990 to 1999, the pre-clinical phases took 3.8 years, the clinical trials 8.6 years and the US approval phase 1.8 years (DiMasi 2001). This overall process can be

(a)

TARGET SUCCESS RATE

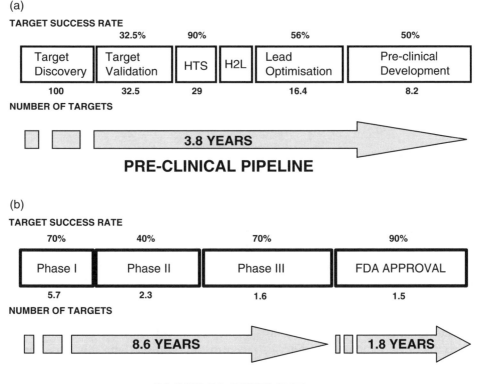

	32.5%		90%		56%		50%

Target Discovery	Target Validation	HTS	H2L	Lead Optimisation	Pre-clinical Development

100   32.5   29   16.4   8.2

NUMBER OF TARGETS

3.8 YEARS

PRE-CLINICAL PIPELINE

(b)

TARGET SUCCESS RATE

70%	40%	70%	90%

Phase I	Phase II	Phase III	FDA APPROVAL

5.7   2.3   1.6   1.5

NUMBER OF TARGETS

8.6 YEARS   **1.8 YEARS**

CLINICAL PIPELINE

Figure 13.1 The drug discovery pipeline is shown schematically starting from target discovery. (a the Pre-clinical pipeline) and going through to clinical trials in man (b the Clinical pipeline). The model shows an estimate of the success rate at each stage of the process and the cumulative number of projects at each stage assuming one starts with 100 projects in Target Discovery. (High-throughput Screening, HTS; Hits to Leads, H2L). The data is adapted from Dimasi (2001) and from Lehman Brothers Report 'The fruits of Genomics' January 2001.

summarized in a flow diagram that defines the major functions involved (Figure 13.1). Indeed, time and failures are two key factors that successful companies work hard to reduce. Careful historical analysis of each step in the process has allowed the calculation of failure rates in moving the project from one step to the next, often called the *attrition rate*. Attrition is so high that only about one in ten compounds entering clinical trials will ever reach the market place and, of these, only one in eight will recoup their investment costs. This high failure rate continues back through every step in the discovery process and one finds an approximately 50 per cent failure rate at each stage. More detailed analysis has also shown that well precedented targets have lower failure rates than unprecedented targets. In this context 'precedented' means the target is well validated in the scientific literature or more importantly drugs against this or similar targets had been through Phase II clinical trials or were already on the market. A concern in the high throughput post-genomic era is that drug discovery pipelines will be filled with more unprecedented or poorly validated targets so increasing attrition rates and ultimately costs. This fact has driven pharmaceutical companies to concentrate most of their efforts on target families that they believe will have a higher chance of success. This leads us to the start of the process – choosing a target.

Target discovery

Most biological and medical research is still carried out in publicly funded institutions and curiosity driven research is still the main source of innovation. Fortunately, much of this work is freely published by the academic community and so provides many of the raw materials for target selection. Reading the literature is clearly not enough in a highly competitive environment, so how do pharmaceutical companies decide on which targets to start drug discovery programmes? Historically researchers studied physiology and pathology and generated testable hypotheses on how to treat or modify certain aspects of the disease. For example, understanding that adrenaline has a central control over heart rate led to the development of adrenergic receptor antagonists (beta blockers) for the control of heart disease. This experimental strategy and the joint efforts of biologists and chemists generated most drugs on the market today. In the 'post-genome' era the way that biology research is conducted has changed. The first and most obvious change is that we now have a list of all of the possible drug targets in the human body. The new challenge is to pick the right ones from this list.

Analysis of all successful chemical medicines to date showed that less than 500 proteins had been targeted by these drugs. This led to the now widely held belief that only a particular set of proteins are able to be acted on by drugs and these proteins are often termed 'druggable' or 'tractable'. Extrapolating this across the whole genome suggested that there might be as many as 5000 to 10 000 of these 'druggable' or 'tractable' targets (Drews 1996; Dews and Ryser 1997). More recent estimates suggest that if one considers only Lipinski compliant agents, only 120 proteins are the targets of all of today's marketed drugs and the total druggable genome is represented by 3051 targets (Hopkins, and Groom, 2002). However, of these 3051 many would not have any influence on disease and perhaps only 600 to 1500 may actually represent valid drug targets. The development of bioinformatics tools allowed researchers to place these known targets into families and it is now accepted that certain protein families are the most favourable targets for small molecule drug discovery. These include some classes of G-protein coupled receptors (GPCRs), ion channels, nuclear receptors and enzymes (Drews and Ryser 1997). It is fair to say that most major pharmaceutical companies now focus most, if not all, of their efforts on these favoured druggable targets. However, bioinformatics can also be used to mine the genome for proteins with a leader sequence. Such proteins may be secreted growth factors or hormones and thus may have therapeutic utility themselves or may be cell surface antigens and so represent potential targets for antibody therapeutics. Many biotechnology companies are exploiting targets such as these for the development of biological drugs. Thus genomics without any deeper biological experimentation can generate a list of target 'candidates'.

However, genomics and bioinformatics alone are still not enough to choose a drug target. There are many hundreds of 'druggable' candidates in the genome and it would not be feasible to start drug discovery programmes on each and every one. Consequently a process of target selection has to be performed and this can be done in a number of ways. One of the simplest ways is to examine levels of expression of a particular protein or its mRNA in healthy versus diseased tissues, techniques known as proteomics and differential gene expression (DGE) respectively. This gives a first clue that a particular protein may be implicated in the disease process although it does not distinguish cause and effect. An alternative approach is to use genetics. Many diseases have a heritable component and studying the inheritance of these traits can identify genes that cause or predispose the individual to that disease. There are many genetic strategies that can be adopted that range from positional cloning of candidate genes as they are inherited through family pedigrees to association genetics which is the study of the linkage of certain genetic loci with disease incidence across large populations. Whereas the link between the gene and the disease can be relatively clear in certain monogenetic diseases such as cystic fibrosis, in more complex multifactorial diseases how a particular allele contributes to the pathology may be difficult to understand or translate into an obvious drug target. Sometimes a great deal of biological research has to be done following the identification of a candidate gene in order to map a particular pathway and to place the candidate into an appropriate context. Invertebrate model organisms such as yeast, nematodes and fruitflies are particularly tractable by genetic means and are useful tools for mapping pathways. Pathway mapping can also be performed by identification of protein–protein interaction networks using techniques such as yeast-two-hybrid screening (Uetz et al. 2000) or by purification of protein complexes and identification of all of the components by proteomics techniques (Bell et al. 2001). Mouse genetics (Zhang 1994) and more recently zebrafish have also been valuable models for identification of disease candidate genes. Knocking genes out, either specifically or by random mutagenesis, followed by careful phenotypic analysis can often lead to the identification of novel drug target candidates.

These various techniques generate a candidate drug target that is hopefully free from predating patent filings and that needs to be validated before progressing along the drug discovery process.

Target validation

Drugs can fail for a variety of reasons, one of which is that the initial target chosen for the drug discovery is inappropriate. Many people in the industry will argue that the only truly validated targets are those for which there is a drug already in man. For the purposes of progressing a candidate target into drug screening, validation has another perhaps less stringent meaning: generating as much confidence as possible that a drug against the target will modify the disease or treat the symptoms. Initially a candidate may have relatively little scientific weight. For example it may be a novel GPCR that is expressed in a particular region of the brain or a novel cell surface protein over-expressed on the surface of a particular cancer cell and this data may have come from mRNA expression profiling. First, one would want to confirm the observation across multiple independent samples, and then confirm that the protein levels are also differentially expressed. At this point one has a correlation between protein expression and some observed physiology or pathophysiology. The challenge is then to show that the expression of this particular target is either causative of the disease or some of the symptoms of the disease. Placing a candidate in the context of a disease requires the use of model systems. These can be cell or tissue based or whole animals but in whichever case they must attempt to be predictive of the situation in man.

There are many available cell lines where some relevant phenotypic readout can be conveniently assayed. Candidate protein targets can be over-expressed in these cells lines by transfection of the corresponding cDNA and the effect on the phenotype measured. Similarly, a candidate target protein can be 'knocked out' by expression of a dominant negative mutant version of the protein or by inhibiting protein expression with antisense oligonucleotides or siRNA. These cell-based techniques can give increased confidence that stimulating or inhibiting a particular target or target pathway may have therapeutic utility and represent a major tool for target validation. These techniques are also scaleable and a number of such 'high-throughput biology' strategies are being developed with the aim of validating targets in a genome wide fashion. One obvious example being genome wide use of siRNA (Gönczy 2000).

Animal models, such as knockout or transgenic mice, are extremely valuable in target validation. The hope is that by modification of the target expression or by expression of a mutant allele, some aspects of the human pathology will be recapitulated in the mouse.

The aim of the target validation step is to generate enough confidence in the target that the expensive and time-consuming process of drug discovery should begin. The better the validation of the biological hypothesis the less the chance of failure further downstream. Thus the output from the target validation step should address the following questions:

- Is there sufficient evidence that the target is implicated in disease pathology?

- Is there evidence that activation or inhibition of the target's activity would modify the disease or provide symptomatic relief?

- Is the target chemically tractable?

- Is the target tractable with a biological agent?

- Is it possible to configure an appropriate screen to identify agents that activate or inhibit the target?

If all of these scientific criteria are met there usually follows a more clinical and commercial analysis at this stage. If one had a drug that acted on this target, how would one configure a clinical trial, how would it be used clinically, what are the current gold standard therapies and how would this compare? It may seem precocious to consider such distant issues so early in the drug discovery process, however experience tells us that if these issues are not clear at the outset, the project may fail later in the process when much more time and money has been spent. This process is usually known as defining the product profile.

Discovery of chemical leads

Having identified a target, performed sufficient biological validation experiments and defined a product profile, one can then start the drug discovery activities. This is the process of finding a small molecule or a biological agent that acts on the target.

High-throughput screening (HTS)

The current practice in most companies for small molecule discovery is to use some form of HTS. Over the past 10 to 15 years there has been a revolution in compound screening technologies largely driven by molecular biology and by engineering solutions so that tens to hundreds of thousands of compounds can be assayed per day. For a successful HTS programme, rapid and reliable assays are an essential prerequisite. Speed is a significant advantage as is reducing the cost of each assay point. To achieve these throughputs the assay must be in a format that is suitable for automation and it must be reliable and robust since the compounds are usually delivered at high concentrations $(1-10\,\mu M)$ in an organic solvent such as DMSO. The quality performance of a screen can be defined by the Z'-factor, which is calculated from the assay signal-to-noise ratio and the signal to background ratio (see Appendix). There are many assay formats mainly based on fluorescence, scintillation proximity assays and luminescence but whatever the platforms there are some basic principles that can be defined. Screens can be run on whole cells, cell or tissue extracts such as membrane preparations or as biochemical assays using recombinant proteins. Running screens on whole cells has some advantages and disadvantages. The target is presented in a 'natural' environment or cellular context and is presumably coupled to its appropriate pathways. The drug has to be active in a cellular context and, consequently must have some favourable physical properties and finally one can screen a whole pathway simultaneously. Any compound acting between the stimulus and the readout should be identified in such an assay (Figure 13.2). The disadvantages are that cell-based assays are more complex and labour intensive to configure and are less robust than biochemical assays. Because there are multiple steps on the pathway where the drug may act these cell-based screens are often called 'black-box' assays. It is usually very difficult to define how the active compound is working and consequently downstream chemical optimization becomes very difficult. There are a variety of readouts used in cell-based assays that range from a natural phenotypic response, such as proliferation, to reporter gene assays. Targets or reporters can also be fused to Green Fluorescent Protein (GFP) and their expression or localization in live cells following stimulation can be measured by microscopic imaging techniques.

The simplest screening format uses a pure recombinant protein or proteins in a biochemical assay where an enzymatic activity or a receptor–ligand interaction is quantified. The advantages are that the components that are being screened against are defined and hopefully one knows how the active compound is working. Quality control of the reagents is also easier to achieve than in cell-based assays and biochemical assays usually have a much better Z'-factor. However, many favoured drug targets are not easy to configure in simple biochemical assay formats. Ion channels and GPCRs need to be in a plasma membrane for them to adopt a correct conformation or subunit composition and consequently such targets are often run in cellular assays or membrane fractions from cells or tissues.

Whatever the screening format the aim is to run a certain number of compounds through the screen and to identify positive compounds or 'hits'. How many compounds and the definition of a 'hit' will vary from company to company and from assay to assay. The majority (often as high as 85 per cent) of HTS 'hits' are weak and have an IC_{50} greater than $10\,\mu M$ (concentration that gives 50 per cent inhibition in an assay) and most fail to make it through hit progression (see Hit Progression). Nevertheless there are a number of common considerations for screening campaigns. The first is what sorts of compound do I run in my screen and where do they come from?

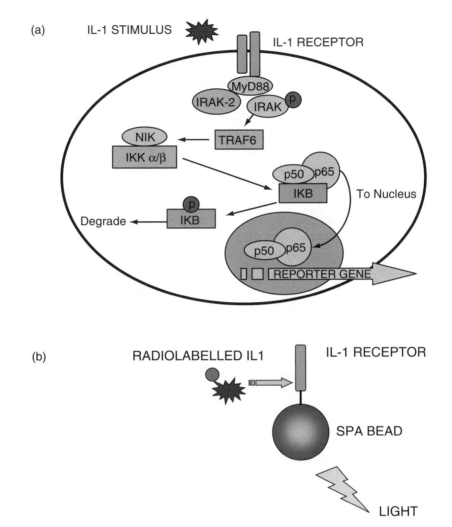

Figure 13.2 (a) An example of a multi-component whole cell screen for IL-1 inhibitors. An IL-1 responsive cell line is engineered to express an IL-1 responsive promoter driving a reporter gene. Stimulation of the cell with IL-1 activates a complex signalling pathway that leads to NF(B (p50/p65) translocating to the nucleus and activating gene expression. Inhibitors of any step in the pathway from receptor binding to gene expression can be identified. An alternative screening strategy is to select one component from the pathway, for example the receptor–ligand binding event, and to screen for inhibitors in a cell-free biochemical assay as shown in (b).

(b) A cell-free receptor binding assay for the discovery of IL-1 receptor antagonists. The IL-1 receptor cDNA is cloned and correctly folded protein is produced in an appropriate expression system. The receptor is attached via a linker to a Scintillation Proximity Assay bead and incubated with radiolabelled IL-1. Binding of the ligand to its receptor causes light to be emitted from the bead that can be measured in a suitable detector. Assays such as these can easily be configured for HTS.

Sources of chemical diversity

Historically, screening was performed with a variety of compounds that included individual chemicals and extracts from natural products such as microbial fermentations and extracts from plants or marine organisms. The natural products offered great chemical diversity, but the active ingredient in the complex mixture is often difficult to identify and once identified may not be amenable to downstream synthetic chemistry. Consequently, natural product screening has become less favoured in most major pharmaceutical companies in recent years. The fine chemicals traditionally came from the proprietary collections that pharmaceutical companies had built up over the years and often contained only a few hundred thousand individual chemical entities. These stored chemical archives represented the compounds and intermediates synthesized by individual medicinal chemists, usually from previous drug discovery programmes. Consequently the historical compound collections were heavily biased towards certain classes of molecules. In the 1990s there was a tremendous drive to increase the numbers of compounds in chemical collections, to increase their diversity and to improve on the 'drug-like' properties. There are a number of synthetic chemistry techniques that have evolved in recent years, which we can collectively call 'combinatorial chemistry', that have in common the ability to generate large and diverse chemical libraries. These advances in high-throughput chemistry have given pharmaceutical companies chemical collections of one to several million entities. There are two broad approaches to chemical library synthesis, one is to generate maximal diversity and the other is targeted design so that the majority of the compounds in the library fulfil some predetermined objective.

Whatever the design objective, in addition to the number of chemical entities in a library their quality is now considered a crucial factor. Christopher Lipinski of Pfizer examined the physical properties of drugs and identified a set of rules that were indicative of the good aqueous solubility and permeability characteristics found in successful drugs. These are now known as the Lipinski Rules and were probably the first systematic attempt to identify drug-like properties in new chemical entities. Most compound collections should be made to conform to Lipinski rules or similar, more advanced criteria that have been used to define good drug-like properties.

Lipinski rules state that poor absorption or permeation is more likely when:

- There are more than 5 H-bond donors (expressed as the sum of OHs and NHs).

- The molecular weight is over 500.

- The CLogP is over 5 (see Appendix).

- There are more than 10 H-bond acceptors (expressed as the sum of Ns and Os).

Rational design

An alternative to random screening of large numbers of compounds is rational drug design. This is based on knowledge of the three-dimensional structure of the target and a mechanistic understanding of how the protein functions. Structural biology techniques such as X-ray crystallography and NMR provide the initial protein data and small molecules that dock with an active site can be designed using sophisticated computational chemistry strategies. These *in silico* designed chemical starting points can be used to generate targeted chemical libraries for conventional screening strategies. In reality the situation is not black and white and most drug discovery programmes use a combination of rational design to direct compound selection and some random diversity-based screening. More recently, high-throughput crystallography techniques have sought to do away with conventional screening altogether (Blundell 2002). Focused chemical libraries can be screened directly by generating hundreds of protein crystals and soaking compounds directly into the crystals prior to visualizing the bound compounds by X-ray crystallography. This highly information-rich screening strategy aims to reduce time and improve the quality of drug leads.

Hit progression

A crucial aspect of any screening strategy, whether it is rational or random, is to have in place an appropriate progression strategy for the initial screening 'hits'. As any compound moves further down the drug discovery pipeline the costs associated with failure start to increase dramatically so careful planning must be in place. The first job is to collect all the screening data and re-test the hits, probably in triplicate (since primary screening is often done as single points). If the re-test is positive a dose response will be performed in order to get an idea of the IC_{50}. If the primary screen was a biochemical assay, one should establish a functional cell based assay for the target and a full cellular dose response will also be performed at this stage.

Secondary assays will then be performed to determine the selectivity of the hit for the target over related targets. For example if the target is a GPCR, selectivity over other family members will be assessed. Finally, a simple test will be performed to assess cytotoxicity of the compound, usually a dye based or bioluminescent assay.

Any compounds that pass all of these triage steps can be considered a real hit and will be examined by chemists for their suitability for chemical optimization to 'leads'. An important final step is a chemical quality control check to ensure that the active chemical entities are still what they are claimed to be in the archive. Compounds held in stores degrade with time or may simply be mis-labelled so it is always important to check before progressing to lead generation.

Hits to leads

The drug screening process by this stage will have generated a number of high quality hits that have some degree of selectivity for the target, hopefully function in a cell-based assay and are not grossly cytotoxic. The aim of the next step is to enhance selectivity and potency and to start to build in more drug-like properties so that the compound can ultimately be tested in an animal model of the disease. Undesirable chemical properties such as reactive or known toxic groups or high plasma protein binding should be eliminated prior to this stage. The project team will examine all the hits to see if there is any obvious structure–activity relationship (SAR). This involves looking for common structures or templates that appear frequently in the screening hits that can be correlated with the activity of the compounds. If this is the case, this information can be used as the basis of targeted library design to generate a new series of compounds related to the original hit. In addition, the primary hit can be used to electronically screen compound collection databases for chemical analogues with related structures. From these two sources, a secondary chemical collection is assembled and re-screened and the hope is that more potent and more selective active compounds can be identified.

This process may be repeated, but usually, if a coherent SAR does not begin to emerge at this stage, the hit series may be abandoned. However, assuming there is a medicinal chemistry strategy, further analogues can be synthesized until an appropriate product profile is achieved in terms of selectivity and potency, a so-called 'lead compound'. It is worth noting that structural biology strategies, particularly X-ray crystallography of drug–target complexes, are extensively used at this stage to help direct the medicinal chemistry process.

Lead molecule optimization

The lead compounds clearly need to be optimized for potency and selectivity against their target in addition to improving their physical properties. However, these are not the only features of a successful drug and in order to reduce late stage failures there is a tendency is to bring predictive *in vitro* studies as early as possible into the drug discovery process. Key features to consider at this stage are Absorption, Distribution, Metabolism, Excretion and Toxicity (ADMET). Oral delivery of drugs is a highly desirable property and consequently it is important to try and develop chemical series that will be absorbed by the intestinal mucosa. There are active and passive mechanisms for drug uptake through the intestinal epithelium and consequently it is difficult to predict simply on the basis of the physical properties of the compound (pKa, lipophilicity and solubility) how it will be absorbed. The most widely used system to measure intestinal absorption is the human CaCo-2 cell line that can be grown as a tight epithelial monolayer on a porous filter. Transport of compounds across the CaCo-2 cell monolayer can be used to predict intestinal uptake of the drug.

Drugs are subject to metabolism within the body. If the drug is absorbed via the oral route it will be transported via the portal vein to the liver where it is subjected to hepatic metabolism and possible elimination via the bile or kidneys. A typical metabolic fate for a drug is oxidation (phase I oxidation) followed by conjugation of the oxidized drug to a polar molecule such as glucose, methionine, sulphate, cysteine or glutathione (phase II conjugation). The key enzymes for phase I metabolism are the many isoforms of the cytochrome P450 family of enzymes and the CYP3A4 isoform metabolizes almost 50 per cent of all known pharmaceuticals. There are also many isoforms of the phase II enzymes and as with the phase I enzymes there are multiple alleles for each isoform. The inheritance of different alleles for these important enzymes can contribute significantly to patient-to-patient variations in drug metabolism. Metabolism of drugs is of central importance for a number of reasons:

- A drug that is rapidly metabolized will have a short half-life in plasma and consequently will require multiple daily doses to achieve the required therapeutic dose. Conversely, a drug that is not metabolized at all may have a very long plasma half-life that may have an influence on safety and toxicity.

- Many drugs are metabolized by the same P450 enzyme isoform and patients taking multiple drugs may suffer drug–drug interactions. One drug interfering with the metabolism of another leading to undesirable effects on drug levels.

- Drugs that are originally non-toxic may be metabolized to toxic products.

For all of these reasons, metabolic studies are performed as early as possible in lead optimization. For this purpose one uses intact hepatocytes or microsomal membranes prepared from liver. The compound is incubated with the microsomes or hepatocytes and the amount of parent compound remaining following incubation is determined by LC-MS. Similarly, assays are conducted to detect potential drug–drug interactions and to examine the ability of the lead compounds to induce P450 expression. Early understanding of metabolism can potentially avoid late stage drug failures and consequently there are many efforts aimed at high-throughput approaches to measure metabolism. Toxicity is also a significant reason for drug failure and this, too, is an area where high-throughput *in vitro* strategies are being developed and can have a significant impact on drug failure rates.

Lead to development candidate

Hopefully at this stage the project has a series of compounds with the desired potency against the target and selectivity against related targets. The compounds should have an acceptable level of metabolic stability, transport across CaCo-2 cells (if the oral route is desirable) and be free from obvious cellular toxicity. Ideally one would also hope to have more than one independent lead series so that there are back-up strategies in place in case of compound failures. At this stage one is ready to test the compounds in animal models of the disease. It is important to realize that animal models rarely, if ever, fully recapitulate the human disease. Nevertheless, in order to build confidence in the therapeutic utility of the target and the chemical series, validation in an animal model is usually indispensable. Before going into the animal model the lead compounds should be evaluated for some basic pharmacokinetic properties in rodents and possibly another species in specific cases. These preliminary studies aim to determine the half-life of the drug in the plasma and the concentrations that can be achieved from various routes of administration. This is a crucial piece of data so that one can be sure that a sensible therapeutic dose of the drug can be achieved in the disease model. Unless one is sure of this the compound may fail in the animal model simply because it has not reached a sufficient level to see a therapeutic effect and a promising target may be abandoned at this stage for the wrong reason.

Assuming the lead compound has fulfilled the desired product profile in animal models it may now be ready to pass into the full development programme. Usually, one would hope to progress more than one chemical class into full development, and ideally one would hope to have three independent candidates. Drugs can fail because the target itself is not valid for that particular indication (mechanism-based failure) or they can fail because of some specific property of the drug itself (compound-based failure). Mechanism-based failures are usually terminal for the project and these should be rare in well-managed projects. Compound-based failures are more unpredictable, which is why having independent back-up chemical series, is important.

Biological candidates

So far we have only discussed small molecule drugs and all of the steps needed to get to a proof of concept study in an animal model. If our drug was a biological agent, for example a monoclonal antibody or a recombinant protein, the process is quite different. By their nature, biological products are often optimized by evolution and the natural form of the protein can be directly used as a therapeutic, for example insulin, growth hormone or Epo. These types of products are also less prone to metabolism and toxicity and the key issues are the ability to manufacture the protein to appropriate standards and in sufficient quantities for testing. Biological therapeutics consequently can progress much more rapidly than small molecules from the phase of target discovery and validation to proof of concept in an animal.

Pre-clinical development

Once a suitable chemical candidate or biological agent has shown acceptable efficacy in an animal model it is ready to pass into the full pre-clinical development process. Many details can be obtained from the Center for Drug Evaluation and Research (*http://www.fda.gov/cder*) and the Center for Biologics Evaluation and Research (*http://www.fda.gov/cber*). The pre-clinical studies must be performed to a prescribed quality standard called Good Laboratory Practice (GLP), although in some cases GLP practices may be adopted earlier in the process. This stage will assess whether the candidate has the correct safety, toxicity and efficacy profile to justify taking it into human studies and whether the candidate can be manufactured and formulated in a cost-effective and reproducible way. The aim is to collect a data package for submission of an Investigational New Drug Application (IND) to the Food and Drugs Administration (FDA) (*http://www.fda.gov/*) which is required before trials of the drug in humans. In the UK the equivalent process is controlled by the Medicines Control Agency (MCA *http://www.mca.gov.uk/home.htm*) and requires an application for a Clinical Trials Exemption (CTX) and the European Agency for the Evaluation of Medicinal Products, the EMEA (*http://www.emea.eu.int*) is responsible for these regulatory affairs across Europe. An IND application is a request for authorization from the FDA to administer an investigational drug or biological product to humans. Such authorization must be secured prior to interstate shipment and administration of any new drug or biological product that is not the subject of an approved New Drug Application (NDA) or Biologics/Product License Application (BLA). Current Federal law requires that a drug be the subject of an approved marketing application before it is transported or distributed across state lines. Because a sponsor will probably want to ship the investigational drug to clinical investigators in many states, it must seek an exemption from that legal requirement. The IND is the means through which the sponsor technically obtains this exemption from the FDA.

The FDA's role in the development of a new drug begins when the drug's sponsor (usually the manufacturer or potential marketer), having screened the new molecule for pharmacological activity and acute toxicity potential in animals, wants to test its diagnostic or therapeutic potential in humans. At that point, the molecule changes in legal status under the Federal Food, Drug, and Cosmetic Act and becomes a new drug subject to specific requirements of the drug regulatory system.

There are three types of INDs:

- An Investigator IND is submitted by a physician who both initiates and conducts an investigation, and under whose immediate direction the investigational drug is administered or dispensed. A physician might submit a research IND to propose studying an unapproved drug, or an approved product for a new indication or in a new patient population.

- An Emergency use IND allows the FDA to authorize use of an experimental drug in an emergency situation that does not allow time for submission of an IND. It is also used for patients who do not meet the criteria of an existing study protocol, or if an approved study protocol does not exist.

- A Treatment IND is submitted for experimental drugs showing promise in clinical testing for serious or immediately life-threatening conditions while the final clinical work is conducted and the FDA review takes place.

The IND application must contain information in three broad areas:

- Animal Pharmacology and Toxicology Studies – Preclinical data to permit an assessment as to whether the product is reasonably safe for initial testing in humans. Also included is any previous experience with the drug in humans (often foreign use).

- Manufacturing Information – Information pertaining to the composition, manufacturer, stability, and controls used for manufacturing the drug substance and the drug product. This information is assessed to ensure that the company can adequately produce and supply consistent batches of the drug.

- Clinical Protocols and Investigator Information – Detailed protocols for proposed clinical studies to assess whether the initial-phase trials will expose subjects to unnecessary risks. Also, information on the qualifications of clinical investigators – professionals (generally physicians) who oversee the administration of the experimental compound – to assess whether they are qualified to fulfil their clinical trial duties. Finally, commitments to obtain informed consent from the research subjects, to obtain review of the study by an institutional review board (IRB) or an ethics committee, and to adhere to the investigational new drug regulations.

Once the IND is submitted, the sponsor must wait 30 calendar days before initiating any clinical trials. During this time, FDA has an opportunity to review the IND for safety to assure that research subjects will not be subjected to unreasonable risk.

In the UK all new active substances, and some established substances for new clinical indications, were previously evaluated in patients under a Clinical Trials Certificate (CTC). Most trials are now conducted under an exemption scheme the Clinical Trial Exemption (CTX). This exemption scheme for clinical trials was introduced in 1981 to speed up the trial of new substances. A manufacturer's licence is not required for the manufacture or assembly of a medicinal product for the sole purpose of a clinical trial. Since 1981 pharmaceutical manufacturers and suppliers may supply medicinal products for the purpose of conducting clinical trials without holding CTCs or marketing authorizations under the provisions of the CTX scheme. The main requirements for this exemption are that suppliers notify the Licensing Authority (LA) of their intention to supply investigators (doctors or dentists who are going to conduct the clinical trial) and support this with a summary of the relevant pharmaceutical data (information on the manufacture and composition of the product) and the pre-clinical safety data (information on the testing of the product in animals).

Other conditions associated with this exemption scheme are that:

- a registered medical practitioner must certify the accuracy of the summary;

- the supplier undertakes to inform the LA of any refusal to permit the trial by an ethics committee;

- the supplier also undertakes to inform the LA of any data or reports which affect the safety of the product.

In order to meet these strict criteria in the USA and Europe the preclinical development programme can be summarized as follows:

Chemistry process development. 2 to 3 kg of drug may be required for phase I clinical trials and consequently the chemical synthesis needs to be scaled up to this level. The synthetic route that was used to make the compound in the research laboratory may not be suitable for pilot plant scale up and new processes may need to be developed taking into account the cost of the materials required and ease with which the chemicals can be bought or made. It must also be established that the quality and purity (>95 per cent) of the end product remain consistent and that the by-products in the final drug are also consistent. If the drug does eventually reach the market chemical synthesis will move from pilot plant to full-scale manufacture and this may require further process development.

Pharmaceutics is involved with the formulation of the drug and the development of analytical chemistry methods to ensure that the chemistry process meets all required quality controls.

Absorption, Distribution, Metabolism and Excretion studies (ADME). Detailed studies are now performed, usually in two species one of which is a rodent the other may be a dog or small primate. These factors contribute what is called the *pharmacokinetic* properties of the drug. One also seeks to confirm the *pharmaco-dynamic* properties of the drug, which we can define as the biochemical and physiological effects of the drug and its mechanism of action.

Carcinogenicity testing is an essential component of the toxicology package and usually involves genotoxicity testing both in the Ames test and the micronucleus test. The *Ames* test measures the effect of the drug on the induction of genetic damage in *Salmonella typhimurium*. The modified microorganism has a mutant gene for ATP synthetase, an enzyme that is required for histidine synthesis. Consequently it cannot grow on histidine deficient media unless a reverse mutation occurs. The *in vivo micronucleus test* is used to screen test articles for clastogenic (chromosome-breaking) and aneugenic (loss of whole chromosome) activity. The cells evaluated in this assay are erythrocyte populations in either the peripheral blood or bone marrow compartment. The test is based on the observation that mitotic cells with chromosome breaks exhibit disturbances in the ana-phase distribution of their chromatin. After telophase, this displaced chromatin can be excluded from the nuclei of the daughter cells and is found in the cytoplasm as a micronucleus (MN). Clastogens or spindle poisons that cause chromosomal damage in stem cells result in the formation of readily detectable micronuclei in the anucleated immature erythrocytes. For chronic carcinogenicity testing rats and mice are fed with high doses of the drug for their whole life and histopatho-logical studies are used to detect increased incidence of cancers.

Toxicity studies require dose range finding studies and are usually performed in the rat and one other non-rodent species (often the dog) to determine the *max-imum tolerated dose* (MTD). These are followed by acute single dose and chronic 28 day repeat dose studies in rat and dog. The test compound should be adminis-tered at doses causing no adverse effect and doses causing major life-threatening effects. Animals should be observed for at least 14 days after drug administration and all mortalities, clinical signs, time of onset, duration and reversibility of toxicity should be recorded. Gross necroscopies should be performed on all animals in the study (Anon 1996).

Chronic toxicity testing is conducted in rodent and non-rodent species and may continue after the drug has entered human clinical trials. Chronic toxicity testing involves carcinogenicity, allergic responses and reproductive and teratogenic effects. An International Conference on Harmonization (ICH) concerning the duration of *chronic toxicity testing* recommended tests of 6 months duration in rodents and 9 months in non-rodents (Anon 1999). The FDA considers 9-month studies in non-rodents acceptable for most drug development programmes, but shorter studies may be equally acceptable in some circumstances and longer studies may be more appropriate in others, as follows:

- Six-month studies may be acceptable for indications of chronic conditions associated with short-term, intermittent drug exposure, such as bacterial infections, migraine, erectile dysfunction, and herpes.

- Six-month studies may be acceptable for drugs intended for indications for life-threatening diseases for which substantial long-term human clinical data are available, such as cancer chemotherapy in advanced disease or in adjuvant use.

- Twelve-month studies may be more appropriate for chronically used drugs to be approved on the basis of short-term clinical trials employing efficacy surrogate markers where safety data from humans are limited to short-term exposure, such as some acquired immunodeficiency syndrome (AIDS) therapies.

- Twelve-month studies may be more appropriate for new molecular entities acting at new molecular targets where post-marketing experience is not available for the pharmacological class. Thus, the therapeutic is the first in a pharmacological class for which there is limited human or animal experience on its long-term toxic potential.

Safety pharmacology studies are defined as those studies that investigate the potential undesirable *pharmacodynamic* effects of a substance on physiological functions in relation to exposure in the therapeutic range and above. The core battery of tests includes cardiovascular, respiratory and central nervous system parameters and other systems may need to be examined in specific cases. Guidelines can be found at *http://www.fda.gov/cber/guidelines.htm and http://www.fda. gov/cder/guidance/index.htm* (Anon 2001).

Development issues for biological drugs

The manufacture of proteins or antibodies by recombinant DNA technology has resulted in a number of blockbuster drugs. Biological drugs face different problems to NCEs, however the toxicology and safety packages required for both types of product are similar. The most notable differences are that biopharmaceutics normally do not require studies in metabolism, genotoxicity and carcinogenicity. Progress has been made on international harmonisation of guidelines for the pre-clinical safety testing of biopharmaceuticals (Anon 1997; Cavagnaro, 2002). Since most biopharmaceuticals are human specific the study design must first determine species cross-reactivity of the active agent. This is usually done by showing biological activity on cell lines from the species in question and then in the animal itself. Toxicity studies are usually acceptable in one species rather than two for NCEs. Toxicity of biopharmaceuticals is usually due to an exaggerated pharmacological response and careful analysis of dose–response curves should be undertaken. Another major factor is the development of antibodies to the protein by the patient and in some cases this gives rise to life-threatening consequences if the antibodies also neutralize the endogenous protein as in the case of Epo (Prabhakar and Muhlfelder 1997). In most cases patient antibodies simply reduce the efficacy of the biopharmaceutical. This is also an important factor to consider in pre-clinical efficacy, safety or toxicity studies. Immunogenicity of biopharmaceuticals is influenced by many factors (Schellekens 2002):

- Variations in sequence from the natural protein;

- Variations in glycosylation patterns;

- Contaminants and process related impurities;

- Modification of the protein due to storage or formulation;

- The dose and length of treatment;

- The immune status of the patient;

- Unknown factors.

Consequently careful physico-chemical analysis of protein therapeutic is essential. Analysis of impurities, heterogeneity, aggregation and degradation must be rigorously performed. Testing for immunogenicity in animal models as a predictor of human immunogenicity is of questionable value as the protein will be a human sequence and consequently the animal will normally recognize it as a foreign antigen. The use of non-human primates may be the closest one can get to pre-clinical testing for human immunogenicity.

Biopharmaceuticals are not subject to metabolism like small molecules but are degraded and cleared from the circulation and the determination of half-life of the product is essential for predicting future regimens in man.

Clinical trials

Once the drug has passed all of the pre-clinical hurdles and an IND has been filed it is now ready to be tested in humans. These trials are conducted in three phases called Phase I, II and III. It is well known that there is great variability between the responses of different individuals to different drugs. These differences manifest themselves as patients who respond to the medicine or not and those that exhibit adverse events following administration. These patient-to-patient variations are to some extent genetically predetermined and in recent years there has been growing interest in the study of pharmacogenomics that aims to correlate genotype which drug response (Lindpaintner 2002; and also see Chapter 1).

Phase I trials usually take about 12 months to complete and involve between 20 and 80 healthy volunteers. The tests are designed to assess the safety profile of the drug including a safe dosage range. The studies also seek to quantify how the drug is absorbed, distributed, metabolized and excreted.

Phase II trials can take about 2 years and are well-controlled and closely monitored clinical studies in approximately 100 to 300 volunteer patients. The studies are designed to obtain preliminary data on the effectiveness of the drug for a particular indication in patients with the disease or condition. This phase also helps determine the common short-term side effects or risks associated with the drug. The clinical trial design may include a number of study arms to compare the new drug with the current best standard of care and the trial may be blinded and placebo controlled.

Phase III lasts approximately 3 years and may involve between 1000 and 3000 patient volunteers. Phase III trials are only initiated after clear evidence of efficacy in Phase II and are intended to confirm the effectiveness in a broader patient sample and for a longer period. The study design should be adequate to allow extrapolation of the risk–benefit relationship for the drug in the general population and allow a suitable physician labelling to be produced.

Following completion of the Phase III clinical trial the sponsor analyses all of the data and if they are satisfied with the efficacy and safety of the drug, files a New Drug Application (NDA) with the FDA. The NDA must contain all of the scientific data that the sponsor has gathered and may contain up to 100 000 pages of information. By law the FDA is allowed 6 months to review the NDA, however in practice this time period is almost always exceeded. The average time for approval of an NCE in 1992 was 30 months, although this time is reducing.

Once the FDA has approved the NDA, the medicine becomes available for doctors to prescribe. The company may continue to supply the FDA additional information including any data on adverse reactions. For some medicines the FDA requires additional Phase IV post-marketing surveillance studies to evaluate the long-term effects of the drug.

Accelerated development and review is a highly specialized mechanism for speeding the development of drugs that promise significant benefit over existing therapy for serious or life-threatening illnesses for which no therapy exists. The rapid review process must be balanced with safeguards to protect the patient and the integrity of the review process and the sponsor must continue testing the drug after approval to clearly demonstrate therapeutic benefit.

An orphan drug is defined in the 1984 amendment of the Orphan Drug Act as a drug intended to treat a condition affecting fewer than 200 000 persons in the USA, or which will not recover development costs, plus a reasonable profit, within 7 years following FDA approval. The incentives include 7 years of exclusive marketing following FDA approval, tax credits for clinical research expense, grant support for the investigation of rare disease treatments and a user fee waiver.

Intellectual property

From this description of drug discovery it should be clear that the process is long, expensive and highly risky. In order for drug developers to recoup their investments it is essential that they protect their inventions with patents so as to stop others for as long as possible from making, using or selling the product. Patents are territorial and a UK patent only gives the holder rights within the UK. There are a number of significant differences between the USA and Europe in patent law but in all territories the law is constantly evolving, particularly in the granting of patents on gene sequences. However, there are some guiding principles that an invention must fulfil:

- The invention must be novel and can not have been made public in any way, anywhere in the world, before the application for a patent is filed.

- The invention must involve an inventive step, which means that when compared with what is already known, it would not be obvious to someone with a good knowledge and experience of the field.

- The invention must be capable of industrial application or use. The concept of usefulness means the invention must take the practical form of an apparatus or device, a product such as a new material or substance or an industrial process or method of operation.

- An invention is NOT patentable if it is a discovery, a scientific theory or mathematical method, an aesthetic creation, a method for performing a mental act, the presentation of information or a computer programe. In addition, it is not possible to get a patent for an invention if it is a new animal or plant variety, a method of treatment of the human or animal body by surgery or therapy or a method of diagnosis.

Inventors need to ensure that their inventions are not disclosed prior to patent filing and that accurate and dated laboratory notebooks are kept, bearing in mind that in the USA priority is given to the first to invent rather than the first to file, which is the case in Europe. In the drug discovery process one must be careful to establish adequate patent protection for the product at the same time as ensuring that one is not infringing other party's patents. Consequently, Intellectual Property is a key component of the drug discovery process.

Detailed information can be obtained from the following sites:

http://www.patent.gov.uk
http://www.uspto.gov
http://www.european-patent-office.org

Summary

This chapter hopes to show that drug discovery is a long, complex process that involves many diverse skills. The process is constantly evolving to meet the changing demands of society and in responses to changes in scientific knowledge and technology. What remains constant is the need for better more cost-effective healthcare.

References

Anon (1996) Guidance for Industry: Single dose acute toxicity testing for pharmaceuticals. *Center for Drug Evaluation and Research http://www.fda.gov/cder/guidance/pt1.pdf.*

Anon (1997) ICH guidance: S6 safety evaluation of biotechnology derived pharmaceuticals. *Center for Drug Evaluation and Research. http://www.fda.gov/cder/guidance/1859fnl.pdf.*

Anon (1999) International Conference on Harmonisation; guidance on the duration of chronic toxicity testing in animals (rodent and nonrodent toxicity testing); availability. Notice. Food and Drug Administration, *Fed Regist* **64**(122): 34259–34260.

Anon (2001) ICH guidance for Industry: S7A Safety Pharmacology Studies for Human Pharmaceuticals. *Center for Biologics Evaluation and Research http://www.fda.gov/cber/guidelines.htm.*

Bell, A.W., Ward, M.A., Blackstock, W.P., Freeman, H.N.M., *et al.* (2001) Proteomics characterization of abundant Golgi membrane proteins. *J Biol Chem* **276**: 5152–5165.

Blundell, T.L., Jhoti, H. and Abell, C. (2002) High-throughput crystallography for lead discovery in drug design. *Nature Rev Drug Discovery* **1**: 45–54.

Cavagnaro, J.A. (2002) Preclinical safety evaluation of biotechnology-derived pharmaceuticals. *Nature Rev Drug Discovery* **1**: 469–475.

DiMasi, J.A. (2001) new drug development in the United States from 1963 to 1999. *Clin Pharmacol & Therapeutics* **69**: 286–296.

Dixon, R.A., Kobilka, B.K., Strader, D.J., Benovic, J.L., *et al.* (1986) Cloning of the gene and the cDNA for mammalian β-adrenergic receptor and homology with rhodopsin. *Nature* **321**: 75–79.

Drews, J. and Ryser, S. (1997) Classic drug targets. *Nature Biotechnol* **15**: 1318–1319.

Drews, J. (1996) Genomic sciences and the medicine of tomorrow. *Nature Biotechnol* **14**: 1516–1518.

Gönczy, P., Echeverri, C., Oegema, K., Coulson, A., *et al* (2000) Functional genomics analysis of cell division in *C. elegans* using RNAi of genes on chromosome III. *Nature* **408**: 331–336.

Hopkins, A.L. and Groom, C.R. (2002) The druggable genome. *Nature Rev Drug Discovery* **1**: 727–730.

Lindpaintner, K. (2002) The impact of pharmacogenetics and pharmacogenomics on drug discovery. *Nature Rev Drug Discovery* **1**: 463–469.

Prabhakar, S.S. and Muhlfelder, T. (1997) Antibodies to recombinant human erythropoietin causing pure red cell aplasia. *Clin Nephrology* **47**: 331–335.

Schellekens, H. (2002) Bioequivalence and immunogenicity of biopharmaceuticals. *Nature Rev Drug Discovery* **1**: 457–462.

Uetz, P., Giot, L., Cagney, G., Mansfield, T.A., *et al.* (2000) A comprehensive analysis of protein–protein interactions in *Saccharomyces cerevisiae*. *Nature* **403**: 623–627.

Zhang, Y., Proenca, R., Maffei, M., Barone, M. and Leopold, L. (1994) Positional cloning of the mouse obese gene and its human homologue. *Nature* **372**:425–432.

Appendix

$$Z'\text{-factor} = 1 - 3\sigma\,\frac{\text{positive control} + 3\sigma \text{ negative control}}{\text{Mean positive control} - \text{Mean negative control}}$$

Where σ represents the standard deviation.

CLogP   calculated LogP

LogP   a measure of lipophilicity of a compound. Its partition coefficient between an organic solvent (usually octanol) and an aqueous buffer

$$\text{LogP} = \frac{\text{concentration of non-ionized drug in octanol}}{\text{Concentration of non-ionized drug in buffer}}$$

Values range from -6 to $+6$, high positive values indicate high lipid solubility and low negative values indicate hydrophilic properties.

14
Structure-based Drug Discovery

Chen-Chen Kan, Kevin Hambly and **Derek A. Debe.**

Overview of drug discovery and development

Drug discovery is an ever-changing and challenging endeavour. Decades of research are required to uncover the molecular mechanisms underlying disease pathologies and identify compounds having therapeutic effects on those disease pathways. Before they can be approved for clinical use, modern therapeutic agents must undergo rigorous characterization to define the physicochemical and stereochemical properties of active chemical ingredients. Regulatory agencies in the USA, Europe and Japan must ensure that all approved drugs meet quality control standards, and that drugs demonstrate sufficient selectivity and pharmacological potency to achieve the desired clinical effects while avoiding undesirable adverse reactions.

Ideal drugs produce their therapeutic effects via interference with a particular biological process through binding and interacting with specific macromolecular drug receptors (or targets). A drug target may be an enzyme, cell surface receptor, DNA, RNA or macromolecule from any other chemical class that is critical for a specific and essential function in a disease pathway. The decoding of human and microbial genomes has yielded a large number of potential drug targets, creating an enormous opportunity for future development of novel therapeutics to treat diseases, especially those diseases with unmet medical needs such as lung, prostate and liver cancer, and Alzheimer's disease.

Molecular Analysis and Genome Discovery edited by Ralph Rapley and Stuart Harbron
© 2004 John Wiley & Sons, Ltd   ISBN 0 471 49847 5 (cased)   ISBN 0 471 49919 6 (pbk)

In this chapter, we will focus our discussion on the use of protein structure information in drug discovery, specifically as it applies to the design of small molecule inhibitors against enzyme targets, which represent about 36 per cent of all drug targets in the human genome (Hopkins and Groom 2002).

Historically, drug discovery has relied heavily on brute-force screening efforts to discover lead compounds, involving the systematic testing of vast inventories of naturally occurring and/or synthetic compound libraries in various functional assays. After determining the chemical structures of initial leads, numerous derivatives would be synthesized by medicinal chemists for further testing. The lead optimization process would then be guided by analysis of structure–activity relationships derived from empirical data obtained from functional assays. At this stage, appropriate modifications are made in order to optimize a compound's stereochemistry and physicochemical properties indicated by the logP (for lipophilicity) and pKa values. Compounds with optimal *in vivo* potency, specificity and pharmacokinetic properties would then be selected for preclinical testing in disease models in animals.

Figure 14.1 illustrates a schematic overview of the drug discovery and development process as it is commonly practised in the bio/pharmaceutical industry. The Pharmaceutical Research and Manufacturers of America (PhRMA) estimate that it costs an average of $800 million to bring a new drug to market; this includes drug discovery and development costs associated with many unsuccessful projects that fail to deliver an approved drug at the end. And, the process takes 10–15 years. PhRMA also estimates that for every newly approved drug, five drug candidates entered clinical testing, 250 entered pre-clinical testing, and 5000–10 000 compounds were tested during lead discovery and optimization. These numbers clearly indicate a very high attrition rate in the drug discovery and development process. New approaches and technologies are urgently needed to improve the overall success rate and increase the efficiency of drug discovery.

Recent advances in structural biology and computing power have led to the development of a structured-based approach to drug development. This approach uses structural information that was not readily attainable in the past to guide and

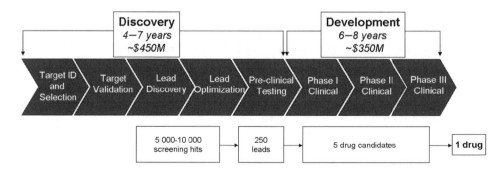

Figure 14.1   Drug discovery and development process

focus drug design efforts, thereby enhancing the efficiency as well as the overall success rate in the process of discovering novel therapeutic agents.

Principles of structure-based drug discovery

Structure-based drug design was pioneered by Goodford and colleagues who utilized the three-dimensional structure of haemoglobin as the basis for ligand design (Beddell *et al*. 1976). Another well-known example that demonstrates how protein structures can provide detailed, atomic-level structural insight to explain drug selectivity was carried out in the early 1980s by Matthews *et al*. (1985), who solved the structure of a ternary complex of chicken dihydrofolate reductase (DHFR) containing both a cofactor (NADPH) and an inhibitor (trimethoprim). The structure of the ternary complex provided an understanding of the structural basis for differential inhibition of bacterial DHFR and vertebrate DHFR by the antibiotic drug trimethoprim. Specifically, a hydrogen bond between the 4-amino group of trimethoprim and the backbone carbonyl of the Val-115 residue of *E. coli* DHFR was found to be absent in the ternary complex of chicken DHFR. Subsequently, information derived from the *E. coli* DHFR ternary structure was used successfully to design a trimethoprim analogue with improved *in vitro* potency (Kuyper *et al*. 1985).

Due to the success of the early studies described above, structure-based drug discovery gained increased acceptance throughout the pharmaceutical industry. Structure-based discovery teams have since been assembled at many different drug discovery organizations, and several companies – such as Agouron Pharmaceuticals, Vertex Pharmaceuticals, and more recently, Sugen, Inc. – have adopted a structure-based approach as their primary drug discovery platform. These structure-based drug discovery teams are highly cross-disciplinary, including researchers from diverse scientific fields, ranging from molecular biology and protein crystallography to computational and medicinal chemistry.

At the onset of a drug discovery project, identifying drug design targets that are suitable for therapeutic intervention requires a sound understanding of the molecular mechanisms underlying those diseases with unmet medical needs. Once targets are selected, DNA sequences encoding the therapeutic targets can be created and cloned into DNA vectors using recombinant DNA technology. In order to facilitate protein crystallographic studies, milligram to gram quantities of these selected drug receptors need to be produced as recombinant proteins using expression hosts such as bacteria strain *E. coli*, insect SF 9 or SF21 cell lines, or Chinese Hamster Ovary (CHO) cell lines. The development of convenient affinity protein chromatography methods, as well as more sensitive and specific biochemical activity assays, have both increased the efficiency of protein purification procedures and the purity of isolated proteins, and have also provided more cost-efficient and convenient methods to ensure that purified recombinant proteins retain their native biochemical activity and conformation. Also, advances in X-ray crystallography equipment, computing power and computational algorithms such as the automated crystallographic structure solution algorithm SOLVE (Terwilliger and Berendzen 1999), have made the high-resolution structure information much more readily available. Computerized Fourier analysis, interactive 3D visualization, and molecular modelling software together now offer drug designers a sophisticated platform to extract, graphically illustrate, manipulate and design the detailed atomic arrangements of ligands in the binding pockets of drug targets. These capabilities enable researchers efficiently to carry out iterative cycles of drug design and optimization (Figure 14.2).

In order to understand the critical interactions between ligand and receptor, the natural substrate and/or cofactor is often used initially to generate a ligand–target binary complex structure or a ternary complex. For the discovery of novel leads, one commonly used approach is to mine chemical databases for compounds that bear similarity to natural ligands, and thus have a higher likelihood of binding to the drug receptor. Hits (or chemical analogues) obtained through data mining are, in some cases, available for purchase or synthetically feasible, thus they may

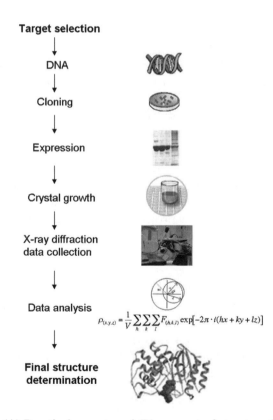

Target selection

DNA

Cloning

Expression

Crystal growth

X-ray diffraction
data collection

Data analysis

$$\rho_{(x,y,z)} = \frac{1}{V} \sum_h \sum_k \sum_l F_{(h,k,l)} \exp[-2\pi \cdot i(hx + ky + lz)]$$

**Final structure
determination**

Figure 14.2 (A) Practical aspect, and (B) concept of structure based drug design

Structure Based Drug Design (SBDD)

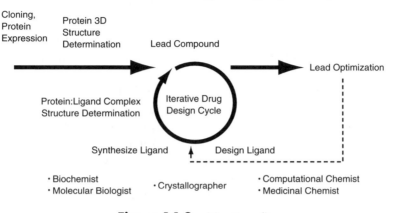

Figure 14.2   (*Continued*)

be experimentally tested immediately. However, database searches often do not provide structurally novel compounds, and they might also be biased toward certain chemical classes. Thus, potential leads obtained through such search efforts may be confined to a limited range of chemical diversity.

An alternative approach is to utilize the process of *de novo* ligand design at the beginning of the iterative drug design cycle. One of the most widely used software tools for structure-based *de novo* ligand design is the energy-based GRID program (Goodford 1985). GRID can be used to create a three-dimensional grating (or grid) for the binding site of the design target. Various probe groups can then be placed at the vertices of the grid and the interaction energy between the probe and specific chemical groups on the protein target can be calculated using empirical energy functions. Grids produced for each probe with an energy value assigned at each vertex then can be used to identify the most favourable locations in which to position particular functional groups for ligand design. These functional groups can then be connected together to form a molecular scaffold. The scaffold can serve as a template for the creation of real compounds through the attachment of additional chemical substituents that have the suitable sizes and chemical properties to fill the extra empty space in and around the active site.

An alternative to the GRID method for drug design is the use of a knowledge-based approach based on analyses of crystal structures of ligand–target binary complexes. Binary complex structures offer information about the geometric arrangement of interactions in the three-dimensional space between bound ligands and the target responsible for binding affinity. Based on this knowledge-based approach, the program LUDI has been widely used computationally to dock small molecular fragments into protein binding pockets (Böhm 1992). This structure-based docking approach requires detailed structural information for the binding site to be surveyed. For example, it is important to identify possible

H-bonding donor/acceptor sites, favourable regions for hydrophobic or electrostatic interactions, and space for van der Waal's interaction. After having surveyed the binding site and the range of preferred geometries for particular types of interactions, chemical fragments are selected based on their physicochemical properties and the ability of the atoms of the fragments to interact with, or fit to, corresponding points in the binding site.

In general, *in silico* search efforts such as GRID and LUDI have yielded small molecule hits with moderate to weak binding affinity. Therefore, empirical approaches have been developed that use X-ray crystallography or nuclear magnetic resonance (NMR) techniques to screen for fragments that can bind to a particular protein target with binding affinity in or below the μM range (Carr and Jhoti 2002). In particular, the study of structure–activity relationships by nuclear magnetic resonance (SAR by NMR method) has attracted much attention. This method compares the NMR spectra of an isolated protein with that of a protein–ligand complex (Shuker *et al.* 1996) to identify small chemical fragments that bind to the protein of interest. Using this method, small 'building block' molecules, which bind to a design target, can be identified. These building blocks are subsequently linked together to create a new molecule that will bind to the target protein with increased affinity. The program CAVEAT was one of the first methods developed to link molecular fragments together to make a new molecule based on specific geometrical relationships between each pair of molecular fragments or linking (Lauri and Bartlett 1986). CAVEAT searches for molecular connectors in a database that contains ring systems extracted from the Cambridge Structure Database, as well as chemical structures from the TRIAD and ILIAD Three-Dimensional Databases of Computed Structures. One limitation to this approach is that the molecular fragments may be farther apart than the longest connector in the database. An alternative program, SPROUT, can be used to connect the fragments by creating a skeleton that can be grown either one atom at a time, or by joining molecular templates commonly found in drug molecules (Gillet *et al.* 1993).

After initial leads are discovered or designed, iterative drug design cycles are carried out based on the results of activity assays and the structure information obtained from binary or ternary complex structures. These iterative cycles are designed to improve the selectivity, potency, ease of synthesis, and pharmacokinetic properties of lead compounds before they enter clinical development. Drugs with binding affinity in the subnanomolar (nM) to picomolar (pM) range that pass stringent selectivity tests can potentially allow a lower dosage to be administered to achieve the desired clinical effect, thus reducing the risk for toxicity and adverse effects that is often observed with higher dosages of drug compounds.

The binding affinity between ligands and drug receptors can be explained by the thermodynamics of molecular interactions. The dissociation constant (Kd) for the binding equilibrium between a ligand, its receptor and the ligand–receptor complex is dictated by $\Delta G = RT \ln K$, where K is the equilibrium constant, R the gas constant $(8.314\,J\,K^{-1}\,mol^{-1})$, T the absolute temperature in kelvins, and $Kd = 1/K$. Optimization of binding affinity of a lead compound usually requires a reduction of 4–8 kcal/mol in the Gibbs free energy (ΔG), which is composed of two opposing terms, enthalpy (ΔH) and entropy (ΔS). Since the Gibbs free energy $\Delta G = \Delta H - T\Delta S$ at absolute temperature T in kelvins, favourable interactions leading to increased binding enthalpy are often offset by a decrease in entropy, and vice versa. Thus, the challenge faced by drug designers is how to balance the two opposing forces of enthalpy and entropy in order to maximize the binding affinity for a given drug to its target.

As it has become clear that the structure-based drug design process is an interdisciplinary endeavour with a high degree of complexity, it can only be successful when knowledge and skill-sets from various scientific fields become integrated via close collaboration. For such collaborations to be productive, the efficient dissemination and exchange of information, data and ideas is essential. Thus far, structure-based approaches have had the greatest impact in the infectious disease area, for example in the aforementioned DHFR inhibitor design, as well as in the design of the influenza neuraminidase inhibitors Relenza and Tamiflu, and novel cephalosporin beta-lactamase inhibitors that combat antibiotic resistance. The greatest excitement for structure-based drug design was generated by the design of HIV-1 protease inhibitors such as Saquinavir, Indinavir, Ritonavir, Nelfinavir and Amprenavir for the treatment of AIDS (Hardman *et al.* 2001). The design of clinically useful inhibitors of HIV protease for modern antiviral therapy has been widely viewed as the first major triumph of the structure-based drug design approach.

Case studies of structure-based drug discovery

Drug design for the HIV-1 protease substrate binding site

HIV protease apo structure

The human immunodeficiency virus protease (HIV PR) has been a target for drug design since the HIV type 1 virus (HIV-1, a member of the retrovirus family) was identified as the aetiological agent for AIDS in 1984. According to the AIDS Epidemic Update published in December 2002 by the Joint United Nations Programme on HIV/AIDS (UNAIDS, http://www.unaids.org/), the number of people living with HIV/AIDS totals 42 million. In the year 2002 alone, there were an additional 5 million people infected with HIV, and AIDS caused 3.1 million deaths worldwide. Because of the rapid emergence of drug resistance against existing anti-HIV therapeutic agents, the AIDS epidemic continues to be the most threatening disease for the human population.

Among the proteins encoded by the HIV-1 genome, the enzymes reverse transcriptase (RT), integrase (IN) and protease (PR) have been considered as suitable structure-based drug design targets. After viral entry into infected cells, reverse transcriptase is required to make a DNA copy of the HIV single-stranded RNA genome. Integrase activity is then required for the provirus DNA to be integrated into the host chromosome. During an active viral replication cycle, the polypeptide precursors p55gag and p160gag-pol are produced by the host's transcription and translation machinery. HIV-1 protease is released auto-proteolytically from the p160gag-pol polypeptide, and PR then cleaves the polypeptide precursor into smaller, functionally active proteins, including RT, IN and additional PR enzymes. A compilation of all proteolytic cleavage site sequences present in p55gag and p160gag-pol polypeptides reveals that Phe and Pro are used most often at the substrate P1 and P1' positions, which flank the scissile peptide bond, while small residues and aliphatic amino acids such as Glu/ Gln are preferred at the P2 and P2' positions.

HIV-1 protease contains the conserved Asp-Thr-Gly aspartic protease sequence motif, and it shares sequence homology with other members in the aspartic protease family. However HIV PR is only 99 amino acids in length, in contrast to mammalian aspartic proteases that are approximately 325 amino acids long. It is interesting to note that the first structure of a retrovirus protease was not HIV protease, but the Rous sarcoma virus (RSV) protease (PDB code: 2RSP, Jaskolski *et al.* 1990). Thus, homology modelling techniques were initially used to build a structural model of HIV-1 PR using 2RSP as a template. Subsequently, in 1989 the crystal structure of HIV-1 PR was determined (PDB code: 2HVP, and 3HVP, Weber *et al.* 1989; Navia *et al.* 1989; Wlodawer *et al.* 1989). The PR crystal structure reveals that the enzyme forms a non-covalent homodimer, with residues from each of the two subunits enclosing the active site at the dimer interface. The active site has a two-fold crystallo-graphic symmetry, with each subunit contributing a catalytic aspartic acid residue (Asp 25 and Asp 25') (Wlodawer and Erickson 1993). The substrate binding pocket resides between the flap domain (residues 47–54) and P1 (or 80s) loop (residues 79 to 83, as shown in Figure 14.3). In addition, residues Gly 27, Asp 29, Gly 48 and Ile 50 from each subunit are also involved in substrate/ ligand binding.

Structure of HIV PR binary complex and inhibitor design

Initially, substrate analogues were designed as inhibitors for HIV PR. Small non-peptidic peptidomimetics were designed based on knowledge of preferred amino acid residues at the substrate binding subsites of HIV PR. These compounds were designed to mimic the shape complementarity and chemical functionality of the natural peptidyl substrates. The first potent inhibitors of HIV PR were a series of

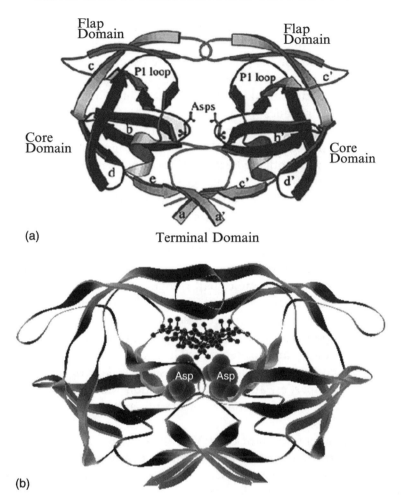

Figure 14.3 (a) apo structure, (b) HIV PR: acetyl-pepstain complex (PDB: 5HVP)

peptide-based diols, which were able to displace the water molecule in the active site and form hydrogen bonds directly with the catalytic aspartate residues. However, these compounds were found to have poor oral biovailability. Subsequently, a series of co-crystal structures of HIV PR binary complexes containing such peptidomimetic inhibitors became available.

The binary structure of HIV protease with bound acetyl-pepstatin at 2.0 Å resolution showed that the inhibitor was bound on one side by the active site aspartic acids, and on the other by a pair of two-fold symmetrical antiparallel beta-hairpin structures known as flaps (PDB 5HVP as shown in Figure 14.3b). The structure of this binary complex shows that a core of 44 amino acids in each subunit, including residues in the active site and at the dimer interface, remains mostly unchanged, as indicated by a 0.39 Å root mean square deviation of alpha carbon positions between complexed and uncomplexed forms of HIV protease. According to the thermal flexibility indicated by B-factor values, and order parameters determined for the protein structure by protein crystallography and NMR methods (Ishima *et al.* 1999), the most static region of the HIV protease appears to be the residues flanking Asp25 in each monomer. The most striking changes occurred at residues 44–57 comprising the so-called flaps of the HIV protease (Fitzgerald *et al.* 1990). In the structure of the HIV protease:inhibitor complexes, a water molecule, water 301 (as shown in Figure 14.3(b)), was found to form hydrogen bonds to both the P2 and P1′ backbone carbonyl groups of acetyl-pepstatin as well as the backbone amino group of Ile 50 and Ile 50′ located in the two flaps. As compared with the apo structure of HIV protease without bound ligand, the binding of substrate or inhibitor induces substantial conformational changes in the flap region of the HIV PR, where a backbone movement as large as 7 Å has been documented (Miller *et al.* 1989). This ligand-induced conformational change manifests the induced-fit mechanism as opposed to the more conventional lock-and-key mechanism for molecular recognition. The ligand-induced conformational changes further highlight the benefit of utilizing the structural information obtained from binary complexes to design novel inhibitors and optimize lead compounds through iterative cycles of drug design, chemical synthesis, inhibition testing and protein crystallography. Once HIV PR inhibitors with *in vitro* potency were obtained, issues such as *in vivo* antiviral activity, pharmacokinetic properties, and cost of production were also addressed based on structural knowledge of inhibitor constraints and the latitude allowed for further modifications.

Nelfinavir is a potent inhibitor of HIV protease with a K_i of 2 nM. An X-ray co-crystal structure of the HIV PR:Nelfinavir complex (Kaldor *et al.* 1997) reveals how the novel thiophenyl ether at the P1 position, and the phenol-amide substituent at the P2 position, interact with the S1 and S2 subsites of HIV-1 protease, respectively. Nelfinavir binds to the HIV protease substrate-binding site in an extended conformation (Figure 14.4). The thiophenyl group binds to the

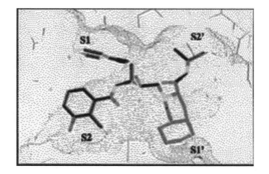

Figure 14.4 View of Nelfinavir in complex with the HIV-1 protease (PDB code:1OHR) (Kaldor *et al.* 1997)

S1 pocket, and the 2-methyl-3-hydroxy benzamide part resides in the S2 pocket. The tert-butylcarboxamide moiety, and the bi-cyclic ring structure occupy the S2' and S1' subsites, respectively. The central hydroxyl group pointing out of the plane binds to the catalytic aspartic acids. A tightly bound water molecule at the bottom of the substrate binding site on the opposite side of the central hydroxyl group relays hydrogen bonds from the two amide carbonyls of Nelfinavir with the Ile50 and Ile50' residues from the flap region, similar to the observation that was made with previous protease–inhibitor complexes. Many clinically successful compounds share a significant degree of similarity both in core structure as well as in substituent groups. It is interesting to note that the root mean square deviation between the backbone atoms of Nelfinavir (AG1343) and Saquinavir (Ro 31-8959) was only 0.49 Å, indicating that these two inhibitors adopt a very similar binding mode in the HIV PR binding pocket.

In addition to Nelfinavir, there are five other FDA-approved HIV protease inhibitors produced by the structure-based drug design approach. They are Amprenavir, Indinavir, Lopinavir, Ritonavir and Saquinavir. The recent clinical advances in the use of potent combinations of anti-retroviral agents, including inhibitors of reverse transcriptase and protease, has made a significant impact on AIDS disease management. New anti-retroviral regimens, the so-called highly active anti-retroviral therapies (HAART), have reduced the morbidity and mortality caused by HIV infection (Palella 1998).

Drug resistances and adaptive ligand design

The remarkable success and clinical benefits of the HIV protease inhibitors and HAART therapy can drive the viral load of HIV-1 to undetectable levels in infected patients. However, drug resistant HIV mutants have quickly emerged, containing mutations in the HIV PR coding sequence. Since HIV protease functions as an obligatory dimer, any mutation in its coding sequence on the genetic level that leads to a single amino acid change appears twice in the functional enzyme. These mutations have been mapped to the active site as well as to other distant regions away from the active site (Condra 1996; Boden and Markowitz 1998). Mutations are observed against all HIV protease inhibitors approved by the FDA to date, and the sequences collected in the HIV mutants database indicates a significant cross-resistance pattern toward various HIV PR inhibitors. Except some mutant forms of HIV protease that arise in response to HARRT treatment remain sensitive to Nelfinavir. Therefore, medicinal chemists face the continuing challenge of designing adaptive ligands as broad-spectrum anti-retroviral therapeutics. Ideally these next-generation inhibitors contain more flexible elements but are still able to maintain high affinity and specificity for all drug-resistant HIV protease mutants instead of just a single HIV protease variant encoded by a specific sequence. However, the introduction of flexible substituents has to be compensated for by favourable interactions (i.e. binding enthalpy), in order to ensure that both affinity and specificity are maintained.

Recently, such an adaptive HIV PR inhibitor, JE-2147, was reported (Reiling 2002). JE-2147 is a potent allophenylnorstatin (Apn)-based tetrapeptide mimetic that works as a transition state analogue and has a K_i of approximately 40 pM and IC50 of 30 nM in a cell-based assay (Figure 14.5). The high-resolution co-crystal structure of the HIV PR:JE-2147 complex has allowed refinement of the anisotropic displacement parameter (ADP) for all atoms. Instead of aniso-tropic motion of atoms indicated by ADP, isotropic B factor represents a spher-ical Gaussian approximation of atomic motion in all directions. Flexible JE-2147 buries 781 Å^2, or 96 per cent, of its solvent accessible surface area, compared with 88 per cent for the other HIV PR inhibitors. The most critical interaction between JE-2147 and HIV PR involves the O-methyl group of the phenyl ring at the P2′ position of the inhibitor, which allows JE-2147 to pack more deeply into the S2 pocket, interacting primarily with Ile47B of the HIV PR dimer. This interaction buries 47.1 Å^2 of solvent-accessible surface area of the P2′ group of JE-2147, compared with only 26 Å^2 of the tert-butyl group at the corresponding position of Nelfinavir. Based on the ADP profiles, the flap region of the JE-2147:HIV PR

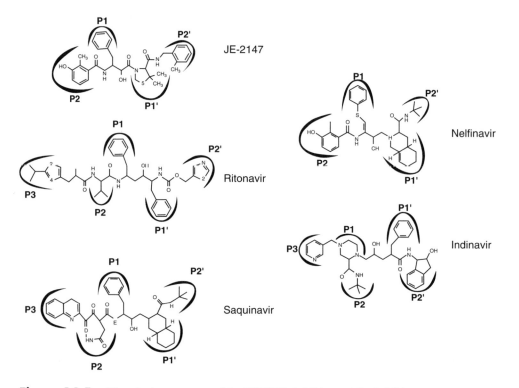

Figure 14.5   Chemical structures of the HIV PR inhibitors. The inhibitor structures are labeled from P3 to P2′ from left to right

complex has a particularly tight conformation with the primary binding flap of monomer A (amino acids 49–52) being 0.2 – 1.2 Å closer to the active site and having a more planar orientation of the side-chain of Ile50A. In this manner, it is speculated that this tighter packing of the flaps may contribute to the more favourable enthalpy of binding of JE-2147 relative to other inhibitors. Results obtained from ADP analysis suggested that JE-2147 binding might be more enthalpically driven than the binding of earlier generations of inhibitors. Based on the high-resolution data obtained at 1.09 Å, and a low average isotropic B-factor for all atoms of 12.9 Å^2, authors also claimed that the JE-2147 complex must have less internal entropy than those other inhibitor complexes.

Nevertheless, mutations of L10F, M46I, I47V and I84V in the HIV PR sequence have arisen in cell culture that render drug resistance toward JE-2147, i.e. a 28-fold increase in IC50. Among all the JE-2147 resistance mutations, I47V appears to have the greatest impact in weakening the inhibitor binding in the S2′ more than the S2 pocket due to the loss of critical interactions with the P2′ moiety as described earlier. This drug resistance profile is unique to JE-2147, being distinctively different from the major mutations of D30N, G48V, I50V, V82A, I84V and L90M that have been observed in proteases resistant to first-generation inhibitors (Yoshimura 1999, and Protease Sequence Database at *http://hivdb.stanford.edu/*). Therefore, JE-2147 represents an extremely potent inhibitor with the potential to broaden the spectrum of inhibitory activity against mutant HIV protease.

In order to gain an understanding of the structural basis for the design of potent broad-spectrum protease inhibitors against both the wild-type and mutant HIV PR, a different approach was taken by Munshi and co-workers. This group carried out a detailed analysis of the crystal structures of binary complexes formed by mixing the wild-type protease or a 9X-mutant of HIV protease (Leu10Val, Lys20Met, Leu24Ile, Ser37Asp, Met46Ile, Ile54Val, Leu63Pro, Ala71Val, and Val82Thr) with one of the following protease inhibitors such as Indinavir, L-756,423, L-739,622 and Saquinavir (Munshi 2000). This effort led to the identification of an alternate binding site for the P1–P3 group of Indinavir and its analogues. This alternate binding site involves the intrinsic conformational flexibility in the main-chain backbone of the 80s loop of the HIV protease. The P1–P3 benzofuran moiety of L-756,423 is inserted between the 80s loop and the flap and binds to the S1–S3 binding pockets. These alternative binding pockets accommodate the bulkier benzofuran group of L-756,423 and the Pro 81 on the 80s loop (where the pyridyl group of Indinavir binds) is moved away by 2.0 Å and 2.8 Å from its original location, in the wild-type and 9X HIV protease complex, respectively (Figure 14.6). In this case, the P1–P3 group of L-756,423 is less flexible and largely more hydrophobic than that of Indinavir, and the inhibitor binds more deeply into the alternative S1–S3 binding pockets. It is the backbone movement of the 80s loop that was exploited for the design of novel and potent inhibitors with K_i in the low nanomolar range. The two regions showing the

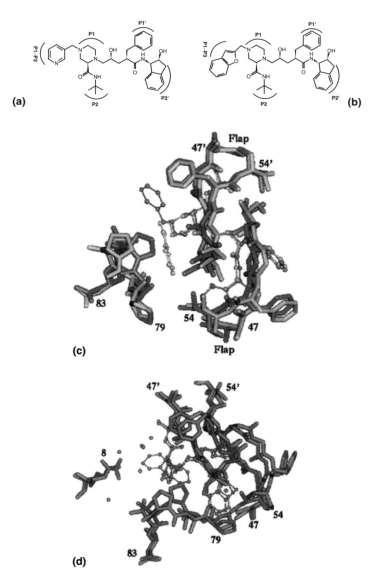

Figure 14.6 Alternative binding sites for the P1-P3 groups of HIV-1 protease inhibitors. ((a) and (b)) Chemical structure of Indinavir and L-756423, (c) Superposition of the crystal structures of wild-type HIV-1 protease complexed with Indinavir and L-756423. The inhibitor structures are represented as ball-and-stick models. The P1-P3 benzofuran of L-756423 is sandwiched between the 80s loop and the flap with Pro81 being pulled away by 2.0 Å., (d) Superposition of the crystal structures of Indinavir complexed with wild type HIV-1 protease, and with the 9X mutant protease. The P1-P3 pyridyl group occupies two different positions. In the 9X structure, Pro81 residue is shifted as far as 2.8 Å. The 80s loop, comprised of residues 79–83, and flap residues 47–54 of the two monomers are depicted in both (c) and (d) (Munshi *et al.* 2000)

greatest dynamic behaviour are residues 37–42, 13–19, and the regions around residue 68 and 81 (Zoete *et al.* 2002).

To date, the Structural Database of HIV Protease (*http://srdata.nist.gov/hivdb/*) has archived more than 272 experimentally determined 3D structures of Human Immunodeficiency Virus 1 (HIV-1), Human Immunodeficiency Virus 2 (HIV-2) and Simian Immunodeficiency Virus (SIV) proteases and their complexes with inhibitors or products of substrate cleavage. In addition to the apo structures of HIV PR available at the Protein Data Bank (PDB), the Structural Database of HIV Protease contains some unique co-crystal structures of binary complexes containing bound ligands such as Indinavir and Nelfinavir. The creation of this database has served as a testimony to the importance of structural information for continuing drug discovery for treating the devastating AIDS epidemic. With easy access to databases via the Internet, structure information has now become readily available to scientists in the field.

Drug design for the cofactor binding site of inosine monophosphate dehydrogenase

Inosine monophosphate dehydrogenase (IMPDH) catalyses the nicotinamide adenine dinucleotide (NAD)-dependent oxidation of inosine-5′-monophosphate (IMP) to xanthosine-5′-monophosphate (XMP), which is then converted into guanosine-5′-monophosphate (GMP) by GMP synthetase. Native IMPDH exists as a homotetramer with subunits of 56–58 kDa in mass (Carr *et al.* 1993). The active-site pocket in each monomer of dimeric IMPDH contains two connected binding pockets: the IMP-binding site and the NAD-binding site (Figure 14.7).

Initially, IMPDH was considered as an attractive cancer drug design target because it catalyses the rate-limiting step in converting IMP to XMP (Crabtree and Henderson 1971), and increased IMPDH activity was observed in rapidly proliferating leukaemia cell lines and solid tumours (Jackson *et al.* 1975; Webber 1983). As it turns out, IMPDH activity is also extremely important for providing the guanosine nucleotide needed to initiate proliferative responses of B- and T-lymphocytes when stimulated by mitogen or antigen (Allison *et al.* 1975).

Mycophenolic acid (MPA) inhibits IMPDH by acting as an uncompetitive inhibitor that replaces the nicotinamide portion of the nicotinamide adenine dinucleotide cofactor and a catalytic water molecule from IMPDH. MPA was first isolated from fermentation culture media of several *Penicillium* species. Mycophenolate mofetil (MMF, CellCept by Hoffman La-Roche) was approved in 1995 by the FDA for prophylaxis of acute allograph organ rejection in patients receiving renal transplants, and more recently was approved for heart transplant recipients. After oral administration, the pro-drug MMF, a morpholinoethyl ester of MPA, is rapidly absorbed and converted to the active metabolite MPA. In

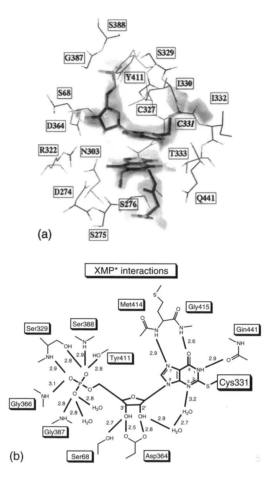

Figure 14.7 (a) Ternary complex structure of the IMP dehydrogenase showing the active site that contains both the substrate and cofactor. (Sintchak *et al.* 1996). Carbon atoms are shown in green, nitrogen in blue, oxygen in red, sulfur in yellow, and phosphorus in purple. There is a covalent bond between the sulfur atom of Cys-331 and the C2 carbon of the hypoxanthine ring. Side chains that interact with substrate or inhibitor to form the active-site pocket are shown using thin bonds. Schematic representations of the XMP*–IMPDH interactions and the MPA–IMPDH interactions are shown in (b) and (c), respectively

addition to MPA's utility as an immunosuppressant, this molecule has also been shown to suppress HIV replication *in vitro*. In addition, MPA has been shown to potentiate the antiviral activity of HIV reverse transcriptase inhibitors such as abacavir (a guanosine analogue), due to the depletion of nucleotide substrates needed for viral replication (Margolis *et al.* 1999). Inability to eradicate HIV and the toxicity associated with HAART has prompted the development of MPA as an alternative anti-HIV therapy (Chapuis *et al.* 2000). The AIDS Clinical Trials Group (ACTG) is considering mycophenolic acid for further clinical studies.

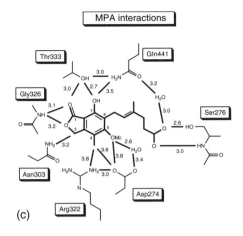

Figure 14.7 (*Continued*)

In this case study, we would like to illustrate how structure information obtained from a ternary complex of IMPDH containing the substrate XMP and an inhibitor with a natural product origin can be used to design better immuno-suppressant drugs, as well as drugs designed to treat other diseases such as hepatitis and AIDS.

The X-ray crystal structure of IMPDH from Chinese hamster in complex with IMP and MPA at 2.6 Å resolution revealed that each of the monomers consists of two domains, with the larger domain folded into an α/β barrel (PDB: 1JR1, Sintchak *et al.* 1996). As described above, the active-site pocket of IMPDH is made up of the IMP-binding site containing the catalytic Cys-331 residue and the NAD-binding site where MPA binds. The catalytic Cys-311 is involved in forming a covalent intermediate with the hypoxanthine ring during oxidation of IMP to XMP. The bicyclic ring system of MPA packs underneath the hypoxanthine ring in this ternary complex structure, thereby trapping this covalent intermediate and preventing the production and release of XMP. Both the substrate–enzyme (XMP-IMPDH) and cofactor mimetic–enzyme (MPA-IMPDH) interactions involve residues from flexible loop regions, including both the active site loop (residues 325–342) and the flap sequence (residues 400–450). Cys-331 forms a covalent bond with the C2 carbon of the hypoxanthine, and Tyr-411 (located on a β strand of the flap region) forms a hydrogen bond with the phosphate moiety of the substrate XMP (Figure 14.7B, C).

Information on the specific interactions that occur between IMPDH and its ligands is useful for rationalizing structure–activity relationships observed for MPA and its analogues. For instance, replacement of the phenolic hydroxyl group by an amino moiety leads to an increase of IC50 from 0.020 μM to 0.154 μM, an eight-fold reduction in potency (Figure 14.8). This phenolic hydroxyl group was identified to be essential for high potency in the inhibition of IMPDH (Nelson *et al.* 1996). The structure indicates that this change from an

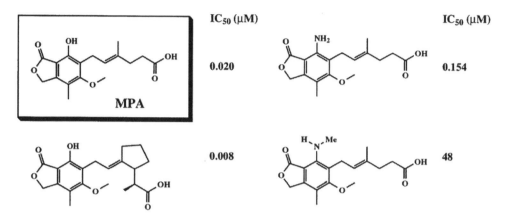

Figure 14.8 Chemical structure of MPA as well as three analogues, and corresponding IC50 values (Sjogren 1995)

OH-group to an NH$_2$-group would cause the loss of one hydrogen bond in the network of MPA's OH-group with IMPDH's OH-of Thr333 and NH$_2$-group of Gln441 (Figure 14.7B). The crystal structure of the ternary complex illustrates the structure–activity relationship for the hydrogen bond formed between the hexenoic acid side-chain of MPA with Ser-276 of IMPDH. Specific interaction between the hexenoic acid side-chain of MPA and IMPDH explains how reduction of this side-chain to an alcohol, or esterification of the carboxylate group, leads to a 50-fold reduction in IMPDH inhibition (D. Armistead unpublished data) thus provides a basis for further design and optimization of novel IMPDH inhibitors.

In spite of the anti-cancer and anti-proliferative activity of mycophenolic acid, the clinical use of mycophenolate mofetil is limited by the observation of toxic gastrointestinal effects, caused by the active metabolite MPA (Allison and Eugui 1993; Shaw *et al.* 1995). Therefore, continuing efforts have been made utilizing structure-based approach to design novel IMPDH inhibitors possessing clinical efficacy and no toxicity. Recently, a novel IMPDH inhibitor, VX-497 (Vertex), was obtained from such an attempt and has entered clinical trials to be evaluated as a novel treatment of hepatitis C and other diseases such as psoriasis (Armistead *et al.* 1998). Similar to the binding of MPA to IMPDH, the phenyl-oxazole moiety of the VX-497 occupies the NAD binding site and packs underneath XMP. In addition, VX-497 makes several new interactions to IMPDH as compared with MPA.

Using IMPDH as the target, drug design efforts have also been made for infectious diseases, because IMPDH enzymes from lower organisms have much less affinity toward MPA and thiazole-4-carboxamide adenine dinucleotide (TAD) than mammalian enzymes. The recently determined 2.2 Å resolution X-ray structure of a *Trichomonas foetus* IMPDH enzyme complexed with the substrate IMP, and the inhibitor beta-methylene-TAD (beta-Me-TAD) allows a greater understanding of the differences in IMPDH selectivity observed between lower eukaryotes and mammals (Gan *et al.* 2002). In this ternary complex structure, the active site loop is ordered and forms hydrogen bonds to the carboxamide of beta-Me-TAD, indicating that the binding of the cofactor NAD to the protozoan IMPDH might promote the nucleophilic attack of IMP by the catalytic Cys residue. Such an interaction between the ligand in the cofactor-binding site and the active site catalytic residue is unique for the protozoan enzyme, and it suggests that the NAD site of the protozoan enzyme may be exploited for designing selective and potent inhibitors for non-mammalian IMPDH enzymes. Exploitation of these structural differences may produce novel IMPDH inhibitors with sufficient selectivity to be developed as anti-infective therapeutic agents for bacterial, fungal and parasitic infections.

In both the HIV PR and IMPDH case, binding of the inhibitor to its respective drug receptor (or target) involves side-chains of amino acid residues residing on flexible loops or flaps. In addition, some mutations causing drug resistance are also located in flexible parts of the protein. Such mutations may not appear to disrupt the local environment directly. Instead, they may exert their effect by causing subtle changes in interaction-networks that could lead to movement of other parts of the ligand-binding pocket. Unfortunately, electron density of amino acid side-chains from flexible regions is often too weak to assign specific orientations to them. In light of protein plasticity and ligand-induced conformational changes observed in some drug design targets, structure-based drug design efforts face significant challenges in the design of potent inhibitors with broad-spectrum activity against both the wild-type enzyme as well as drug-resistant variants to deal with the rapid emergence of drug resistance. In addition to static high-resolution protein structure, more detailed kinetics studies to determine association and dissociation constants for ligand binding often provide additional insightful information for the discovery of selective and potent lead compounds for drug development.

Structure-based drug discovery in the post-genomics era

The successful sequencing of the human genome – a landmark achievement in the field of genetics – now poses a major challenge to the field of structural biology. In the coming years, public and private structural genomics efforts employing costly experimental techniques will attempt to furnish three-dimensional structure information for many of the proteins encoded by the human genome. However, the automated high-throughput process of experimental structure determination has had an overall low success rate of less than 1 per cent, with pilot structural genomics projects producing only 50 structures from an initial set of approximately 5000 targets (Service 2002). The public and private high-throughput structural genomics initiatives expect the throughput of protein structure determination to continue to improve as scientists gain more experience and bring additional technologies into the process. However, it is important to note that proteins are more variable in terms of stability and surface chemistry and their behaviour and performance in an automated process is much less predictable. In addition, it may become obvious that multiple protein expression and purification processes are needed in order to handle those proteins that are less soluble or more prone to undergo conformational changes that create physical heterogeneity in response to sub-optimal storage conditions.

Despite the challenges of high-throughput experimental three-dimensional structure determination, the use of protein structure information in drug discovery has substantially increased over the past decade. As more structures of drug targets have been determined, and as structure-based drug design methods have garnered increasing acceptance, protein structure determination has now become an integral component of drug discovery research.

Structural proteomics

High-throughput crystallography

Recent technical advances in laboratory automation have created a great deal of interest in automating the protein crystallographic process as a way to decrease the cost and time associated with experimental determination of protein structures. Companies such as San Diego, CA-based Syrrx and Structural GenomiX have invested tens of millions of dollars in state-of-the-art robotics platforms aimed at the miniaturization and automation of expression, purification and crystallization of proteins for structure determination. Unfortunately, these automated procedures often compromise the quality of protein crystals for an increase in speed and throughput. In order to obtain three-dimensional structural coordinates from lower quality crystals that are smaller in size or mosaic in morphology, scientists must use extremely intense and focused X-ray beams for collecting diffraction data. The high-powered radiation sources necessary for this type of analysis can only be produced at synchrotron beamline facilities, of which there are only five in the USA. Also, sometimes in order to obtain high quality diffracting crystals it is required to genetically engineer changes into a protein, making these cases not amenable to high throughput approach (Chen *et al.* 1996; Jenkins *et al.* 1996).

One of the primary concerns for the high-throughput crystallography approach is that the experimental determination of ALL human protein crystal structures would require decades, and control experiments have shown that only about 10–15 per cent of all proteins are amenable to crystallization for structure determination (Burley *et al.* 1999). In addition, nearly half of all known drug targets are membrane proteins, which are notoriously difficult to crystallize and are thus not readily amenable to the high-throughput crystallography approach. Therefore, it remains to be seen whether or not the enormous investment in high-throughput crystallography will translate into substantial benefits for the discovery of new medicine.

Computational structural proteomics

One practical alternative to high-throughput X-ray crystallography is large-scale protein structure modelling. As new structures continue to be generated at an increased speed, and more complete libraries of known protein folds are revealed, computational modelling methods will play a key role in filling in the gaps in structure space by providing protein structure models for all orthologues and paralogues of a given known structure. Standard comparative modelling methods generally require that a query protein sequence shares at least 30 per cent sequence identity (and preferably >40 per cent) with a known structure in order to be accurately modelled. Given the current state-of-the-art in comparative modelling, it has been estimated that ∼ 16 000 novel and optimally selected structures would need to be experimentally solved in order to allow 90 per cent of all known protein sequences to be modelled (Vitkup *et al.* 2001). Even with the technological improvements that have been made in high-throughput crystallography, it would still take tens of millions of dollars and well over a decade experimentally to determine this number of protein structures. Thus, improving the comparative modelling approach – such as lowering the sequence identity threshold necessary for accurate model building, and improving the ability to model membrane proteins – represents a more feasible route to achieving a 'comprehensive' view of the structural proteome in an efficient manner.

Current challenges for drug discovery

Genomics projects have produced an explosion in the number of potential new drug targets, thus the opportunity to develop novel therapeutic compounds is now greater than ever before. Recent estimates have placed the new number of potential human drug targets at around 3000 to 10 000 (Drews 2000). Also, the number of targets under investigation at the average large pharmaceutical company has been forecasted to increase four-fold over the next 5 years, from approximately 50 to nearly 200, with the proportion of novel, untested target classes increasing from 30 per cent to 70 per cent (Lehman Brothers, McKinsey & Co. 2001). The explosion in the number of novel targets creates a serious challenge for drug discovery researchers, as the industry struggles to translate vast amounts of uncharted genomic information into novel drugs.

To deal effectively with the increased costs and risks that accompany drug development in novel target space, drug discovery organizations are finding that they need to shift their emphasis from target identification toward target prioritization and validation by effective means. The prioritization of those targets that have higher benefit-to-risk ratios will be the key to successful drug discovery by reducing the downstream attrition rates of drug development. Improved structure-based analysis and enhanced understanding of target–drug interactions at the three-dimensional structure level can aid in this prioritization process. Through critical evaluation of physicochemical properties of binding cavities within three-dimensional protein structures, only those novel proteins containing druggable binding sites can be selected as drug discovery targets for unmet medical needs with a higher degree of confidence.

Companies such as Eidogen in Pasadena, CA, USA, Inpharmatica in the United Kingdom (*http://www.eidogen.com/*, and *http://www.inpharmatica.co.uk*) and others have developed proprietary structural informatics platforms and databases containing both experimentally determined protein structures and computation models. For example, Eidogen's web-based Target Informatics Platform™ (TIP™) is a database of more than 500 000 high-quality protein structures, including those determined using a proprietary algorithm capable of building accurate structural models from templates sharing as little as 20 per cent sequence identity. All structures in the TIP™ database have also been annotated with a proprietary algorithm that identifies and characterizes small-molecule binding sites within protein structures with a high degree of confidence. Eidogen's algorithm is different from, and significantly superior to the publicly available program PASS (Brady and Stouten 2000). The computational structure determination and binding site analysis capabilities of TIP™ are combined with a suite of PC-based visualization and binding site comparison tools. Comprehensive structural content and detailed binding site analyses provide researchers with the necessary tools critically to evaluate a novel target's susceptibility to drug action more completely and efficiently. The ability to carry out accurate, confident, *in silico* annotation of small molecule binding sites could provide significant value to drug discovery by providing biologists and chemists with a hypothesis and/or an understanding of how a drug might bind to, and modulate the activity of, its respective target(s). Computational binding site annotation is also capable of opening up new drug discovery opportunities, such as in the identification of novel drug sites on previously known target proteins. Table 14.1 lists several test cases where TIP™ was used successfully to identify novel binding sites *in silico* with no *a priori* knowledge of the experimentally determined location of these previously unidentified, non-active site regions of drug targets.

The discovery of novel therapeutic drugs using conventional medicinal chemistry and modern high-throughput screening techniques (HTS) has proven to be rather inefficient. Nevertheless, high-throughput screening of large compound libraries continues to produce lead compounds for over 50 per cent of the drug discovery projects in big Pharma companies. To improve the efficiency and cost-effectiveness of the drug discovery process, co-crystal structures of proteins complexed with initial weak binding 'hits' from HTS allow rational design of

Table 14.1 Recently published cases of novel druggable sites identified in validated targets

Target	Disease	Description of site	Reference
Glycogen phosphorylase	Diabetes	Allosteric CP32626-binding site	Oikonomakos *et al.* (2000)
Lymphocyte function-associated antigen-1 (LFA-1)	Inflammation/ Hypertension	Allosteric statin-binding site	Weitz-Schmidt *et al.* (2001)
P38 MAP kinase	Inflammation/ Cancer	Allosteric BIRB796-binding site	Pargellis *et al.* (2002)
Fructose-1,6-bisphosphatase	Diabetes	Allosteric anilinoquinazoline-binding site	Wright *et al.* (2002)

directed or focused combinatorial libraries. This allows the quick exploration of new binding space, due to ligand-induced side-chain re-orientation or main chain movement, revealed by these co-crystal structures as was described in the HIV protease case study. In addition, co-crystal structures containing weak binding 'hits' provide critical protein:ligand interaction information that enables medicinal chemists to select and to focus on only those hits with the specific structural and chemical features having the highest probability of being elaborated into tight binding drug leads. The iterative cycle of structure-based drug design can then be used to optimize a compound's potency, selectivity profile and pharmacokinetic properties during the lead optimization phase of a drug discovery project.

As we progress into the post-genomics era, efforts made in high-throughput crystallography and computational structural proteomics will undoubtedly provide a more comprehensive view of the structural space occupied by drug design targets of various protein families, such as protein kinases, proteases, phosphodiesterases and nuclear receptors. More importantly, the combined experimental and computational efforts will enable scientists to enumerate and compare subtle but important structural and chemical differences in the ligand binding sites of proteins within a given protein family. Such an enabling knowledge base is already emerging from the efforts made in structural genomics by scientists in both the academic and industrial sectors. As drug discovery becomes more and more of an integrated process, it is anticipated that protein structures will be utilized to guide the design of directed and focused combinatorial chemical libraries to enhance the success rate of HTS and reduce the cost. In the near future, we may expect to see protein structural information and compound structure-activity data to become more highly correlated, thus creating a multi-dimensional map of the chemical space that is occupied by drug targets and their cognate ligands. Structure-based drug discovery will continue to be carried out in a higher-throughput manner, although it is important to note that the future success of structure-based drug discovery will be dependent on not only the *quantity* but also the *quality* of the data that is deposited into centralized databases. Despite tremendous technology improvements, drug discovery remains a significantly challenging endeavour. Its success continues to require critical thinking, effective exchange of knowledge at the interface between all disciplines of science and medicine, and sometimes serendipity because many *unknowns* remain in life sciences.

Acknowledgements

I wish to acknowledge the support from Keck Graduate Institute of my efforts in writing this manuscript. I would like to thank my co-authors for their collaborative efforts and contributions, and Ms Lisa Marlette at Eidogen for her administrative assistance. Specially, I would also like to thank Drs David. Matthews and Siegfried Reich for their critical review of this manuscript, and my other former colleagues at Pfizer La Jolla Laboratories for their contribution in the structure-based drug discovery of Nelfinavir.

References

Allison A. C., Hovi, T., Watts, R. W. E. and Webster, A. D. B. (1975) Immunological observations on patients with Lesch-Nyhan syndrome and on the role of de novo purine synthesis in lymphocyte transformation. *Lancet* **ii**: 1179–1182.

Allison, A. C. and Eugui, E. M. (1993) Immunosuppressive and other effects of mycophenolic acid and an ester prodrug, mycophenolate mofetil. *Immunol Rev* **136**: 5–28.

Allison, A. C., Hovi, T., Watts, R. W. E. and Webster, A. D. B. (1977) The role of *de novo* purine synthesis in lymphocyte transformation. *Ciba Foundation Symp* **48**: 207.

Artis, D. R., Elworthy, T. R., Hawley, R. C., Loughhead, D. G., *et al.* (1995) 5-substituted derivatives of mycophenolic acid, World patent WO 95/22538.

Beddell, C. R., Goodford, P. J., Norrington, F. E., Wilkinson S. and Wootton, R. (1976) Compounds designed to fit a site of known structure in human haemoglobin. *Br J Pharmacol* **57**: 201–209.

Boden, D. and Markowitz, M. (1998) Resistance to human immunodeficiency virus type 1 protease inhibitors. *Antimicrob Agents Chemother* **42**: 2775–2783.

Böhm, H.-J. (1992) LUDI-rule-based automatic design of new substituents for enzyme inhibitor leads. *J Computer-Aided Molecular Design* **6**: 593–606.

Brady G. P. and Stouten P. F. (2000) Fast prediction and visualization of protein binding pockets with PASS. *J Comput-Aided Mol Des* **14**: 383–401.

Burley, S. K., Almo, S. C., Bonanno, J. B., Capel, M., *et al.* (1999) Structural genomics: beyond the human genome project. *Nature Gen* **23**(2): 151–157.

Carr, R. and Jhoti, H. (2002) Structure-based screening using either x-ray crystallography or NMR. *Drug Discovery Today* **7**: 522–527.

Carr, S. F., Papp, E., Wu, J. C. and Natsumeda, Y. (1993) Characterization of human type I and type II IMP dehydrogenase. *J Biol Chem* **268**: 27286–27290.

Chapuis, A. G., Rizzardi, G. P., D'Agostino, C., Attinger, A., *et al.* (2000) Effects of mycophenolic acid on human immunodeficiency virus infection in-vitro and in vivo. *Nature Med* **6**(7): 762–768.

Chen, P., Tsuge, H., Almassy, R. J., Gribskov, C. L., *et al.* (1996) Structure of the human cytomegalovirus protease catalytic domain reveals a novel serine protease fold and catalytic triad. *Cell* **86**(5): 835–843.

Class, S. (2002) Pharma Overview. *Chem. & Eng. News*, December 2 issue: 39–49.

Condra, J. H., Holder, D. J., Schleif, W. A., Blahy, O. M., *et al.* (1996) Genetic correlates of in vivo viral resistance to indinavir, a human immunodeficiency virus type 1 protease inhibitor. *J Virol* **70**:8270–8276.

Crabtree, G. W. and Henderson, J. F. (1971) Rate-limiting steps in the interconversion of purine bionucleotides in Ehrlich ascites tumor cells in-vitro. *Cancer Res* **31**: 985–991.

Drews J. (2000) Drug discovery: a historical perspective. *Science* **287**: 1960–1964.

Fitzgerald, P.M., McKeever, B.M., VanMiddlesworth, J.F., Springer, J.P., *et al.* (1990) Crystallographic analysis of a complex between human immunodeficiency virus type 1 protease and acetyl-pepstatin at 2.0-6A resolution. *J Biol Chem* **265**: 14209–14219.

Franklin, T. J. and Cooke, J. M. (1969) The inhibition of nucleic acid synthesis by mycophenolic acid. *Biochem J* **113**: 515–524.

Gan, L., Petsko, G. A. and Hedstrom, L. (2002) Crystal structure of a ternary complex of *Tritrichomonas foetus* inosine 5'-monophosphate dehydrogenase: NAD+ orients the active site loop for catalysis. *Biochemistry* **41**(44): 13309–13317.

Gillet, V. J., Johnson, A. P., Mata, P. Sik, S. and Williams, P. (1993) SPROUT – A program for structure generation. *J Comp Aided Mol Design* **7**: 127–153.

Goodford, P. J. (1985) A computational procedure for determining energetically favorable binding sites on biologically important macromolecules. *J Med Chem* **28**: 849–857.

Goodman & Gilman's The Pharmacological Basis of Therapeutics, 10th edition, 2001, eds. Hardman J. G., Limbird, L. E. & Gilman A. G., McGraw-Hill Medical Publishing Division.

Holmes, E. W., Pehlke, D. M., and Kelley, W. N. (1974) Human IMP dehydrogenase: kinetics and regulatory properties. *Biochim Biophys Acta* **364**: 209–217.

Hopkins, A. L. and Groom, C. R. (2002) The druggable genome. *Nature* **1**: 727–730.

Ishima, R., Freedberg, D. I., Wang, Y. X., Louis, J. M. and Torchia, D.A. (1999) Flap opening and dimer-interface flexibility in the free and inhibitor-bound HIV protease, and their implications for function. *Structure Fold Des* **7**:1047–1055.

Jackson, R. C. B., Weber, G. and Harris, H. P. (1975) IMP dehydrogenase, an enzyme linked with proliferation and malignancy. *Nature* **256**: 331–333.

Jaskolski, M., Miller, M., Rao, J. K., Leis, J. and Wlodawer, A. (1990) Structure of the aspartic protease from Rous sarcoma retrovirus refined at 2-Å resolution. *Biochemistry* **29**: 5889–5898.

Jenkins, T. M., Engelman, A., Ghirlando, R. and Craigie, R. (1996) A soluble active mutant of HIV-1 integrase: involvement of both the core and carboxyl-terminal domains in multimerization. *J Bio Chem* **271**(13): 7712–7718.

Kaldor, S. W., Kalish, V. J., Davies, J. F. 2nd, Shetty, B. V., *et al.* (1997) Viracept (nelfinavir mesylate, AG1343): a potent, orally bioavailable inhibitor of HIV-1 protease. *J Med Chem* **40**: 3979–3985.

Kuyper, L. F., Roth, B., Baccanari, D. P., Ferone, R., *et al.* (1985) Receptor-based design of dihydrofolate reductase inhibitors: comparison of crystallographically determined enzyme binding with enzyme affinity in a series of carboxy-substituted trimethoprim analogues. *J Med Chem* **28**: 303–311.

Lauri, G. and Bartlett, P. A. (1994) A program to facilitate the design of organic molecules. *J Comput-aided Mol Design* **8**: 51–66.

Lehman Brothers, McKinsey & Co. (2001) *The Fruits of Genomics: Drug Pipelines Face Indigestion Until the New Biology Ripens.*

Margolis, D. *et al.* (1999) Abacavir and mycophenolic acid, an inhibitor of inosine monophosphate dehydrogenase, have profound and synergistic anti-HIV activity. *J Acquir Immune Defic Syndr* **21**:362–370.

Matthews, D. A., Bolin, J. T., Burridge, J. M., Filman, D. J., *et al.* (1985) Dihydrofolate reductase. The stereochemistry of inhibitor selectivity. *J Biol Chem* **260**: 392–399.

Miller, M., Schneider, J., Sathyanarayana, B. K., Toth, M. V., *et al.* (1989) Structure of complex of synthetic HIV-1 protease with a substrate-based inhibitor at 2.3 A resolution. *Science* **246**: 1149–1152.

Munshi, S., Chen, Z., Yan, Y., Li, Y. *et al.* (2000) Alternate binding site for the P1-P3 group of a class of potent HIV-1 protease inhibitors as a result of concerted structural change in 80'S loop. *Acta Crystallogr D Biol Crystallogr* **56**(4): 381–388.

Navia, M. A., Fitzgerald, P. M., McKeever, B. M., Leu, C. T., *et al.* (1989) Three-dimensional structure of aspartyl protease from human immunodeficiency virus HIV-1. *Nature* **337**: 615–620.

Nelson, P. H., Carr, S. F., Devens, B. H., Eugui, E. M., *et al.* (1996) Structure–activity relationships for inhibition of inosine monophosphate dehydrogenase by nuclear variants of mycophenolic acid. *J Med Chem* **39**(21): 4181–4196.

Oikonomakos, N. G., Skamnaki, V. T., Tsitsanou, K. E., Gavalas, N. G. and Johnson, L. N. (2000) A new allosteric site in glycogen phosphorylase b as a target for drug interactions. *Structure Fold Des* **8**(6): 575–584.

Palella, F. J. Jr., Delaney, K. M., Moorman, A. C., Loveless, M. O., *et al.* (1998) Declining morbidity and mortality among patients with advanced human immunodeficiency virus infection. *N Eng J Med* **338**: 853–860.

Pargellis, C., Tong, L., Churchill, L., Cirillo, P. F., *et al.* (2002) Inhibition of p38 MAP kinase by utilizing a novel allosteric binding site. *Nat Struct Biol* **9**(4): 268–272.

Prosise, G. L., Wu, J. Z. and Luecke, H. (2002) Crystal structure of *Tritrichomonas foetus* inosine monophosphate dehydrogenase in complex with inhibitor ribavirin monophosphate reveals a catalysis-dependent ion-binding site. *J Biol Chem* **277**(52):50654–50659.

Reiling, K. K., Endres, N. F., Dauber, D. S., Craik, C. S. and Stroud, R. M. (2002) Anisotropic dynamics of the JE-2147-HIV protease complex: drug resistance and thermodynamic binding mode examined in a 1.09 A structure. *Biochemistry* **41**: 4582–4594.

Service, R. F. (2002) Structural genomics. Tapping DNA for structures produces a trickle. *Science* **298**: 948–950.

Shaw, L. M., Sollinger, H. W., Halloran, P., Morris, R. E., *et al.* (1995) Mycophenolate mofetil: a report of the consensus panel. *Therapeut Drug Monitor* **17**: 690–699.

Shuker S. B., Hajduk, P. J., Meadows, R. P. and Fesik, S. W. (1996) Discovering high-affinity ligands for proteins: SAR by NMR. *Science* **274**: 1531–1534.

Sintchak, M. D., Fleming, M. A., Futer, O., Raybuck, S. A., *et al.* (1996) Structure and mechanism of inosine monophosphate dehydrogenase in complex with immunosuppressant mycophenolic acid. *Cell* **85**: 921–930.

Sjogren, E. B. (1995) 4-amino derivatives of mycophenolic acid with immunosuppressant activity. World patent WO 95/22535.

Snyder, F. F., Henderson, J. F. and Cook, D. A. (1972) Inhibition of purine metabolism: computer assisted analysis of drug effects. *Biochem Pharmacol* **21**:2351–2357.

Terwilliger, T. C. and Berendzen J. (1999) Automated MAD and MIR structure solution. *Acta Crystallographica* **D55**: 849–861.

Todd, M. J., Luque, I., Todd, M. J., Milutinovich, M., *et al.* (2000) Thermodynamic basis of resistance to HIV-1 protease inhibition: calorimetric analysis of the V82F/I84V active site resistant mutant, Biochemistry, 39: 11876–11883.

Velazquez-Campoy, A., Kiso, Y. and Freire, E. (2002) The binding energetics of first- and second-generation HIV-1 protease inhibitors: implications for drug design. *Arch Biochem Biophys* **390**:169–175.

Vitkup, D., Melamud, E., Moult, J. and Sander, C. (2002) Completeness in structural genomics. *Nat Struct Biol* **8**(6): 559–566.

Weber, I. T. (1990) Evaluation of homology modeling of HIV protease, *Proteins* **7**:172–184.

Webber, G. (1983) Biochemical strategy of cancer cells and the design of chemotherapy. *Cancer Res* **43**: 3466–3492.

Weber, I. T., Miller, M., Jaskolski, M., Leis, J., *et al.* (1989) Molecular modeling of the HIV-1 protease and its substrate binding site. *Science* **243**: 928–931.

Weitz-Schmidt, G., Welzenbach, K., Brinkmann, V., Kamata, T. *et al.* (2001) Statins selectively inhibit leukocyte function antigen-1 by binding to a novel regulatory integrin site. *Nat Med* **7**(6):687–692.

Wlodawer, A., Miller, M., Jaskolski, M., Sathyanarayana, B. K., *et al.* (1989) Conserved folding in retroviral proteases: crystal structure of a synthetic HIV-1 protease. *Science* **245**: 616–621.

Wright, S. W., Carlo, A. A., Carty, M. D., Danley, D. E., *et al.* (2002) Anilinoquinazoline inhibitors of fructose 1,6-bisphosphatase bind at a novel allosteric site: synthesis, in vitro characterization, and X-ray crystallography. *J Med Chem* **45**(18):3865–3877.

Yoshimura, K., Kato, R., Yusa, K., Kavlick, M. F., *et al.* (1999) JE-2147: a dipeptide protease inhibitor (PI) that potently inhibits multi-PI-resistant HIV-1. *Proc Natl Acad Sci USA*, **96**:8675–8680.

Zoete, V., Michielin, O. and Karplus, M. (2002) Relation between sequence and structure of HIV-1 protease inhibitor complexes: a model system for the analysis of protein flexibility. *J Mol Biol* **315**:21–52.

15
Protein Interaction-targeted Drug Discovery

Gary Hudes, Sanjay Menon and **Erica A. Golemis**

Introduction

The definition of the complete genomic sequence for multiple organisms at the end of the second millennium is widely acknowledged to have constituted a scientific milestone. At present, major efforts are underway to build from this enormous DNA sequence resource to a completely defined complement of genes and proteins, and to define physical and functional interactions between the proteins. For biologists in general, among the many great changes in scientific thinking and practice accompanying the accomplishment of these feats, the most fundamental benefit will be the first rigorous 'definition of field' of relevant players for living systems, comparable in import to the definition of a periodic table of elements for chemists, and moving understanding of organismal function to a new level of clarity. For clinicians and those interested in drug discovery in particular, the promise of such clarity is the prospect that a completely defined biological system will unmask critical control points impacted by disease, allowing precise engineering of therapeutic agents to target the point of pathological disturbance, improving efficacy and reducing side effects.

Achievement of these ultimate therapeutic goals remains some distance away. Nevertheless, in contrast to past practice in which drugs were first screened for

Molecular Analysis and Genome Discovery edited by Ralph Rapley and Stuart Harbron
© 2004 John Wiley & Sons, Ltd   ISBN 0 471 49847 5 (cased)   ISBN 0 471 49919 6 (pbk)

general efficacy in *in vivo* systems, and mode of function defined well after the fact of developing a usable agent, current drug discovery efforts are now generally organized against individual protein targets, with demonstration of efficacy *in vivo* following at least preliminary establishment of mode of action. Already, there are some striking examples of the success of such protein-targeted strategies. These include the clinical potency of STI571 (also termed Gleevec or imatinib mesylate) against BCR-ABL, c-Kit, and other related kinases transforming therapy of chronic myelogenous leukaemia (CML) and gastrointestinal stromal tumours (GIST) (Druker 2002). At this point in time, one of the most critical areas of debate must be how generally such successes are likely to be achieved, and strategically, how to maximize the chances of finding a good drug.

Within the larger context of protein-targeted drug discovery, one area of considerable interest has been the question of utilizing Protein–Protein interactions (PPIs) as targets for drug intervention. In the broadest sense, all drugs directed against a protein target can be considered to target PPIs, insofar as no protein works in complete isolation from other cellular proteins; thus, drugs interfering with the function of a protein are likely to interfere in some way with that protein's association with the rest of the cellular machinery. Although accurate at one level, the breadth of such a definition tends to rob it of specific meaning: more usefully, a PPI should be taken to describe an association between two or more proteins of some duration that has functional import for a given disease process. The interest of targeting a PPI lies in the facts that important signalling proteins tend to interact with many different partners under differing growth circumstances; and secondly, that complete elimination of the function of some important signalling proteins (for example, by knockout or siRNA) frequently is deleterious to the viability of a cell or organism. Hence, beyond the specificity of action gained by targeting a drug to a protein of interest, it is to be hoped that disease-relevant signalling axes dependent on particular PPIs might be disrupted, while benign or essential PPIs might be maintained. The idea is attractive: but is it practicable?

PPIs: classes, functional relationships, and regulation

Proteins interact in many different ways. Consideration of the different classes of interactions suggests some issues related to targeting an interaction with a small molecule agent, which makes a brief review of interaction modes useful (see Figure 15.1 for examples).

At the simplest level, a functionally significant PPI defines the stable association of a protein to form a homodimer with itself or heterodimer with a second protein, such that the dimer has novel properties not possessed by the protein as an isolated monomer. Examples of this class of PPI would include assembly of a transcription factor (e.g. dimerization of MyoD with itself or other helix-

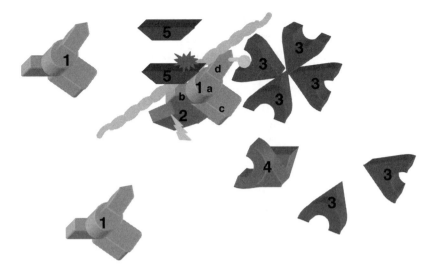

Figure 15.1 Examples of different types of protein interaction. In a hypothetical situation, proteins a,b,c,d define subunits of a first complex, 1 that predominantly exists as a heterotetramer. Protein 2 is stably covalently modified by addition of a SUMO moiety that targets it to a localization in which it forms a stable complex with 1. Following transient modification of the 1d subunit by phosphorylation (pale blue circle), it forms a complex with 3, inducing homotetramerization of protein 3: the 3 homotetramer protects 1d from dephosphorylation by protein 4. In the presence of DNA, complex 1 additionally binds to another protein 5, resulting in activation of 5.

loop-helix transcription factors (Weintraub *et al.* 1991)), or creation of an active cyclin-dependent kinase (CDK)-cyclin pair (Meijer *et al.* 1989; Draetta *et al.* 1989). At a more complicated level, proteins may self-oligomerize to a higher level (e.g. in assembly of a homo-octomeric porphobilinogen synthase, required for synthesis of tetrapyrroles such as heme (Kervinen *et al.* 2000)); or in the extensive oligomers represented by the chaperone-mediated assembly of tubulin into microtubules (Lopez-Fanarraga *et al.* 2001)). Proteins may engage in simultaneous interactions with two or more partner proteins to form a functional complex (e.g. the T cell receptor (Malissen and Malissen 1996)). In these more complex interactions, each protein may form moderate to high affinity interactions with each of its partners. Alternatively, one element of a complex may act as a 'scaffolding' protein, promoting the association of two components that would normally interact weakly (e.g. KSR regulation of interactions between Ras, Raf and MEK (Dhillon and Kolch 2002)). In an extension of this idea, in some cases one component of a PPI actively guides or stabilizes the conformation of an interactive partner, governing both its activity and interaction capability: for example, the Hsp90 complex assembles with chaperones structurally fragile proteins such as the c-Raf-1 oncoprotein, Akt, ErbB2 and others (Neckers 2002).

Conversely, some functionally important interactions have elements of extreme transience. Classically, kinase interactions with and modifications of substrates have been considered transient, as have been most interactions involving the modification by one protein of a second. In the light of current knowledge of signal transduction systems, however, it is important not to assume that all elements of such modification reactions are transient. For instance, upon cellular attachment, integrin ligation triggers the activation of a complex involving focal adhesion kinase (FAK), Src, and a large docking protein, p130Cas, that is substrate for both FAK and Src (O'Neill *et al.* 2000). Although the phosphorylation of p130Cas by FAK and Src is in each case the product of a 'transient' catalytic reaction, nevertheless the association of the proteins is of some duration and sufficient stability to be detectable by co-immunoprecipitation and other approaches, because of additional interactions between p130Cas and other, non-catalytic domains of FAK and Src (Ruest *et al.* 2001). Other interactions may also possess the character of both transience and stability but in a different way, as, for example, when a cellular 'motor' protein such as myosin or kinesin progresses along a filamentous actin or tubulin-based substrate. For these proteins, individual contacts between any individual surface of the motor, or monomeric module of the filament, are extremely brief, but the overall durability of interaction with the filament is high. At the other end of the scale, some protein interactions become ultimately durable, through culmination in the formation of covalent bonds between two proteins: an example of this class of interaction would be the attachment of a small ubiquitin-related modifier (SUMO) peptide to an appropriate substrate (Alarcon-Vargas and Ronai 2002). Finally, in some cases, interactions are functionally important, but of low general specificity by design: for example, the nuclear export machinery promotes the rapid transit to the cytoplasm of numerous different factors, including proteins and mRNAs, and hence must be able to sort appropriate substrates based on very limited recognition motifs (Weis 2002).

An appreciable number of PPIs have a fundamental requirement for a component that is not a protein. Among transcription factors and DNA repair complexes, the presence of DNA is necessary as a nucleating and stabilizing factor for many interactions (Volker *et al.* 2001). Other interactions depend on microenvironments contributed by cellular membranes: for example, lipid rafts (Alonso and Millan 2001) promote the assembly at the cell surface of important signalling complexes, including those dependent on the T-cell receptor. Naturally produced peptidyl or other small molecule (e.g. hormonal) ligands that bind and form stable associations with protein receptors are critical regulators of PPIs in many different cellular processes of relevance to disease. The binding of these small molecules to their receptors can contribute directly to PPIs, by contributing to a dimerization interface; or indirectly, by creating or stabilizing a conformational state in their receptor that is conducive to interaction with other partners.

These many different kinds of PPI pose different kinds of challenges for the design of inhibitors. For some classes of interaction, such as WW domains or SH3 domains, or kinases, the recognition motif is relatively small, comprised of 4–6 contiguously linked amino acids. Other proteins, such as those involved in obligate dimerizations, or in large complexes, are likely to have extensive cell surface contacts between partners contributing to their high affinities. Previous metastudies of crystal structures of dimerized molecules have estimated a range for interface surface area interactions of $\sim$350 – 4900 Å^2 (average of $\sim$800 Å^2) per protein monomer, which are involved in such interactions ((Janin *et al.* 1988; Jones and Thornton 1995; Lo Conte *et al.* 1999), and extensive discussion in (Stites 1997)). This roughly corresponds to surface contacts of $\sim$14–25 amino acids per protein monomer, with these contacts involving $\sim$7–30 per cent of the total surface area of the protein (highest for homodimers and proteins involved in larger complexes) with enrichment of hydrophobic residues. Current analyses of the characteristics of 'good drugs' have suggested that optimal agents will be of low molecular weight ($<$500 Da), and neither extremely hydrophobic or hydrophilic (Lipinski *et al.* 2001). These values would at first glance suggest it is easier to target the discrete interactions between, for example, an SH3 domain and its polyproline substrate, should such interactions drive PPIs on a clinically important signalling pathway. Perhaps, or alternatively, the limited nature of such an interaction may lead to significant problems in achieving specificity in a drug agent. Conversely, a number of studies have demonstrated that in spite of the large interfaces involved in some PPIs, a limited number of 'hot spots', comprising only a small number of amino acids, provide the major binding energy between proteins; efforts to identify such hot spots as targets for intelligent drug design are ongoing (DeLano 2002). At present, given the limited number of drugs known to be targeted to specific proteins, the general potential of drugs for modulating PPIs remains a point of speculation. Some excellent recent extended discussions of this topic have been published (Toogood 2002; Cochran 2000; 2001).

Protein interactions in disease-related signalling: categories of disruption

In thinking of PPIs in the context of disease, it is important conceptually to distinguish between cases in which changes in PPIs are causative of a disease state, as opposed to considering cases where a given PPI is essentially unaltered in disease induction, but essential for allowing expression of a diseased phenotype (e.g. required for cell viability). In the former case, which is particularly relevant to genetically linked, clonal diseases such as cancer, one or more PPIs are qualitatively different in a diseased versus a normal cell, and hence provide a potentially unique target that could be disrupted without interfering with the biology of normal cells. In contrast, in the latter case, drugs targeting a PPI will not discriminate between diseased and normal cells, but rather affect all cells in which the PPI occurs, raising issues discussed in greater detail in the following section. For diseases in which changes in PPIs are thought to be causative of disease progression, a number of different types of change have been observed or proposed. These include:

1. Overexpression of PPI participants. It has been proposed that enhancement of the levels of specific genes and their protein products in cancer may represent a novel class of tumor promoting change, with the proteins thus elevated operationally regarded as activated oncogenes although unaltered in sequence (Sager 1997). Microarray profiling to compare normal versus diseased cells and tissues has as one major goal the identification of genes which are differentially expressed in disease, with the idea that genes showing increased expression promote the disease, while those with reduced expression may normally act to hinder disease progress (see Chapter 9). While correlation of expression by no means constitutes proof of causation, it is clear that a number of proteins associated with proliferation, cell migration, angiogenesis, resistance to apoptosis and other processes tend to have enhanced expression in cancer cells, with highest levels observed in the most malignant populations (Sager 1997; Clark *et al.* 2000). This increased abundance, typically from 2–10 times that in normal cells, can drive the formation of complexes between these proteins and their interactors. Qualitatively, however, interaction with positive and negative regulators of their activity would be equally favoured.

Much work is currently devoted to the identification of such upregulated factors. However, an important caution on selecting such proteins as drug discovery targets is provided by recent results of combined high-throughput genetic/microarray analyses in yeast, which indicate that the set of genes induced following a stimulus is largely non-congruent with the set of genes known to have a functional role in the yeast cellular response to the same stimulus (Birrell *et al.* 2001; Giaever *et al.* 2002). For human disease, this field is in its early stages, but may well follow the yeast paradigm: hence, in picking targets, it will be important to develop search criteria that rigorously emphasize protein causality as well as correlative expression in a disease.

2. Loss of PPI participants. For many tumour suppressors (NF2, TSC1 and TSC2, Rb, p53), oncogenesis follows upon lost or reduced function induced either by complete loss of protein expression (due to deletion, or due to a destabilizing mutation that promotes protein degradation), or by point mutation that eliminates a key activity of the protein. Loss of a protein obviously will disrupt signalling related to PPIs in which the protein is involved. In the case of inactivating point mutations, in some cases the reason for lost function again relates to loss of one or more specific PPIs, or trapping of the protein in an inactive conformation. The former situation is not likely to be easily redressed through application of a small molecule agent, making such proteins poor targets for directed drug design: however, there is some progress in correcting the latter class of defect. For example, loss of p53 function occurs commonly through a combination of gene deletion, followed by mutation of the remaining p53 allele. Most p53 mutations are missense mutations that code for proteins with unstable folding and loss of DNA binding, resulting in loss of transcriptional activating function. In what could be described as the molecular equivalent of a 'tummy tuck', a series of small molecules were designed to gird the altered DNA binding domain, thereby restoring a conformation conducive to DNA binding and restoring transcriptional activity of the protein (Foster *et al.* 1999). A similar concept could be applied to a PPI, in which a loss of function, i.e. a specific PPI, caused by a missense mutaton, is restored by a small molecule that binds the mutant protein and changes conformation to enhance binding with the partner protein.

3. *Missense mutations: regulation of PPIs* There are many cases in which disease results from a single mutational change in a protein, either pre-existing as a hereditary predisposing factor (e.g. germline mutations in BRCA1 or BRCA2), or arising clonally in a tumour. These mutations can have significant effects on an extensive set of PPIs. As one example, a well documented subclass of oncogenic lesion is a mutation that generates a constitutively active form of a protein that normally oscillates between active and inactive states. Proteins thus targeted include GTPases (e.g. Ras), kinases (BCR-ABL), transmembrane growth factor receptors (EGFR), and others. Activation might occur from a single amino acid change in a catalytic site (e.g. the G12V transition inducing constitutive Ras activation), or derive from a gross physical re-arrangement, such as the chromosomal translocation creating BCR-Abl. Such mutations can have profound impacts on multiple PPIs. The mutated protein itself changes its pattern of interaction with its normal regulatory proteins. Only GDP-bound Ras binds activating GDP-GTP exchange factors (GEF) such as SOS; G12V-Ras is solely liganded with GTP, selecting strongly against such an interaction. Conversely, the interaction of G12V-Ras with negative regulatory factors such as p190Ras-GAP should be unaffected, or potentially enhanced (sequestering the GAP from interactions with other possible substrates); however, this interaction is no longer productive, with Ras remaining bound to GTP. Finally, G12V-Ras is much more likely to exist in a complex with its effectors, such as Raf, PI3K and Ral-GDS. The ability to target proteins with such precisely focused small molecule agents is not likely to be immediately forthcoming on a global basis, although it is a goal of considerable interest (Xu and Luo 2002).

Selecting targets and screens

Clearly, there are many different classes of protein interactions involved in the control of cellular function, and there are many different ways in which these interactions may be disrupted in disease. To focus on a single disease class, cancer, well in excess of 100 genes have been described as oncogenes or tumour suppressors. Based on current global proteomic studies of protein interactions in the model organisms *S. cerevisiae* (Gavin *et al.* 2002; Ito *et al.* 2001; Uetz *et al.* 2000) or *C. elegans* (Boulton *et al.* 2002), an average protein is involved in more than 10 PPIs. Assuming such interaction properties also pertain to human proteins, at least 1000 interactions can be suggested as directly linked to cancer causing proteins, and might theoretically comprise useful targets for drug discovery. In a condition of unlimited resources, many of these might be productively explored.

However, drug discovery and development are inherently extremely expensive processes, fundamentally constraining the number of targets that can be addressed. In designing a protein-targeted or PPI-targeted drug discovery programme, it is important to develop robust criteria to guide choice among the many potential targets. Given drug discovery largely occurs in the for-profit sector, a significant set of criteria will not be strictly scientific. Instead, criteria such as size of affected population with a particular involved protein- or PPI-dependent pathology; chronic or acute nature of disease; the availability of other treatments for the disease that might compromise market share; and patent protection restrictions on addressing the protein or pathway involved, are going to be important determining factors in target selection. Beyond these pragmatic issues, it is also important to try to establish scientifically based target selection criteria to maximize the chances of developing a small molecule agent that will be able to effectively modulate the PPI. A number of factors that would contribute to such criteria are discussed below.

Extracellular versus intracellular interactions

One of the most fundamental divisions between classes of PPI is whether they occur intracellularly or within the extracellular space. There are many reasons to favour extracellular targets. For drugs to be permeable within cells, it is generally thought that they must not exceed a relatively low molecular weight ($\sim$500Da), have low charge, and not be extremely hydrophilic or hydrophobic (Lipinski *et al.* 2001). These parameters greatly restrict the potential chemical space that can be explored searching for a bioactive agent; the restrictions on size in particular also limit the potential binding energy of a given compound. Further, entire classes of therapeutic agents, including peptides and monoclonal antibody derivatives tend to be non-cell-permeable, eliminating these agents from consideration.

The preceding points reflect the qualities of the drugs; from the perspective of targets, it is clear that there are many good candidates relative to disease that are expressed in the extracellular space. Integrins engage in dialogue with the extracellular matrix, and cadherins mediate cell–cell interactions: these and associated proteins are important in processes ranging from angiogenesis through metastasis to wound healing (Hajra and Fearon 2002; Rust *et al.* 2002). There are many transmembrane receptor proteins that play critical roles in cellular proliferation control, that are fundamental elements of disease-dependent autocrine loops (Zwick *et al.* 2002). Further, many of these proteins normally engage in interactions not only with other large proteins, but with small peptides of defined sequence. The sequence information from these ligands can be used to guide the chemistry of drug discovery, while the small size of the binding site also contributes to a good outlook in designing blocking agents. As discussed below, some of the most successful examples of PPI-blocking drugs are targeted at these classes of proteins.

Interaction affinity and protein abundance

Proteins are large and tend to interact with high affinity. Drugs are small and as a rule capable of much lower affinity interactions with their targets. As has been discussed in detail by others (Cochran 2000), as long as it is possible to achieve a situation where a small molecule agent is present in sufficient excess over its PPI targets, the difference in relative affinity in the interactions should not pose a critical problem for effectively targeting a PPI. In practical terms, because of drug formulation and drug developmental issues, a 'good drug' is generally defined as one that has a sub-micromolar (mid-nanomolar or better) potency, while few drugs in current use are delivered with higher than low micromolar concentrations. By this reasoning, if a drug must be used at 100-fold excess relevant to its target PPI, then the target of a drug with 100 nM affinity for its binding should be present at 1 nM concentration or less (Cochran 2000)).

Given that the typical mRNA is estimated to be present at frequencies of 1–20 copies per cell, this is likely to include a significant fraction of cellular proteins. However, choosing a particularly low abundance protein target would seem to be a straightforward way of optimizing drug target ratio. Nevertheless, there are a number of additional factors that caution against an overly simplistic application of such a rule. For instance, while *in vitro* biochemical studies of proteins are frequently considered the gold standard describing the interaction properties of a given protein, in light of modern cell biological studies concerning protein intracellular movement versus activity, it is likely that a significant rethinking of the dominance of *in vitro* studies would be useful. An early study illuminating this point is the work of Miyamoto *et al.* (1995) studying the localization of intracellular effector proteins following the activation of integrin-dependent signalling cascades by placement of a bead with an integrin ligand on the surface of a cell (Miyamoto *et al.* 1995). Of note, although the majority of tracked effectors, including RhoA, Rac1, Ras, Raf, MEK, ERK and JNK were diffusely spread through the cytoplasm prior to activation, within minutes following integrin stimulation, all had relocalized proximal to the integrin ligand bead, creating an extremely high relative local concentration. Further, many of the proteins had become tyrosine phosphorylated, creating additional binding/interaction sites for phospho-tyrosine binding motifs such as SH2 domains, and thus completely altering the likely PPI profile. In the past several years, it has become clear that rapid dynamic changes in protein localization and physical modification state such as described above may represent the norm in many types of signalling interaction, including the interactions of many proteins associated with disease such as PTEN, PI3K, JAK-STAT, SREBP and others. This reality greatly complicates any effort to choose targets based on biochemically based estimates of interaction affinity and intracellular concentration, as values pertaining to PPIs prior to delivery of an activation stimulus are likely to differ sharply from values relevant to PPIs occurring under conditions of pathway activation. These studies emphasize the importance of incorporating significant cell biological analysis into any *in vitro*-based drug screening strategy.

Protein family or unique protein: *essential or non-essential*

Based on the past several years of advances in genomics and proteomics, for any proposed PPI disease target, it is possible to mine public databases to identify extensive information regarding the two interacting proteins (for pros and cons of use of such databases, see discussion in Edwards *et al.* 2002). Within humans, it is possible to determine whether each of the proteins is completely unique in sequence, or whether it possesses specific functional domains which are highly conserved within a group (e.g. the catalytic domains of kinases), or whether it is a member of a protein family (e.g. H-Ras, K-Ras, and N-Ras). It is possible with little effort to construct an alignment displaying orthologues of the proteins from multiple species, allowing determination of highly conserved versus divergent sequence elements within the coding sequence, allowing prediction of important structural or functional groups. For proteins with orthologues in lower eukaryotes such as yeast (*S. cerevisiae*), and increasingly *D. melanogaster* and *C. elegans*, groups have developed compilations of systematic knockout (Winzeler *et al.* 1999) or deletion mapping data, and protein interaction maps (Gavin *et al.* 2002; Ito *et al.* 2001; Uetz *et al.* 2000) that can be used to gain further insights into protein functionality. These resources from model systems can now be readily supplemented by assessment of the effect of transiently deleting individual proteins in human cells through use of techniques such as siRNA-mediated mRNA degradation (Tuschl *et al.* 1999). Taken in sum, together with additional structural or mutational data available for proteins in the PPI of interest, as well as developing resources to describe the expression profile of individual mRNAs and proteins throughout the body, it is possible to amass a detailed portrait of the protein's sphere of activities and interacting partners which would provide some context for estimating whether efficiently and specifically disrupting the PPI using a drug is likely to be tolerated or extremely deleterious, and to have limited (specific) or pleiotropic effects.

In an ideal situation, a PPI-dependent signalling pathway would be essential at the site of disease (for instance, in a tumour), but partly or entirely dispensable in non-affected tissues. Classically, chemotherapeutic strategies have almost uniformly been based on the difference in proliferation index between cancerous cells and most normal body cells, with side effects limited to toxicity in blood cell production and gut lining. A next generation PPI-based therapy might seek, for example, to identify PPIs required for proliferation in the tumour cells, but not required (because of redundancy, or cell type differences) in haematopoiesis or intestinal epithelial cell replenishment. It may or may not be desirable to attempt to limit specificity of a PPI-targeting drug to a single family member from a group. A recent analysis of STI571 efficacy demonstrates advantages of both a highly specific target (e.g., BCR-ABL is only present and required in affected tumour cells) and some lack of specificity in the developed drug (e.g. the STI571 inhibitor designed to target BCR-ABL also shows some activity against the structurally related kinases c-kit and PDGF, extending the range of tumours against which the agent is active (Druker 2002)). No matter the degree of specificity sought in a screen design for PPI-targeted inhibitors, the use of informatic resources should increasingly aid in prediction and interpretation of mode of action.

Indirect modulation of PPIs

Finally, although the most obvious way to think about a PPI-targeted drug is to identify a drug that directly impedes the physical interaction of the two protein components, it is also conceivable to achieve the same functional effect by indirect means. For example, when the Ras protein interacts with and activates its effector Raf-1 to promote subsequent activation of the downstream effector MEK1, the activation process requires anchorage of Ras to the plasma membrane via farnesylated sequences at the carboxy-terminal tail of Ras. The primary rationale in developing farnesyltransferase inhibitors (FTIs), even though in practice the action of these compounds turned out to be more complex, was to impede the addition of the farnesyl groups to Ras, thus indirectly interfering with a productive Ras-Raf interaction and activation. Similarly, an agent that did not interfere with the interaction of Ras with Raf, but impaired the ability of the Ras-Raf-1 complex to activate MEK1, would simulate a PPI inhibitor. There are many examples of key PPIs for which the occurrence and subsequent transmission of biological signal are dependent upon the initial modification of one or both PPI participant proteins. While the main purpose of this review is to focus on issues for drugs directly targeting PPIs, indirect means should also be considered if a PPI is sufficiently important in a disease process.

Screening

There are many different approaches to screening for PPI-targeted drugs. Some involve computational docking simulations that can be performed using available protein target crystal or PPI co-crystal structures in conjunction with various chemical structures to probe the interaction of a chemical inhibitor with its proein binding domain (Bissantz *et al.* 2000). Others involve direct screening for disruption of a protein interaction using methodologies such as fluorescence polarization (Zhang *et al.* 2002), or the two-hybrid system (Kato-Stankiewicz *et al.* 2002). Other approaches are functionally based, and seek to identify reversion of PPI-induced signalling cascades, or otherwise functionally assess the consequence of drug treatment in regulating the PPI (Peterson *et al.* 2001). Potentially, small molecule PPI inhibitor leads may be identified via high-throughput screening of synthetic combinatorial chemistry libraries or natural product mixtures, virtual screening approaches, peptidomimetic or proteomimetic-based strategies, or structure-based or related rational drug design methods (Toogood 2002). The leads resulting from such screens can be optimized for potency and target selectivity through chemical modification based on the results of classical structure–activity relationship analysis. Numerous additional steps are required to ascertain stability, solubility, pharmacokinetics, bioavailability, metabolism and toxicology, and antitumour activity in animal models, during which it may be necessary further to modify the chemical structure of the inhibitor such that the above characteristics are favourable. These methodologies are beyond the scope of this chapter.

Progress in cancer treatment

Efforts to obtain small molecule inhibitors of PPIs have surged in the past several years. Given the length of the drug discovery and drug development processes, (see Chapter 13) at present few of these agents have progressed to evaluation in clinical trials: a listing of some promising agents at varying stages of development is presented in Figure 15.2. However, although a large stable of PPI targeted drugs in clinical trials is some distance away, other agents such as humanized antibodies and bioactive peptides and peptidomimetics designed to target individual proteins or PPIs, and inhibit specific signalling pathways, are currently under active evalu-

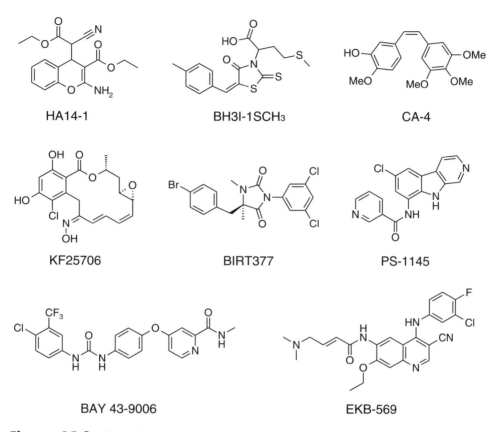

Figure 15.2 Examples of protein-and PPI-targeted small molecule inhibitors. HA14-1 and BH3I1SCH$_3$ inhibit binding of Bcl-2 family members (Huang *et al.* 2002; Lugovskoy *et al.* 2002); CA-4 is a tubulin polymerization inhibitor (Wang *et al.* 2002b); KF25706 and other radicicol derivatives are Hsp90 ligands (Agatsuma *et al.* 2002); BIRT377 is an allosteric LFA-1 antagonist (Last-Barney *et al.* 2001); PS-1145 is an NF-kB inhibitor (Ritzeler *et al.* 2001); BAY 43-9006 is a Raf kinase inhibitor (Lyons *et al.* 2001); and EKB-569 is an irreversible inhibitor of the EGFR kinase (Wissner *et al.* 2003)

ation. Results of some of these studies are discussed here as illustrative of the types of issues expected to be encountered with PPI-directed drugs.

Clinical trial design for inhibitors of PPI: detecting activity of non-cytotoxic drugs

Limited success achieved with cytotoxic agents using non-selective approaches of blocking global DNA and protein synthesis to induce temporary tumour regression, but at the cost of disabling toxicity. To what degree can therapeutic effects be expected with single agents that selectively target a single PPI, and how can we detect clinically important effects if not by the traditional benchmark of tumour regression? A realistic expectation of molecular targeted therapy is that inhibition of one key protein, one PPI, or even one signalling module may be achieved with minimal host toxicity, but without noticeable tumour regression. In animal models, most of these agents do not reduce tumour size, but can delay tumour growth or reduce the number of metastatic lesions that develop after tumour implantation.

Because a significant delay in tumour growth would still be considered beneficial to patients with cancer, clinical trials of so-called 'cytostatic agents' must be designed to detect benefit in ways other than by reduction in tumour size (Korn *et al.* 2001). If such benefit is to be measured as a delay in time to tumour progression, then a critical issue will be variability of the natural history of the disease untreated or after prior interventions. Several tumours, such as renal and prostatic carcinomas, are characterized by highly variable rates of tumour progression, so that enrichment procedures based on known prognostic factors may be required to restrict enrolment in a given trial to those patients with the most predictable courses (Korn *et al.* 2001). Although randomized trials utilizing a no treatment control arm may be attractive for determining the effect of treatment on survival or time to tumour progression, these studies are large and expensive. Thus, preliminary evidence of efficacy is usually sought in smaller phase II, non-randomized trials using a meaningful benchmark, such as a doubling of the expected time to progression, to signal success and to justify larger, definitive randomized trials.

Of practical concern in the clinical evaluation of targeted therapies is the interpretation of negative studies. If there is evidence that the intended molecular targeting has occurred in human tissues, then a lack of clinical response could reflect the underlying molecular heterogeneity of human tumours, even those of the same organ and histologic type. Identification of molecular markers or signatures of response from normal or tumour tissue in preclinical and clinical studies would help define the disease settings for which a given anti-tumour agent would have the greatest chance to provide benefit. However, these correlations have not yet been made for the majority of targeted therapies. Consequently, many clinical trials of targeted therapies are vastly undersized and underpowered to detect efficacy, assuming that a small proportion of patients with a given histologic tumour type are susceptible to a single agent (Betensky *et al.* 2002).

Examples of PPI inhibitors in clinical trials

Antibodies that block ligand-receptor binding and receptor–protein interactions

Overexpression and activation of oncoproteins is a hallmark of malignancy, thus providing a logical starting point for therapeutic targeting of PPI. In human breast cancer, amplification of *ERBB2* and overexpession of the ErbB2 protein contributes to tumour progression of the disease in 25 per cent of cases (Slamon *et al.* 1989). Among the first agents to interfere with PPI in the clinic was transtuzamab or Herceptin, a humanized monoclonal antibody that binds the ErbB2 receptor extracellular domain with high affinity (Carter *et al.* 1992). Although a specific ligand has not been identified for this receptor, when overexpressed, ErbB2 spontaneously dimerizes leading to activaton of the cytoplamic kinase domain and autophosphorylation. Moreover, ErbB2 heterodimerizes with the other three members of the epidermal growth factor receptor (EGF) family, and is the preferred partner for pairing after, for example, binding of EGF or TGFα to the ErbB1 (EGF) receptor (Olayioye *et al.* 2000). Binding of Herceptin to ErbB2 inhibits dimerization, receptor autophosphorylation, and interaction of receptor cytoplasmic domains with SH2 and PTB motifs of downstream signalling proteins. Approval of Herceptin was based primarily on a large, controlled clinical trial that demonstrated a greater proportion of responders and longer survival after treatment with herceptin plus chemotherapy compared with chemotherapy alone (Slamon *et al.* 2001). Herceptin in combination with paclitaxel has become a standard treatment for patients with metastatic breast cancer that overexpresses *ERBB2*.

A number of critical issues that have emerged from the clinical experience with herceptin will likely recur with other molecular targeted therapies. Of greatest importance was the observation that benefit was restricted to patients with tumours expressing the ErbB2 protein as determined in tumour tissue by strong immunostaining, or *ERBB2* gene amplification, by FISH. A second important finding was the association of Herceptin treatment with cardiomyopathy and congestive heart failure in 4.7 per cent of patients (Cobleigh *et al.* 1999), and the potentiation of anthracycline-associated cardiomyopathy by Herceptin in 28 per cent of patients (Slamon *et al.* 2001). A plausible mechanism for Herceptin cardiomyopathy emerges if the function of ErbB2 is considered. ErbB2 is essential for heart development, and ErbB2 knock-out yields a lethal phenotype in mice, with death at embryonic day 10.5 due to the inability of cardiac trebeculae to maintain blood flow (Lee *et al.* 1995). Of greater importance in the clinic, ErbB2 is protective against the development of dilated cardiomyopathy, as shown in ErbB2-deficient conditional mutant mice, a model in which deletion of the ErbB2 receptor tyrosine kinase is restricted to the heart. In this model, all animals develop dilated cardiomyopathy, and cardiac myocytes exhibit increased sensitivity to anthracyclines (Crone *et al.* 2002). Thus, therapy with an agent specifically targeted against an oncoprotein implicated in tumour progression may also disrupt PPI of the same oncoprotein in normal tissues. Fortunately, careful monitoring of cardiac function in patients receiving Herceptin, and avoidance of concurrent anthracycline therapy usually prevents clinically significant cardiomyopathy. Still, the unexpected cardiotoxicity associated with Herceptin should alert us to the existence of essential tissue PPIs of other oncogenic signalling proteins.

Small molecule inhibitors of receptor tyrosine kinases

Other antibodies and small molecule inhibitors have been developed to block PPI of the epidermal growth factor receptor (EGFr) (Herbst *et al.* 2002; Hidalgo *et al.* 2001), the vascular endothelial growth factor (VEGF), fibroblast growth factor (FGF), and platelet derived growth factors, c-abl kinase, and Flt1 receptors (Druker *et al.* 2001). The small molecule inhibitors compete with ATP binding in the cytoplasmic portion of the receptor protein, blocking tyrosine autophosphorylation and thus interaction with SH2 and SH3 domains of other proteins. As noted above, the clinical success of cAbl and cKit blockade by Gleevec in CML and GISTs, respectively (Druker *et al.* 2001; Demetri *et al.* 2002), parallels the central role played by these proteins in each disease. By contrast, the modest clinical effects of EGF inhibition are most likely a reflection of the more limited and variable contribution of EGFr activation to tumour growth and survival. Thus, the success of PPI targeted therapies should be commensurate to what is known about the molecular pathology of a given tumour type. In the case of CML, constitutive activation of the mutant c-Abl kinase is pathogenetic, and therapy aimed at this molecular lesion can induce disease remission.

Problems in the development of the EGFr inhibitor ZD1839 (Iressa) are instructive. This agent demonstrated modest but reproducible antitumour activity when used as a second or third line therapy of patients with non-small cell lung carcinoma, a population that could be considered highly selected based on their survival despite failure of prior chemotherapy. By contrast, two large, international randomized studies failed to show a survival advantage for the addition of Iressa to standard chemotherapy compared with chemotherapy alone as first-line treatment of patients with metastatic lung cancer. How can these results be reconciled? One possible explanation is that the population treated in the initial clinical trials was 'enriched' with a higher proportion of patients with EGF/EGFr-driven tumours. Another reason is that the modest effects observed in earlier trials were too small to affect survival, a much higher standard of drug efficacy. Unfortunately, tumour expression of EGFr has not correlated with response to treatment, and markers of successful treatment outcome with ZD1839 have not yet been identified.

Interfering with PPI by altering protein localization: Farnesyltransferase inhibitors

Many proteins require post-translational lipidation for membrane localization and function. The enzymes that catalyse transfer of 15 carbon farnesyl and 20-carbon geranylgeranyl lipid groups to proteins have been targeted with small molecule inhibitors. First to be evaluated in the clinic were inhibitors of farnesyltransferase (FTase), a heterodimer that transfers a 15-carbon farnesyl group to

Ras, Rap and other GTP-binding proteins. Initial interest in farnesyltransferase inhibitors (FTIs) was driven by the possibility of blocking the function of Ras proteins, which are mutated in 25 per cent of human tumours. Although this approach was greeted with enthusiasm based on the ability of FTIs to reverse malignancy in H- and N-Ras mutant transgenic mice, a number of problems were encountered in more genetically complex models. Although all ras isoforms can be farnesylated, K-Ras and N-Ras are preferentially geranylgeranylated by the enzyme geranylgeranyltransferase-I and can function in the absence of FTase. The alternative prenylation of these isoforms, coupled with the fact that K-Ras is the isoform most commonly mutated in human tumours, has called into question the likelihood that Ras function is significantly affected by FTIs. Further, in addition to Ras, over 100 proteins undergo farnesylation, and the key farnesylated protein(s) mediating FTI action have yet to be indentified.

Neurofibromatosis type 1 (NF-1) is an autosomal dominant, deforming disease affecting 1 in 4000 births and characterized by multiple neurofibromas, or peripheral nerve sheath tumours. In contrast to the complex molecular pathology accompanying Ras mutation in solid tumours, the molecular lesions underlying NF-1 consist of loss of function mutations of neurofibromin, a Ras GTPase activating protein (GAP) encoded by the *NF1* gene on chromosome 17 (Shen *et al.* 1996). Mutation of neurofibromin with loss of Ras GAP function increases the level of Ras signalling (Cichowski and Jacks 2001). Blocking Ras membrane localization and function by inhibiting farnesylation with an FTI can circumvent the loss of neurofibromin-Ras interaction that negatively regulates Ras (Yan *et al.* 1995), a therapeutic approach that is undergoing clinical evaluation.

Direct interference with PPI interaction with small peptides or small molecules

Another variation on PPI-targeted therapy with potential for a high degree of specificity is the use of peptides or small molecules to directly disrupt protein interactions. This approach is facilitated by knowledge of the amino acid sequence and surface characteristics of domains involved in the binding of a protein with one or more partners. Illustrating this concept is the disruption of the interaction between the extracellular matrix protein fibronectin and cellular integrin fibronectin receptors. Like the full-length protein, the fibronectin-derived pentapeptide Pro-His-Ser-Arg-Asn (PHSRN) interacts with the third amino-terminal domain of the α 5 chain of the α5β1 integrin fibronectin receptor (Burrows *et al.* 1999) and stimulates basement membrane invasion of DU145 prostate carcinoma cells. Substituting a cysteine for the arginine results in a pentapeptide competitive inhibitor, PHSCN, that blocks fibronectin-dependent tumour cell invasion *in vitro* and *in vivo*. The PHSCN inhibitor slows the rate of tumour growth at the site of implantation and dramatically reduces the number of lung metastases that develop in nude mice innoculated with DU145 tumour cells (Livant *et al.* 2000). Inhibition of the fibronectin-$\alpha5\beta1$ integrin interaction with a similar peptide is being tested in a Phase I clinical trial. This approach is an attractive one for drug development because it incorporates knowledge of protein sequence, structure, and function that can quickly lead to peptide inhibitors for initial proof of concept that a specific PPI should be targeted.

Perspective on early clinical trial results: desirability of combination therapies

With the exceptions STI-571 (Gleevec) in CML and GIST, the results of the first generation clinical trials of PPI inhibitors have been modest. However, a number of important lessons have been learned in the clinic that will insure greater success of future trials. Expectations that a single agent can have widespread activity in all tumours, or even a single 'type' of cancer have been tempered by appreciation of molecular heterogeneity for even a single histologic type of cancer, which predicts that tumours with specific molecular signatures will respond differently to a given inhibitor. Further, considering that many genetic mutations and altered proteins contribute to the unrestrained growth and invasiveness of most solid tumours, it is unlikely that a single therapy will reverse the malignant phenotype. As shown in experimental models (Wang *et al.* 2002a), inhibition of more than one receptor or signalling pathway frequently produces greater antitumor effects. There may be value in targeting multiple receptors at the cell periphery, as shown experimentally for dual inhibition of the EGF and ErbB2 receptors, which are known to heterodimerize (Normanno *et al.* 2002; Ye *et al.* 1999). The most effective combinations may require simultaneous inhibition of multiple signalling pathways as shown by Wang *et al.* in studies using a three-dimensional reconstituted basement membrane assay that preserves cell adhesion signalling. Inhibition of MAPK, PI3K and cell adhesion signalling was required to induce cell death or phenotypic reversion in multiple invasive breast carcinoma cell lines (Wang *et al.* 2002a). Hence, even though clinical trials of individual PPI- or pathway-targeting agents may not yield dramatic effects, the gradual development of a panel of agents that can be used in combination to target complementary oncogenic pathways is a source of considerable hope.

Issues in the development of drug combinations include determination of the optimal duration and order of drug administraton. The sequence in which agents are administered may be especially important when using an inhibitor of a signalling pathway with a cytotoxic agent. Cytotoxicity may be magnified – both against tumour and host – resulting from modification of the stress response, or by perturbation of cell cycle control. Conversely, cell cycle effects of one agent can interfere with the antitumor effects of a second agent, as exemplified by combinations of S-phase inhibitors and antimitotic agents (Shah and Schwartz

2001) In these combinations, exposure to the S-phase active agent preceding the antimitotic drug arrests cells in S-phase and reduces the population of G2-M phase cells that are vulnerable to taxanes and other antimitotic agents. Consequently, the net cell kill is less than additive. Conversely, additive or greater cytotoxicity results when cells are exposed to the antimitotic agent first (Bahadori *et al.* 2001; Koutcher *et al.* 2000). The combined effects of an inhibitor of PPI and a cytotoxic agent may be difficult to predict if each individual drug has more than one target or cellular effect. Since many PPIs affect both cell proliferation and survival, the interaction with a cytotoxic agent could be complex. Furthermore, the combined effects, like those of each single agent, may depend on the context of altered cell signalling. Thus, a tumour comprised of cells with mutations in *RAS* or *PTEN* may not be responsive to combined EGFr and ErbB2r blockade because the downstream MAPK and AKT pathways are constitutively active even without input from growth factor receptor signals.

Summary

PPI-directed drug discovery is still in its early days. Above all, this chapter seeks to emphasize the dependence of this endeavour on interactive research that combines insights from chemistry, structural biology, biochemistry, cell biology and signal transduction, genomics/proteomics, and clinical research. Each of these fields has separately made enormous advances in information resources and sophistication of analysis in the past decade, with these advances fuelling both the rationale for and ability to develop targeted drugs. It is to be hoped that the next decade sees the fruit of this work begin to transform disease treatment in the clinic.

References

Agatsuma, T., Ogawa, H., Akasaka, K., Asai, A., *et al.* (2002) Halohydrin and oxime derivatives of radicicol: synthesis and antitumor activities. *Bioorg Med Chem* **10**: 3445–3454.

Alarcon-Vargas, D. and Ronai, Z. (2002) SUMO in Cancer-Wrestlers Wanted. *Cancer Biol Ther* **1**: 237–242.

Alonso, M.A. and Millan, J. (2001) The role of lipid rafts in signalling and membrane trafficking in T lymphocytes. *J Cell Sci* **114**: 3957–3965.

Bahadori, H.R., Green, M.R. and Catapano, C.V. (2001) Synergistic interaction between topotecan and microtubule-interfering agents. *Cancer Chemother Pharmacol* **48**: 188–196.

Betensky, R.A., Louis, D.N. and Cairncross, J.G. (2002) Influence of unrecognized molecular heterogeneity on randomized clinical trials. *J Clin Oncol* **20**: 2495–2499.

Birrell, G.W., Giaever, G., Chu, A.M., Davis, R.W. and Brown, J.M. (2001) A genome-wide screen in *Saccharomyces cerevisiae* for genes affecting UV radiation sensitivity. *Proc Natl Acad Sci USA* **98**: 12608–12613.

Bissantz, C., Folkers, G. and Rognan, D. (2000) Protein-based virtual screening of chemical databases. 1. Evaluation of different docking/scoring combinations. *J Med Chem* **43**: 4759–4767.

Boulton, S.J., Gartner, A., Reboul, J., Vaglio, P., *et al.* (2002) Combined functional genomic maps of the *C. elegans* DNA damage response. *Science* **295**: 127–131.

Burrows, L., Clark, K., Mould, A.P. and Humphries, M.J. (1999) Fine mapping of inhibitory anti-alpha5 monoclonal antibody epitopes that differentially affect integrin–ligand binding. *Biochem J* **344**(2): 527–533.

Carter, P., Presta, L., Gorman, C.M., Ridgway, J.B., *et al.* (1992) Humanization of an anti-p185HER2 antibody for human cancer therapy. *Proc Natl Acad Sci USA* **89**: 4285–4289.

Cichowski, K. and Jacks, T. (2001) NF1 tumor suppressor gene function: narrowing the GAP. *Cell* **104**: 593–604.

Clark, E.A., Golub, T.R., Lander, E.S. and Hynes, R.O. (2000) Genomic analysis of metastasis reveals an essential role for RhoC. *Nature* **406**: 532–535.

Cobleigh, M.A., Vogel, C.L., Tripathy, D., Robert, N.J., *et al.* (1999) Multinational study of the efficacy and safety of humanized anti-HER2 monoclonal antibody in women who have HER2-overexpressing metastatic breast cancer that has progressed after chemotherapy for metastatic disease. *J Clin Oncol* **17**: 2639–2648.

Cochran, A.G. (2000) Antagonists of protein-protein interactions. *Chem Biol* **7**: R85–R94.

Cochran, A.G. (2001) Protein–protein interfaces: mimics and inhibitors. *Curr Opin Chem Biol* **5**: 654–659.

Crone, S.A., Zhao, Y.Y., Fan, L., Gu, Y., *et al.* (2002) ErbB2 is essential in the prevention of dilated cardiomyopathy. *Nat Med* **8**: 459–465.

DeLano, W.L. (2002) Unraveling hot spots in binding interfaces: progress and challenges. *Curr Opin Struct Biol* **12**: 14–20.

Demetri, G.D., von Mehren, M., Blanke, C.D., Van den Abbeele, A.D., *et al.* (2002) Efficacy and safety of imatinib mesylate in advanced gastrointestinal stromal tumors. *N Engl J Med* **347**: 472–480.

Dhillon, A.S. and Kolch, W. (2002) Untying the regulation of the Raf-1 kinase. *Arch Biochem Biophys* **404**: 3–9.

Draetta, G., Luca, R., Westendorf, J., Brizuela, L., *et al.* (1989) cdc2 protein kinase is complexed with both cyclin A and B: evidence for proteolytic inactivation of MPF. *Cell* **56**: 829–838.

Druker, B.J. (2002) Perspectives on the development of a molecular targeted agent. *Cancer Cell* **1**: 31–36.

Druker, B.J., Talpaz, M., Resta, D.J., Peng, B., *et al.* (2001) Efficacy and safety of a specific inhibitor of the BCR-ABL tyrosine kinase in chronic myeloid leukemia. *N Engl J Med* **344**: 1031–1037.

Edwards, A.M., Kus, B., Jansen, R., Greenbaum, D., *et al.* (2002) Bridging structural biology and genomics: assessing protein interaction data with known complexes. *Trends Genet* **18**: 529–536.

Foster, B.A., Coffey, H.A., Morin, M.J. and Rastinejad, F. (1999) Pharmacological rescue of mutant p53 conformation and function. *Science* **286**: 2507–2510.

Gavin, A.C., Bosche, M., Krause, R., Grandi, P., *et al.* (2002) Functional organization of the yeast proteome by systematic analysis of protein complexes. *Nature* **415**: 141–147.

Giaever, G., Chu, A.M., Ni, L., Connelly, C., Riles, L., *et al.* (2002) Functional profiling of the *Saccharomyces cerevisiae* genome. *Nature* **418**: 387–391.

Hajra, K.M. and Fearon, E.R. (2002) Cadherin and catenin alterations in human cancer. *Genes Chromosomes Cancer* **34**: 255–268.

Herbst, R.S., Maddox, A.M., Rothenberg, M.L., Small, E.J., *et al.* (2002) Selective oral epidermal growth factor receptor tyrosine kinase inhibitor ZD1839 is generally well-tolerated and has activity in non-small-cell lung cancer and other solid tumors: results of a phase I trial. *J Clin Oncol* **20**: 3815–3825.

Hidalgo, M., Siu, L.L., Nemunaitis, J., Rizzo, J., *et al.* (2001) Phase I and pharmacologic study of OSI-774, an epidermal growth factor receptor tyrosine kinase inhibitor, in patients with advanced solid malignancies. *J Clin Oncol* **19**: 3267–3279.

Huang, Z., Liu, D., Han, X., Zhang, Z. and Wang, J. (2002) Small molecule inhibitors of Bcl-2 proteins. *In* US6492389.

Ito, T., Chiba, T., Ozawa, R., Yoshida, M., Hattori, M. and Sakaki, Y. (2001) A comprehensive two-hybrid analysis to explore the yeast protein interactome. *Proc Natl Acad Sci USA* **988**: 4569–4574.

Janin, J., Miller, S. and Chothia, C. (1988) Surface, subunit interfaces and interior of oligomeric proteins. *J Mol Biol* **204**: 155–164.

Jones, S. and Thornton, J.M. (1995) Protein–protein interactions: a review of protein dimer structures. *Prog Biophys Mol Biol* **63**: 31–65.

Kato-Stankiewicz, J., Hakimi, I., Zhi, G. Zhang, J., *et al.* (2002) Inhibitors of Ras/Raf-1 interaction identified by two-hybrid screening revert Ras-dependent transformation phenotypes in human cancer cells. *Proc Natl Acad Sci USA* **99**: 14398–14403.

Kervinen, J., Dunbrack, R.L. Jr., Litwin, S., Martins, J., *et al.* (2000) Porphobilinogen synthase from pea: expression from an artificial gene, kinetic characterization, and novel implications for subunit interactions. *Biochemistry* **39**: 9018–9029.

Korn, E.L., Arbuck, S.G., Pluda, J.M., Simon, R., *et al.* (2001) Clinical trial designs for cytostatic agents: are new approaches needed? *J Clin Oncol* **19**: 265–272.

Koutcher, J.A., Motwani, M., Zakian, K.L., Li, X.K., *et al.* (2000) The in vivo effect of bryostatin-1 on paclitaxel-induced tumor growth, mitotic entry, and blood flow. *Clin Cancer Res* **6**: 1498–1507.

Last-Barney, K., Davidson, W., Cardozo, M., Frye, L.L., *et al.* (2001) Binding site elucidation of hydantoin-based antagonists of LFA-1 using multidisciplinary technologies: evidence for the allosteric inhibition of a protein–protein interaction. *J Am Chem Soc* **123**: 5643–5650.

Lee, K.F., Simon, H., Chen, H., Bates, B., Hung, M.C. and Hauser, C. (1995) Requirement for neuregulin receptor erbB2 in neural and cardiac development. *Nature* **378**: 394–398.

Lipinski, C.A., Lombardo, F., Dominy, B.W. and Feeney, P.J. (2001) Experimental and computational approaches to estimate solubility and permeability in drug discovery and development settings. *Adv Drug Deliv Rev* **46**: 3–26.

Livant, D.L., Brabec, R.K., Pienta, K.J., Allen, D.L., *et al.* (2000) Anti-invasive, anti-tumorigenic, and antimetastatic activities of the PHSCN sequence in prostate carcinoma. *Cancer Res* **60**: 309–320.

Lo Conte, L., Chothia, C. and Janin, J. (1999) The atomic structure of protein–protein recognition sites. *J Mol Biol* **285**: 2177–2198.

Lopez-Fanarraga, M., Avila, J., Guasch, A., Coll, M. and Zabala, J.C. (2001) Review: postchaperonin tubulin folding cofactors and their role in microtubule dynamics. *J Struct Biol* **135**: 219–229.

Lugovskoy, A.A., Degterev, A.I. Fahmy, A.F., Zhou, P., *et al.* (2002) A novel approach for characterizing protein ligand complexes: molecular basis for specificity of small-molecule Bcl-2 inhibitors. *J Am Chem Soc* **124**: 1234–1240.

Lyons, J.F., Wilhelm, S., Hibner, B. and Bollag, G. (2001) Discovery of a novel Raf kinase inhibitor. *Endocr Relat Cancer* **8**: 219–225.

Malissen, B. and Malissen, M. (1996) Functions of TCR and pre-TCR subunits: lessons from gene ablation. *Curr Opin Immunol* **8**: 383–393.

Meijer, L., Arion, D., Golsteyn, R., Pines, J., *et al.* (1989) Cyclin is a component of the sea urchin egg M-phase specific histone H1 kinase. *EMBO J* **8**: 2275–2282.

Miyamoto, S., Teramoto, H., Coso, O.A., Gutkind, J.S., *et al.* (1995) Integrin function: Molecular hierarchies of cytoskeletal and signalling molecules. *J Cell Biol* **131**: 791–805.

Neckers, L. (2002) Hsp90 inhibitors as novel cancer chemotherapeutic agents. *Trends Mol Med* **8**: S55–S61.

Normanno, N., Campiglio, M., De, L.A., Somenzi, G., *et al.* (2002) Cooperative inhibitory effect of ZD1839 (Iressa) in combination with trastuzumab (Herceptin) on human breast cancer cell growth. *Ann Oncol* **13**: 65–72.

O'Neill, G.M., Fashena, S.J. and Golemis, E.A. (2000) Integrin signaling: a new Cas(t) of characters enters the stage. *Trends Cell Biol* **10**: 111–119.

Olayioye, M.A., Neve, R.M., Lane, H.A. and Hynes, N.E. (2000) The ErbB signaling network: receptor heterodimerization in development and cancer. *Embo J* **19**: 3159–3167.

Peterson, J.R., Lokey, R.S., Mitchison, T.J. and Kirschner, M.W. (2001) A chemical inhibitor of N-WASP reveals a new mechanism for targeting protein interactions. *Proc Natl Acad Sci USA* **98**: 10624–10629.

Ritzeler, O., Stilz, H.U., Neises, B., Jaehne, G. and Habermann, J. (2001) Substituted indoles for modulation NF-κB activity. *In* WO0130774.

Ruest, P.J., Shin, N.Y., Polte, T.R., Zhang, X. and Hanks, S.K. (2001) Mechanisms of cas substrate domain tyrosine phosphorylation by fak and src. *Mol Cell Biol* **21**: 7641–7652.

Rust, W.L., Carper, S.W. and Plopper, G.E. (2002) The promise of integrins as effective targets for anticancer agents. *J Biomed Biotechnol* **2**: 124–130.

Sager, R. (1997) Expression genetics in cancer: shifting the focus from DNA to RNA. *Proc Natl Acad Sci USA* **94**: 952–955.

Shah, M.A. and Schwartz, G.K. (2001) Cell cycle-mediated drug resistance: an emerging concept in cancer therapy. *Clin Cancer Res* **7**: 2168–2181.

Shen, M.H., Harper, P.S. and Upadhyaya, M. (1996) Molecular genetics of neurofibromatosis type 1 (NF1). *J Med Genet* **33**: 2–17.

Slamon, D.J., Godolphin, W., Jones, L.A., Holt, J.A., *et al.* (1989) Studies of the HER-2/neu proto-oncogene in human breast and ovarian cancer. *Science* **244**: 707–712.

Slamon, D.J., Leyland-Jones, B., Shak, S., Fuchs, H., *et al.* (2001) Use of chemotherapy plus a monoclonal antibody against HER2 for metastatic breast cancer that overexpresses HER2. *N Engl J Med* **344**: 783–792.

Stites, W.E. (1997) Protein–protein interactions: interface structure, binding thermodynamics, and mutational analysis. *Chem Rev* **97**: 1233–1250.

Toogood, P.L. (2002) Inhibition of protein–protein association by small molecules: approaches and progress. *J Med Chem* **45**: 1543–1558.

Tuschl, T., Zamore, P.D., Lehmann, R., Bartel, D.P. and Sharp, P.A. (1999) Targeted mRNA degradation by double-stranded RNA in vitro. *Genes Dev* **13**: 3191–3197.

Uetz, P., Giot, L., Cagney, G., Mansfield, T.A., *et al.* (2000) A comprehensive analysis of protein–protein interactions in *Saccharomyces cerevisiae. Nature* **403**: 623–627.

Volker, M., Mone, M.J., Karmakar, P., van Hoffen, A., *et al.* (2001) Sequential assembly of the nucleotide excision repair factors in vivo. *Mol Cell* **8**: 213–224.

Wang, F., Hansen, R.K., Radisky, D., Yoneda, T., *et al.* (2002a) Phenotypic reversion or death of cancer cells by altering signaling pathways in three-dimensional contexts. *J Natl Cancer Inst* **94**: 1494–1503.

Wang, L., Woods, K.W., Li, Q., Barr, K.J., *et al.* (2002b) Potent, orally active heterocycle-based combretastatin A-4 analogues: synthesis, structure–activity relationship, pharmacokinetics, and in vivo antitumor activity evaluation. *J Med Chem* **45**: 1697–11711.

Weintraub, H., Davis, R., Tapscott, S., Thayer, M., *et al.* (1991) The myoD gene family: nodal point during specification of the muscle cell lineage. *Science* **251**: 761–766.

Weis, K. (2002) Nucleocytoplasmic transport: cargo trafficking across the border. *Curr Opin Cell Biol* **14**: 328–335.

Winzeler, E.A., Shoemaker, D.D., Astromoff, A., Liang, H., *et al.* (1999) Functional characterization of the S. cerevisiae genome by gene deletion and parallel analysis. *Science* **285**: 901–906.

Wissner, A., Overbeek, E., Reich, M.F., Floyd, M.B., *et al.* (2003) Synthesis and structure–activity relationships of 6,7-disubstituted 4-anilinoquinoline-3-carbonitriles. The design of an orally active, irreversible inhibitor of the tyrosine kinase activity of the epidermal growth factor receptor (EGFR) and the human epidermal growth factor receptor-2 (HER-2). *J Med Chem* **46**: 49–63.

Xu, C.W. and Luo, Z. (2002) Inactivation of Ras function by allele-specific peptide aptamers. *Oncogene* **21**: 5753–5757.

Yan, N., Ricca, C., Fletcher, J., Glover, T., *et al.* (1995) Farnesyltransferase inhibitors block the neurofibromatosis type I (NF1) malignant phenotype. *Cancer Res* **55**: 3569–3575.

Ye, D., Mendelsohn, J. and Fan, Z. (1999) Augmentation of a humanized anti-HER2 mAb 4D5 induced growth inhibition by a human-mouse chimeric anti-EGF receptor mAb C225. *Oncogene* **18**: 731–738.

Zhang, H., Nimmer, P., Rosenberg, S.H., Ng, S.C. and Joseph, M. (2002) Development of a high-throughput fluorescence polarization assay for Bcl-x(L). *Anal Biochem* **307**: 70–75.

Zwick, E., Bange, J. and Ullrich, A. (2002) Receptor tyrosine kinases as targets for anticancer drugs. *Trends Mol Med* **8**: 17–23.

16

Overview of Quantitative Structure–Activity Relationships (QSAR)

David A. Winkler

It is clear that the observable (macroscopic) properties of matter are connected in some, usually complex and non-linear way, with their atomic and molecular (microscopic) properties. The ability to model these relationships is very import-ant in many areas of science, particularly those concerned with the biological properties of molecules. Clearly a single residue change in a peptide sequence, a single atomic change in a small organic molecule, or even a change in the shape of a molecule induced by isomerism can profoundly affect the biological properties of these entities.

Modelling structure–activity relationships

The relationships between molecular structure and biological activities (struc-ture–activity relationships, or SARs) can be studied by very computationally intensive techniques such as molecular mechanics, molecular dynamics, or *ab initio* molecular orbital methods. Methods such as molecular docking and tran-sition state analogue design employ such techniques to understand relationships between molecular structure and biological properties in considerable detail. As well as being computationally demanding, they also often require considerable structural information about the protein target, or the biochemical reaction catalysed by an enzyme. Usually the relationships between structure and activity are so complex that many assumptions and simplifications must

Molecular Analysis and Genome Discovery edited by Ralph Rapley and Stuart Harbron
© 2004 John Wiley & Sons, Ltd   ISBN 0 471 49847 5 (cased)   ISBN 0 471 49919 6 (pbk)

be used in order to make the calculations tractable, and these methods are mainly suitable for modelling *in vitro* rather than *in vivo* data. The application of such techniques to the study of biological structure–activity relationships has been the subject of several recent reviews and books (e.g. Lipkowitz and Boyd 1990; Gubernator and Bohm 1998; Marrone and McCammon 1997 Klebe 2000).

An alternative way of finding complex structure–activity relationships is the subject of this chapter. QSAR (quantitative structure–activity relationships) involves finding an empirical relationship between the molecular properties of a series of molecules of differing structure, and their observed biological properties. This approach is essentially that of pattern recognition and is essentially 'model free', i.e. does not rely on an underlying model for the biological process. Such methods can be useful even when considerable molecular information is available about the biological target and how small molecules (e.g. drugs, or crop protection chemicals) interact with it, but are essential in the majority of cases when this information is not available.

QSAR was first developed by Hansch and Fujita in the 1960s (Hansch and Fujita 1964) and has since been widely used in discovering and optimizing new biologically-active leads, such as drug candidates. The early methods used by Hansch and Fujita have undergone considerable refinement over time. New techniques such as neural networks, fuzzy sets, support vector machines, and genetic algorithms have begun to replace linear statistical methods such as multiple linear regression (MLR) in many cases. In recent times QSAR methods have gained additional prominence because of their ability to generate predictive models for toxicological, physicochemical, and absorption, distribution, metabolism and excretion (ADME) properties. The pharmaceutical industry has recognized that such models are essential to build 'drug-like' behaviour and 'developability' into bioactive leads, resulting in fewer, very costly, downstream failures in the drug development process. This chapter introduces QSAR methods with an emphasis on several important, recent developments in the field. There are many excellent general reviews and textbooks on older, traditional methods of carrying out QSAR that the reader can consult (Grover *et al.* 2000; Martin 1981; Testa 2000; Fujita 1997; Boyd 1990). There are also many websites describing QSAR, molecular descriptors and their roles in bioactive discovery and development.

QSAR methods that deal with classification rather than quantitative model building (so-called qSAR methods) have been dealt with only briefly in this introductory chapter. Readers are referred to recent papers on classification methods for more in depth discussions of qSAR (e.g. Jürgen Bajorath 2001; Brown and Martin 1996).

The QSAR method

The QSAR method involves recognition that a molecule (be it a small organic compound, peptide, protein etc.) can be viewed as a three-dimensional distribution of molecular properties. At a basic level the most important of these properties are steric (e.g. molecular shape and volume), electronic (e.g. electric charge and electrostatic potential), and lipophilic properties (how polar or non-polar molecules are, usually represented by the log of the octanol–water partition coefficient (log P)). Scientists are used to visualizing mainly steric properties of molecules. However, molecules appear quite different when viewed in electrostatic, or lipophilic 'space', as Figure 16.1 illustrates. The relationship between molecular properties and a biological response of interest (for example, an LD_{50}, K_i, IC_{50} etc.) is usually non-linear, complex and unknown. This is also usually the case when the method is applied to the modelling of physicochemical or materials properties (Quantitative Structure–Property Relationships or QSPRs). This chapter will be concerned mainly with the application of QSAR to the quantitative modelling of biological properties.

Implementing the QSAR method involves carrying out a number of key steps:

- converting molecular structures into mathematical descriptors which encapsulate the key properties of the molecules relevant to the activity or property being modelled;

- selecting the most informative descriptors from a larger set of possible, relevant descriptors;

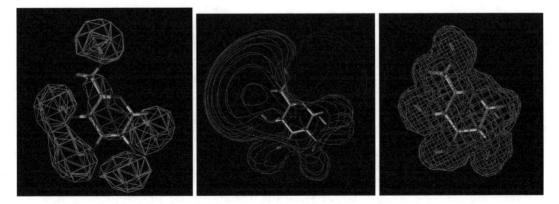

Figure 16.1 A representation of the same molecule in steric space (left), electrostatic space (centre), and lipophilic space (right)

- finding a model describing the relationship between the molecular descriptors and the biological properties (a mapping), preferably using a 'model free' mapping system in which no assumptions are needed as to the functional form of the structure–activity relationship;

- validating the prediction ability of the model, that is, how well it will generalize to new molecules not present in the data set used to generate the model (the training set).

The method is most easily understood by considering an example of a small molecule (e.g. a drug) interacting with a protein receptor or enzyme active site (Jordan *et al.* 2001). Figure 16.2 describes such an interaction. The enzyme is trihydroxynaphthalene reductase and the ligand is a nitroindanone. The ligand contains a number of structural features that interact with complementary features in the protein. In the example shown, the most important complementary interactions are between the hydrogen bond acceptor (the carbonyl group) on the ligand and the hydrogen bond donors on key active site residues serine164 and tyrosine178. There is also an important lipophilic interaction between the indane ring and residues lining the active site, especially tyrosine223, which can π-stack parallel to the aromatic ring in the ligand. These complementary groups interact in a way that is energetically favourable. Although ligands are constantly binding to, and leaving, the binding sites on proteins (i.e. it is an equilibrium process) good ligands spend a lot more time bound than free. The relationship betwen the binding free energy (ΔG) and the logarithm of the equilibrium constant, K_i, relating free and bound ligand concentrations, is described by a simple equation.

$$\Delta G = -RTlnK_i$$

where R is the gas constant and T is the temperature. Consequently, when we carry out QSAR studies we use (whenever possible) free energy related biological responses, or something approximating them. The above example shows a three-dimensional representation of interactions between ligand and protein. In QSAR we use descriptors that capture the essential features of these interactions in lower dimensional space and with less detail. Because the molecular properties of the molecules used in a QSAR study generally relate to the energy of interaction with the protein, we usually seek a relationship between the logarithm of the biological response BR (e.g. IC_{50}, LD_{50}, ED_{90}) and the molecular descriptors. Such a relationship can be expressed (Hansch and Fujita 1964) in the most general form as:

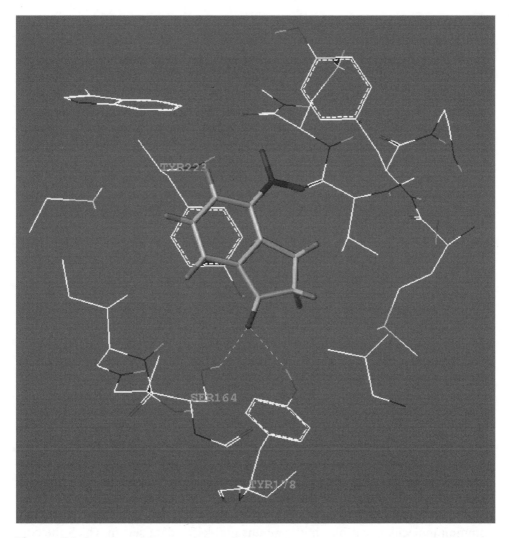

Figure 16.2 Nitroindane bound to the active site of trihydroxynaphthalene reductase showing molecular interactions between ligand and protein. The most important of these are the hydrogen bond interactions with serine 164 and tyrosine 178, and the π-stacking interaction with tyrosine 223

$$\log(\text{BR}) = f(x_1, \ x_2, \ \ldots x_N)$$

where f is usually an unknown, complex, nonlinear function, and $x_1, \ldots x_N$ are molecular descriptors. Building of a QSAR model via the four steps outlined above, involves finding the best form of function f.

Generation of descriptors

There are a myriad of methods for generating molecular descriptors. Packages such as Dragon (Todeschini and Consonni, 2000; (2001) are able to generate almost 1000 descriptors, while methods such as CoMFA (Cramer *et al.* 1988) (discussed below) generate many thousands. Molecular descriptors can be of diverse types. We have chosen to categorize them into fragment descriptors, involving properties of sections of molecules, and whole molecule descriptors, based on the properties of the intact molecule.

Fragment descriptors

The very earliest descriptors used in QSAR were of this type (Hansch and Fujita 1964; Hansch and Hoekman 1995). QSAR was performed using 'substituent constants' like hydrophobic constants π, molar refractivity MR, Hammett constants σ, and several other, less well-known constants. The hydrophobic constant π is related to the difference between the log of the partition coefficient of an unsubstituted molecule and that of a molecule with given substituent. It measures polar/non-polar character of the substituent. MR is a measure of size and polarizability of substituents and is derived from the Lorentz–Lorenz equation. It is a function of refractive index, density and molecular weight. The Hammett constant is a measure of electron withdrawing or electron-donating properties of substituents and was derived originally from the differences between the log of the ionization constant of benzoic acid and that of substituted benzoic acids. An alternative approach is the Free Wilson method (Free and Wilson 1964) where indicator variables were used to show whether a particular chemical group was present (1) or absent (0) at a given position on a chemical structure. Both of these methods worked well for fairly homologous series of molecules where there was a common molecular scaffold, but were unsuitable for data sets in which the core structure was different. Compilations of these and related substituents constants are available in the literature (Hansch and Hoekman 1995).

The recent explosion in the number of molecular descriptors is partly due to ease by which they may be generated by computational means, such as molecular orbital methods (Warne and Nicholson 2000; Karelson *et al.* 1996; Carbo-Dorca *et al.* 2000; recent examples where this has been done on a grand scale may be found in papers by Beck *et al.* 1998). There has also been a focus on developing fragment descriptors that are very computationally efficient. The reason is that rapid searching for leads in large chemical libraries (databases of real chemical compounds) or virtual libraries (databases of chemically reasonable molecules which have not yet been synthesized) require efficient, information-rich descriptors. Surprisingly simple descriptors can yield useful models. For example, molecules may represented simply by counting the numbers of atoms of specific elemental type, with specific numbers of connections (a measure of the hybridization) (Burden 1996; Winkler *et al.* 1998). Although simple, this representation is adequate to encode not only physicochemical parameters, such as lipophilicity and molar refractivity, but also biological activity (e.g. GABA$_A$ receptor activity of benzodiazepines). Subtle, higher level information is captured by these simple descriptors (Brown and Martin 1997).

A current trend is to employ fragment descriptors based on important molecular entities such as hydrophobes (e.g. aromatic rings), hydrogen bond donors (e.g. amines), hydrogen bond acceptors (e.g. carbonyls), positive charges (e.g. NH_4^+), and negative charges (e.g. PO_3^-). The rationale for this approach was first described by Andrews *et al.* (1984). Other fingerprint and general fragment based methods such as molecular holograms (Tong *et al.* 1998; Winkler 1998) generalize this approach of breaking molecules into fragments. Another important class of fragment-based descriptors, the van der Waals Surface Area descriptors (VSA), have been reported by Labute (2000) to have attributes that makes them widely applicable QSAR descriptors. VSA descriptors are derived by adding together the VDW surface area contributions of atoms exhibiting a given property (chosen from steric, electrostatic and lipophilic properties) within a given binned property range. Linear combinations of VSA descriptors correlate well with most other commonly used descriptors.

Fragment-based descriptors have advantages of being computationally efficient and are usually independent of molecular conformation or 3D structure. However, they are usually not as informative, being relatively poor at accounting for effects of stereochemistry in activity.

Whole molecule descriptors

These descriptors typically capture information on molecular size and lipophilicity via properties such as the molecular weight or molecular volume and log of the octanol–water partition coefficient (log P). Log P in particular is found to be important in many QSAR models (Leo and Hansch 1999). The relationship between log P and some biological responses is often an inverted parabola, in which a maximum in the biological response occurs at some optimum log P value. The explanation for this relationship is that it describes the partitioning of drug molecules into biological membranes. Hansch *et al.* (2001) recently reviewed classes of QSAR problems for which hydrophobicity was not an important factor.

Another important class of descriptors is based on treating molecules as molecular graphs. Topological indices (Balaban 2001; Devillers 2000) are derived from molecular graphs in which atoms are the vertices, and bonds the edges. Figure 16.3 shows the conversion of a molecular structure into a molecular graph.

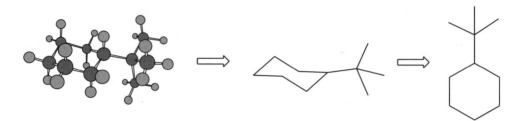

Figure 16.3 Relationship between a molecule and its molecular graph. Three-dimensional structure (left), two-dimensional, hydrogen suppressed structure (centre), and hydrogen-suppressed molecular graph (right)

It is possible to characterize molecular graphs using a number of indices. The most well known of these are the Randic indices (Kier and Hall 1995), and the Kier and Hall electrotopological indices (Hall and Kier 1995). In essence, these indices describe molecules in terms of connected paths through the hydrogen-suppressed molecular graph. For instance, the zeroth and first order Randic indices are:

$$^0\chi = \sum_{i=1}^{n} \delta_i^{-1/2}$$

$$^1\chi = \sum_{s=1}^{Ne} (\delta_i \delta_j)^{-1/2}$$

where the δ_i represent the hydrogen-suppressed valence of vertex (atom) i, n is the number of vertices (atoms) and N_e the number of edges (bonds). Higher order indices are calculated from progressively longer paths in the molecule. The Kier and Hall E-state indices extended this idea to include the effects of bond orders and electronegativity of atoms, and the influence of more distant atoms in a molecule (the 'field') (Kier and Hall 1999).

Recently, eigenvalues (graph invariants) of molecular matrices derived from chemical graphs have shown promise as descriptors for QSAR (Burden 1997; Stanton 1999; Randíc et al. 2001) and for molecular diversity purposes (e.g. characterization of chemical libraries and databases, and for design of optimally-diverse combinatorial libraries) (Pearlman and Smith 1998). Modified adjacency matrices describe how atoms in a molecule are connected. They provide a means of combining the molecular properties with topological information encoding the way a molecule is connected. Figure 16.4 shows an example of a modified adjacency matrix. Several recent studies (e.g. Stanton 1999) have shown that they are effective QSAR descriptors that encode molecular information not provided by other descriptors.

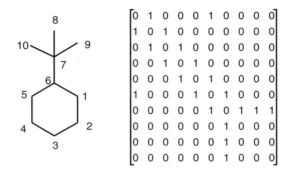

$$\begin{bmatrix} 0 & 1 & 0 & 0 & 0 & 1 & 0 & 0 & 0 & 0 \\ 1 & 0 & 1 & 0 & 0 & 0 & 0 & 0 & 0 & 0 \\ 0 & 1 & 0 & 1 & 0 & 0 & 0 & 0 & 0 & 0 \\ 0 & 0 & 1 & 0 & 1 & 0 & 0 & 0 & 0 & 0 \\ 0 & 0 & 0 & 1 & 0 & 1 & 0 & 0 & 0 & 0 \\ 1 & 0 & 0 & 0 & 1 & 0 & 1 & 0 & 0 & 0 \\ 0 & 0 & 0 & 0 & 0 & 1 & 0 & 1 & 1 & 1 \\ 0 & 0 & 0 & 0 & 0 & 0 & 1 & 0 & 0 & 0 \\ 0 & 0 & 0 & 0 & 0 & 0 & 1 & 0 & 0 & 0 \\ 0 & 0 & 0 & 0 & 0 & 0 & 1 & 0 & 0 & 0 \end{bmatrix}$$

Figure 16.4   Conversion of a 10-atom molecule into a 10×10 adjacency matrix. Off-diagonal elements are 1 if the two atoms are bonded, 0 if not. Eigenvalues of modified adjacency matrices are useful QSAR descriptors

The descriptors discussed so far encode two-dimensional representations of molecules (e.g. molecular graphs). Several widely used molecular descriptors types are calculated from three-dimensional properties of molecules. The CoMFA (comparative molecular field analysis) method (Cramer *et al.* 1988) surrounds the three-dimensional structure of a molecule by an array of grid points. A probe atom is positioned at each grid point outside of the molecule and the steric, electrostatic and (optionally) lipophilic fields calculated at that point. This process generates a molecular field representation of molecules in a training data set. The large number of descriptors this method generates must be dealt with by special regression techniques such as partial least squares (PLS) analysis. Field-based methods have been very useful in finding QSAR models. However, all three-dimensional QSAR methods suffer from the problems of requiring a valid 3D structure (assumptions must be made concerning the biologically active conformer or shape), and an alignment rule to superimpose molecules in the training set in a meaningful way.

Many other types of whole molecule descriptors have been devised ranging from those describing molecular shape, through molecular similarity, to molecular autocorrelation functions. Several recent papers have compared the efficacy of several types of molecular descriptors (Dearden and Ghafourian 1999; Estrada and Molina 2001).

Whole molecule descriptors are the most information dense and usually the most informative. QSAR models derived from 3D molecular descriptors are often more easily interpreted than are those from fragment descriptors or 2D whole molecule descriptors (e.g. topological indices). They are able to account for effects of conformation and isomerism on activity, but this flexibility comes at a computational cost.

The development of new descriptors is an active research area in the relatively few major groups who work on discovery of new QSAR and chemometrics methods. There is still considerable scope to discover more information rich, more generally applicable molecular descriptors than the myriad of those now available.

Descriptor selection

To build a good QSAR model, a minimal set of information-rich descriptors is required. The large number of possible indices creates several problems for the modeller (Topliss and Edwards 1979; Manallack and Livingston 1992):

- many descriptors do not contain molecular information relevant to the problem;

- many descriptors are linearly dependent (contain essentially the same information as other descriptors);

- use of poor descriptors in QSAR yields poor and misleading models;

- using too many descriptors in the model, even if they contain relevant information, can result in overfitting of the model, and loss of ability of the model to generalize to unseen molecules;

- many methods of screening this large pool of potential descriptors for relevant ones can lead to chance correlations (correlations which arise by chance because so many descriptors have been tried in models). In other words, if a large number of random numbers are generated as potential descriptors (which clearly do not contain any useful molecular information), and various subsets of these are used to build models, apparently significant models can arise by chance.

These factors have all led, at various times, to poor models being published in the literature. It is important to consider these points when building a QSAR model, and to employ methods that we discuss in more depth (*vide infra*) that avoid these pitfalls. Variable selection involves choosing descriptors containing relevant information, either by experience, variable reduction methods, or intelligent selection methods.

Knowledge of the biological process being modelled can provide insight into the types of descriptor that are likely to be important. For instance, in modelling toxicity due to uncoupling of mitochondrial oxidative phosphorylation, the mechanism of uncoupling suggests that lipophilic weakly acidic molecules are likely to be most effective. Consequently, an experienced modeller would choose descriptors correlating with these molecular properties.

One of the earliest methods of variable selection was stepwise regression. This was integrated with the model building process and involved stepwise addition (or backwards elimination) of descriptors according to a statistical test, to find the best model. Another, widely used variable reduction method is principal components analysis (PCA). This involves creating a smaller set of new orthogonal descriptors from linear combinations of the original descriptors and using these to generate QSAR models. Many of the papers, reviews and textbooks on QSAR referred to in this paper contain descriptions of PCA (e.g. Eriksson *et al.* 2001).

Most research into variable selection methods is in the area of genetic algorithms (Yasri and Hatsough 2001; Hou *et al.* 1999; Waller and Bradley 1999; Kimura and Funatsu 1999). These methods start with a pool of possible descriptors (often having undergone some rational preprocessing) and take various combinations of these, chosen according to a selection operator. These combinations constitute a population of possible descriptor sets that are evolved under genetic selection rules to find the set that generates the best QSAR model. Evolutionary pressure is applied using a fitness function (often a measure of the validity of the QSAR model), and population members who are less fit (produce bad models) are eliminated.

An alternative approach is to use Bayesian statistics to rank input descriptors according to relevance to the model, effectively removing uninformative descriptors. This method, known as automatic relevance determination (ARD) (Burden *et al.* 2000), has the advantage of being a non-linear process with a sound statistical basis. Choosing descriptor sets for a non-linear QSAR model using a linear method such as PCA is not optimal. The reader is referred to recent reviews (Zheng and Tropsha 2000; Bajorath 2001) that address the variable selection methods and issues in greater detail.

Structure–activity mapping

Many methods have been used to map molecular descriptors to biological activities to produce quantitative models (QSARs). The majority are regression methods, of which multiple linear regression is the simplest and the one first used (Hansch and Fujita 1964). Regression methods attempt to fit a specific function with free parameters to a set of data. This is usually achieved by a gradient descent method such as least squares, that finds the best set of free parameters which minimize the sum of the squares of the errors between the measured values of the dependent variables, and those calculated by the fitted function. Some QSAR problems have relatively linear response surfaces that can be modelled successfully by linear regression methods. For example, Abraham's group (Zhao *et al.* 2001) found simple linear relationships between some free energy-related molecular descriptors and the intestinal absorption of a wide range of drugs.

$$\%Abs = 93 - 0.386IS - 21.3A - 19.0B + 14.6V$$

$$n = 180, \ r^2 = 0.74, S = 15\%$$

where per cent Abs is the percent absorption of the drug in the intestine, A and B are the hydrogen bond acidity and basicity, V is related to the volume of the molecule, IS is the percent insoluble drug in water, n is the number of compounds in the model, r^2 is the square of the correlation coefficient, and S is the standard error of the model.

However, most QSAR problems involve at least some degree of non-linearity. Initially, this was tackled using bilinear, exponential, power law, or polynomial regression to find relationship. However, this required some subjectivity by the researcher in choosing the functional form of the relationship used in the model (Constans and Hirst 2000) and created problems because the true, complex nature of the relationship was often not found. Frequently many models needed to be created until the 'best' one emerged, due to the subjective nature of the choice of functional relationship. Methods such as genetic programming have been used to automate this process by evolving different functional forms, using different descriptors, to find the optimum models. This must be done very carefully to avoid problems with chance correlations (seemingly good models with no predictive power) (Topliss and Edwards 1979).

In the last decade, neural networks have emerged as the most useful way to overcome many of these shortcomings (see for example, Salt *et al.* 1992; Tetko *et al.* 1993; Burns and Whitesides 1993; Gasteiger and Zupan 1993; Aoyama *et al.* 1990; Maggiora *et al.* 1992). Neural networks (NNs) are computer-based mathematical models developed to have analogous functions to idealized simple biological nervous systems (illustrated in Figure 16.5). They are regression methods that automatically learn the functional relationships between molecular structure and properties without any subjective input from the researcher. They consist of layers of processing elements that are analogous to the nerve cells (neurons). These are interconnected to form a network that is, in essence, a parallel computer (Rumelhart and McClelland 1986) even though neural networks are most likely to be simulated in software on non-parallel computers such as PCs or workstations.

The parallel processing nature of the neural networks give them the characteristics of speed, reliability and generalization. The speed occurs because many bits of information can be input and analysed simultaneously. Reliability occurs because the networks can produce reasonable results even when some input data are missing or inaccurate. Generalization is the ability of the network to estimate reasonable results when faced with new data outside its normal range of experience. They are often capable of producing better generalizations than conventional methods, since they are able to model a complicated, non-linear activity surface better.

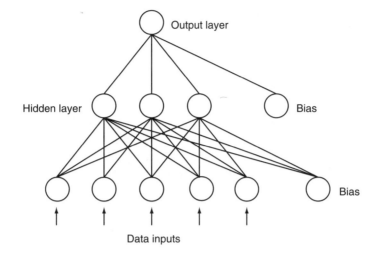

Figure 16.5 A schematic of a typical back-propagation neural network. The input neurodes distribute the data across the hidden layer. The inputs to the hidden layer are weighted, with weights being set during training. The hidden layer neurodes carry out a non-linear transformation of the weighted, summed inputs and distribute their outputs to the output layer. The output layer carries out a weighted sum to provide the predicted output value

Neural networks applied to QSAR problems are trained to build internal functions relating patterns of inputs to outputs (supervised training). After learning these relationships they are able to classify patterns and make decisions or predictions based upon a new patterns of inputs. Another mode of training is unsupervised (the network sees only the inputs, not the corresponding outputs). Such networks can classify a set of input patterns into a specified number of classes and are used for clustering and projection problems.

Structure of an artificial neural network

The basic unit in the neural net is the processing element or neuron/neurode that typically has many data inputs, each modified by a gain factor called a weight (Figure 16.5). The inputs are summed and modified by an internal transfer function, which may be a threshold (step) function, a linear, or non-linear (typically sigmoidal) function. The neurode output signal is then provided as an input to other neurodes.

The practical effect of the layers of interconnected neurons is to form a network like structure where the connection weights between the neurons each hold a part of the knowledge of how the network will reach a decision from a given input. Adjustment of these connection weights is therefore considered to be training the network. The most frequently used neural network for biomedical applications, including QSAR studies (Salt *et al.* 1992), is a supervised type called the back-propagation neural network. This has an excellent ability to predict and classify data. In the back propagation algorithm the adjustment of the weights is accomplished by determining the error in the output neuron in the training phase, then propagating this error back through the network as a means of correcting the weights.

From a functional point of view the dispersal of the knowledge over the whole of the network is considered to one of the main reasons why the networks are excellent at pattern recognition. By analysing the distribution of connection weights it has been possible to show that after training, various parts of the network specialize in detecting specific parts of a pattern to be recognized (Rumelhart and McClelland 1986). Thus a complex pattern can be recognized because various parts of the network each learn to recognize a smaller, simpler part of the whole.

The non-linear output from each neuron is considered to be another important feature of neural networks. Back-propagation neural networks are 'universal approximators', that is, in principle they can approximate any continuous function to arbitrary accuracy (Hornik *et al.* 1988). Consequently, neural networks are useful in predicting results from complex data interactions involving multiple variables where the inter-relationships among them are poorly understood or fuzzy. Back-propagation neural networks, in particular, have been very successful in modelling complex structure–activity relationships as they overcome many of the shortcomings of other methods of structure–activity modelling.

However, neural networks left some structure–activity mapping shortcomings unanswered and introduced additional problems (Manallack and Livingston 1992; 1999). Like most other regression methods, neural networks can over-fit the data. They can also be over-trained, a situation where they get progressively better at predicting the behaviour of the training set, but worse at predicting the behaviour of molecules not used in training. Finding the optimum architecture for the neural net is also a subjective, time-consuming problem. When neural nets are trained several times using the same training data, and random initial weights in the neural nets, they do not always train to the same model, because they can find local minima in the biological response surface.

One solution is to use a special kind of back-propagation neural net, the Bayesian regularized neural net, to build structure–activity models (MacKay 1992; MacKay 1995; Burden and Winkler 1999). This is a very robust regression method which is resistant to overtraining, automatically optimizes the neural net architecture, and can use all of the available data as cross-validation is not necessary. It produces a single mathematically optimum model. Because it uses neural networks the method is a very good generalized mapping methods that can automatically learn quite complex structure–property relationships. It is also capable of learning a number of models simultaneously. For example, a Bayesian neural network has created a single toxicity model containing four distinct sub-models explaining different mechanisms of toxicity (Burden and Winkler 2000).

Alternative methods of mapping molecular structure to biological activity include recursive partitioning (Rusinko *et al.* 1999), Gaussian processes (Burden 2001), radial basis functions (RBF) (Mayer-Bäse and Watzel 1998), and other types of neural networks (Burden 1998).

The Support Vector Machine (SVM), a classification algorithm widely used in pattern recognition, is showing considerable promise in finding good qSAR (classification) models in a very computationally efficient manner (Burbridge *et al.* 2001). The SVM chooses from increasingly complex hypothesis spaces those that can simultaneously minimize training and generalization errors. Burbridge and co-workers compared SVM with back-propagation neural networks, RBF and recursive partitioning and found they were superior to the latter two methods and as effective at the neural net (although quicker to train).

Validation and testing

It is important to know how well a QSAR model can generalize, that is, how well it can predict the biological activity of a new molecule outside of the training set. It is important to identify whether a structure–activity modelling technique has over-fitted the data, whether a neural net has been over-trained, or that chance correlations are present. Several methods have been developed to estimate the validity or predictivity of the derived structure–property model. The most common method is 'leave-one-out' cross-validation (Wold 1991). This involves leaving each molecule out of the training set in turn, then creating a model using the remainder of the training set. The property of the omitted molecule is predicted using the model derived from all of the other molecules. This method is not a very rigorous test of the predictivity of the model and suffers from several disadvantages: the time to carry out cross-validation increases as the square of the size of the training set; the method produces N final models (each corresponding to one of the training set molecules being left out) and it is not clear which is the 'best' model; the molecules left out and predicted are not truly independent as they occur in the training set N–1 times.

A better method is to remove a percentage of the training set into a test set (Tetko *et al.* 1995). The structure–property model is derived using the reduced training set, and the properties of the test set predicted using this model. This is a more rigorous test of the quality of the structure–property model but again suffers from problems: not all of the available data can be used to make the model as some must be held back for the test set; it is not clear how the test set is best selected from the training set, e.g. randomly, using cluster analysis etc.

As Bayesian regularized neural nets do not strictly require a validation set, they overcome the validation bottleneck inherent in other regression methods. Being able to use all of the data to generate the model is an advantage, especially where data sets are small and generation of additional data impossible or expensive. Several seminal papers and reviews provide clear descriptions of the methods and pitfalls of validating QSAR models (Livingston *et al.* 1997; Manallack and Livingston 1992).

Applications and conclusions

QSAR has been applied extensively and successfully over several decades to discover and optimize new bioactive agents such as pharmaceuticals, veterinary drugs and crop protection agents. It has recently grown in prominence as a method for modelling the important drug-like properties of molecules. Such properties are called ADMET properties (Absorption, Distribution, Metabolism, Excretion and Toxicity). Failure to optimize these properties is the major cause of expensive late failure of drug candidates. Examples of recent work in this important area includes toxicity prediction (Cronin 2000, Freidig and Hermens 2001) physicochemical properties prediction (e.g. water solubility, lipophilicity) (Beck *et al.* 2000; Xing and Glen 2002), gastrointestinal absorption (Zhao *et al.* 2001; Agatonovic-Kustrin *et al.* 2001; Benigni *et al.* 2000), activity of peptides (Brusic *et al.* 2001), data mining, drug metabolism (Lewis 2000), and prediction of other pharmacokinetic and ADME properties. Recent reviews (e.g. Podlogar and Ferguson 2000; Vedani and Dobler 2000) summarize work in a number of these areas and Winkler and Burden (2002) have summarized the application of neural networks to combinatorial discovery. The journal *Quantitative Structure–Activity Relationships* contains abstracts of QSAR studies from other journals in each issue.

It is clear that the number of potential applications for structure–property modelling, in the most general case, are extensive and growing daily. Improved molecular descriptors, based on a better understanding of which molecular attributes are most important for a given property being modelled, and increasing use of genetic and artificial intelligence methods will raise QSAR to even greater levels of usefulness than the current high level. A basic understanding of QSAR concepts is essential for most people, across a diverse range of skills, who work in molecular design.

References

Agatonovic-Kustrin, S., Beresford, R., Yusof, A. and Pauzi M. (2001) Theoretically-derived molecular descriptors important in human intestinal absorption, *J Pharm Biomed Anal* **25**: 227–237.

Andrews, P.R., Craik, D.J. and Martin, J.L. (1984) Functional group contributions to drug-receptor interactions, *J Med Chem* **27**: 1648–1657.

Aoyama, T., Suzuki, Y. and Ichikawa, H. (1990) Neural networks applied to quantitative structure–activity relationships, *J Med Chem* **33**: 2583–2590.

Bajorath, J. (2001), Selected concepts and investigations in compound classification, molecular descriptor analysis, and virtual screening, *J Chem Inf Comput Sci* **41**: 233–245.

Balaban, A.T. (2001) A personal view about topological indices for QSAR/QSPR, *QSPR/QSAR Stud. Mol. Descriptors*, 1–30.

Beck, B., Horn, A., Carpenter, J.E. and Clark, T. (1998), Enhanced 3D-databases: a fully electrostatic database of AM1-optimized structures, *J. Chem. Inf. Comput. Sci.* **38**: 1214–1217.

Beck, B., Breindl, A. and Clark, T. (2000) QM/NN QSPR models with error estimation: vapor pressure and logP, *J Chem Inf Comput Sci* **40**: 1046–1051.

Benigni, R., Giuliani, A., Franke, R. and Gruska, A. (2000) Quantitative structure–activity relationships of mutagenic and carcinogenic aromatic amines. *Chem Rev* **100**: 3697–3714.

Boyd, D.B. (1990) 'Successes of computer-assisted molecular design,' In Lipkowitz K, Boyd DB, ed., *Reviews in Computational Chemistry*, Vol. 1, VCH, New York pp. 355–371.

Brown, R. D. and Martin, Y. C. (1996) Use of structure–activity data to compare structure-based clustering methods and descriptors for use in compound selection. *J Chem Inf Comput Sci* **36**: 572–584.

Brown, R.D. and Martin, Y.C. (1997) The information content of 2D and 3D structural descriptors relevant to ligand-receptor binding. *J Chem Inf Comput Sci* **37**: 1–9.

Brusic, V., Bucci, K., Schönbach, C., Petrovsky, N., *et al.* (2001) Efficient discovery of immune response targets by cyclical refinement of QSAR models of peptide binding. *J Mol Graph Mod* **19**: 405–411.

Burbidge, R., Trotter, M., Buxton, B. and Holden, S. (2001) Drug design by machine learning: support vector machines for pharmaceutical data analysis. *Comput Chem* **26**: 5–14.

Burden, F.R. (1996) Using artificial neural networks to predict biological activity from simple molecular structural considerations. *Quant Struct-Activ Relat* **15**: 7–11.

Burden, F.R. (1997) A chemically intuitive molecular index based on eigenvalues of a modified adjacency matrix. *Quant Struct-Act Relat* **16**: 309–314.

Burden, F.R. (1998) Holographic neural networks as non-linear discriminants for chemical applications. *J Chem Inf Comput Sci* **38**: 47–53.

Burden, F.R. (2001) Quantitative structure–activity relationship studies using Gaussian processes. *J Chem Inf Comput Sci* **41**: 830–835.

Burden, F.R. and Winkler D.A. (1999) Robust QSAR models using Bayesian regularized neural networks. *J Med Chem* **42**: 3183–3187.

Burden, F.R. and Winkler, D.A. (2000) A QSAR Model for the acute toxicity of substituted benzenes towards *Tetrahymena pyriformis* using Bayesian regularized neural networks. *Chem Res Toxicol* **13**: 436–440.

Burden, F.R., Ford, M., Whitley, D. and Winkler, D.A. (2000) The use of automatic relevance determination in QSAR studies using Bayesian neural nets. *J Chem Inf Comput Sci* **40**: 1423–1430.

Burns, J.A. and Whitesides, G.M. (1993) Feed-forward neural networks in chemistry: mathematical systems for classification and pattern recognition. *Chem Rev* **93**: 2583–2601.

Carbo-Dorca, R., Amat, L., Besalu, E., Girones, X. and Robert. D. (2000) Quantum mechanical origin of QSAR: theory and applications. *Theochem* **504**: 181–228.

Constans, P. and Hirst, J.D. (2000) Nonparametric regression applied to quantitative structure–activity relationships. *J Chem Inf Comput Sci* **40**: 452–459.

Cramer, R.D., Patterson, D.E. and Bunce, J.D. (1988) Comparative Molecular Field Analysis (CoMFA). 1. Effect of shape on binding of steroids to carrier proteins. *J Am Chem Soc* **110**: 5959–5967.

Cronin, M.T.D. (2000) Computational methods for the prediction of drug toxicity. *Curr Opin Drug Discovery Dev* **3**: 292–297.

Dearden, J.C. and Ghafourian, T. (1999) Hydrogen bonding parameters for QSAR: comparison of indicator variables, hydrogen bond counts, molecular orbital and other parameters. *J Chem Inf Comput Sci* **39**: 231–235.

Devillers, J. (2000) New trends in (Q)SAR modeling with topological Indices. *Curr Opin Drug Discovery Dev* **3**: 275–279.

Eriksson, L., Johansson, E., Kettaneh-Wold, N. and Wold, S. (2001) *Multi- and Megavariate Data Analysis*, Umetrics Academy, Umea.

Estrada, E. and Molina, E. (2001) 'QSPR/QSAR by Graph Theoretical Descriptors Beyond the Frontiers,' in M. V. Diudea, ed., *QSPR/QSAR Studies by Molecular Descriptors,* Nova Science, Huntington, New York, pp. 83–107.

Free, S.M. and Wilson, J.W. (1964) A mathematical contribution to structure–activity studies. *J Med Chem* **7**: 395–399.

Freidig, A.P. and Hermens, J.L.M. (2001) Narcosis and chemical reactivity QSARs for acute fish toxicity. *Quant Struct–Act Relat* **19**: 547–553.

Fujita, T. (1997) Recent success stories leading to commercializable bioactive compounds with the aid of traditional QSAR procedures. *Quant Struc–Act Relat* **16**: 107–112.

Gasteiger, J. and Zupan, J. (1993) Neural networks in chemistry. *Angew Chem Int Ed Engl* **32**: 503–527.

Grover, M., Singh, B., Bakshi, M. and Singh S. (2000) Quantitative structure–property relationships in pharmaceutical research – Part 1. *Pharm Sci Technol Today* **3**: 28–35.

Gubernator, K. and Bohm, H.-J. (eds) (1998) *Structure-based Ligand Design*, Vol. 6, Wiley, Weinheim, p. 147.

Hall, L.H. and Kier, L.B. (1995) Electrotopological state indices for atom types: a novel combination of electronic, topological and valence state information. *J Chem Inf Comput Sci* **35**: 1039–1045.

Hansch, C. and Fujita, T. (1964) ρ-σ-π Analysis. A method for the correlation of biological activity and chemical structure. *J Am Chem Soc* **86**: 1616.

Hansch, C., Leo, L. and Hoekman, D. (1995), in Heller SR, ed., *Exploring QSAR. Hydrophobic, Electronic, and Steric Constants*, ACS, Washington.

Hansch, C., Kurup, A., Garg, R. and Gao, H. (2001) Chem-bioinformatics and QSAR: a review of QSAR lacking positive hydrophobic terms. *Chem Rev* **101**: 619–672.

Hou, T.J., Wang, J.M., Liao, N. and Xu, X.J. (1999) Applications of genetic algorithms on the structure–activity relationship analysis of some cinnamamides. *J Chem Inf Comput Sci* **39**: 775–781.

Jordan, D., Basarab, G., Liao, D.-I., Johnson, W.M.P., *et al.* (2001) Structure-based design of inhibitors of the rice blast fungal enzyme trihydroxynaphthalene reductase (3HNR). *J Mol Graphics Modelling* **19**: 434–447.

Jürgen Bajorath, J. (2001) Selected concepts and investigations in compound classification, molecular descriptor analysis, and virtual screening. *J Chem Inf Comput Sci* **41**: 233–245.

Hornik, K., Stinchcombe, M. and White, H. (1988) Multilayer feedforward networks are universal approximators. *Neural Networks* **2**: 359–366.

Karelson, M., Lobanov, V.S. and Katritzky, A.R. (1996) Quantum-chemical descriptors in QSAR/QSPR studies. *Chem Rev* **96**: 1027–1043.

Kier, L.B. and Hall, L.H. (1995) 'The molecular connectivity chi indexes and kappa shape indexes in structure–property modelling,' in Lipkowitz K.B. and Boyd, D.B., ed., *Reviews in Computational Chemistry*, Volume 2. Verlag Chemie, New York p. 367–422.

Kier, L.B. and Hall, L.H. (1999) *Molecular Structure Descriptions: The Electrotopological State*, Academic Press, San Diego.

Kimura T. and Funatsu, K. (1999) GA strategy for variable selection in QSAR studies: application of GA-based region selection to a 3D-QSAR study of acetylcholinesterase inhibitors. *J Chem Inf Comput Sci* **39**: 112–120.

Klebe, G. (2000) Recent developments in structure-based drug design. *J Mol Med* **78**: 269–281.

Labute, P. (2000) A widely applicable set of molecular descriptors. *J Mol Graph Mod* **18**: 464–477.

Leo, A.J. and Hansch, C. (1999) Role of hydrophobic effects in mechanistic QSAR. *Perspect Drug Discovery Des* **17**: 1–25.

Lewis, D.F.V. (2000) Structural characteristics of human P450s involved in drug metabolism: QSARs and lipophilicity profiles. *Toxicol* 197–203.

Lipkowitz, K. B. and Boyd, D.B. (eds) (1990) *Reviews in Computational Chemistry*, Verlag-Chemie, New York 1990–2003. Vols. 1–18.

Livingstone, D.J., Manallack, D.T. and Tetko, I.V. (1997) Data modelling with neural networks: advantages and limitations. *J Comput-Aided Mol Des* **11**: 135–142.

MacKay, D.J.C. (1992) A practical Bayesian framework for backprop networks. *Neural Computat* **4**: 415–447.

Mackay, D.J.C. (1995) Probable networks and plausible predictions – a review of practical Bayesian methods for supervised neural networks. *Comput Neural Sys* **6**: 469–505.

Maggiora, G.M., Elrod, D.W. and Trenary, R.G. (1992) Computational neural nets as model-free mapping devices. *J Chem Inf Comput Sci* **32**: 732–741.

Manallack, D.T. and Livingston, D.J. (1992) Artificial neural networks: applications and chance effects for QSAR data analysis. *Med Chem Res* **2**: 181–190.

Manallack, D.T. and Livingston, D.J. (1999) Neural networks in drug discovery: have they lived up to their promise? *Eur J Med Chem* **34**: 195–208.

Marrone, T. J., Briggs, J. M. and McCammon, J. A. (1997) Structure-based drug design: computational advances. *Ann Rev Pharmacol Toxicol* **37**: 71–90.

Martin, Y.C. (1981) A practitioner's perspective of the role of quantitative structure–activity analysis in medicinal chemistry. *J Med Chem* **24**: 229–237.

Mayer-Bäse, A. and Watzel, R. (1998) Transformation radial basis neural network for relevant feature selection. *Patt Recog Lett* **19**: 1301–1306.

Pearlman, R.S. and Smith, K.M. (1998) 'Novel software tools for chemical diversity,' in Kubinyi H, Folkers G, and Martin YC, ed.. *3D QSAR in Drug Design*, Volume 2. Kluwer/ESCOM, London p. 339–353.

Podlogar, B.L. and Ferguson, D.M. (2000) QSAR and CoMFA: A perspective on the practical application to drug discovery. *Drug Des Disc* **17**: 4–12.

Randíc, M., Vracko, M. and Novic, M. (2001) 'Eigenvalues as molecular descriptors,' in M. V. Diudea, ed., *QSPR/QSAR Studies by Molecular Descriptors,* Nova Science, Huntington, New York, p. 147–211.

Rumelhart, D.E. and McClelland, J.L. (1986) *Parallel Distributed Processing: Explorations in the Microstructure of Cognition,* MIT Press, Cambridge, Vols. I and II.

Rusinko, A., Farmen, M.W., Lambert, C.G., Brown, P.L. and Young, S.S. (1999) 'Analysis of a large structure/biological activity data set using recursive partitioning. *J Chem Inf Comput Sci* **39**: 1017–1026.

Salt, D.W., Yildiz, N., Livingston, D.J. and Tinsley, C.J. (1992) The use of artificial neural networks in QSAR. *Pestic Sci* **36**: 161–170.

Stanton, D.T. (1999) Evaluation and use of BCUT descriptors in QSAR and QSPR studies. *J Chem Inf Comput Sci* **39**: 11–20.

Testa, B. (2000) Structure–activity-relationships – challenges and context. *Pharm News* **7**: 13–22.

Tetko, I.V., Luik, A.I. and Poda, G.I. (1993) Applications of neural networks in structure–activity relationships of a small number of molecules. *J Med Chem* **36**: 811–814.

Tetko, I.V., Livingstone, D.J. and Luik, A.I (1995) Neural network studies: 1. Comparison of overfitting and overtraining. *J Chem Inf Comput Sci* **35**: 826–833.

Todeschini, R. and Consonni, V. (2000) *Handbook of Molecular Descriptors* Wiley-VCH, Weinheim p. 667.

Todeschini, R. and Consonni, V. (2001) *Dragon, rel. 1.12 for Windows,* Milano, Italy, http://www.disat.unimib.it/chm/.

Tong, W., Lowis, D.R., Perkins, R., Chen, Y. *et al.* (1998) Evaluation of quantitative structure–activity relationship methods for large-scale prediction of chemicals binding to the estrogen receptor. *J Chem Inf Comput Sci* **38**: 669–677.

Topliss, J.G. and Edwards, R.P. (1979) Chance factors in studies of quantitative structure–activity relationships. *J Med Chem* **22**: 1238–1244.

Vedani, A. and Dobler, M. (2000) Multi-dimensional QSAR in drug research: predicting binding affinities, toxicity and pharmacokinetic parameters. *Prog Drug Res* **55**: 105–135.

Waller, C.L. and Bradley, M.P. (1999) Development and validation of a novel variable selection technique with application to multidimensional quantitative structure-activity relationship studies. *J Chem Inf Comput Sci* **39**: 345–355.

Warne, M.A. and Nicholson, J.K. (2000) Quantitative structure–activity relationships (QSARs) in environmental research. Part II. Molecular orbital approaches to property calculations. *Prog Environ Sci* **2**: 31–52.

Winkler, D.A. (1998) Holographic QSAR of benzodiazepines. *Quant Struct-Activ Relat* **17**: 224.

Winkler, D.A. and Burden, F.R. (2002) 'Application of neural networks to large dataset QSAR, virtual screening and library design,' in Bellavance-English, L. ed., *Combinatorial Chemistry Methods and Protocols,* Humana Press, New York.

Winkler, D.A., Burden, F.R. and Watkins, A.J.R. (1998) Atomistic topological indices applied to benzodiazepines using various regression methods. *Quant Struct–Activ Relat* **17**: 14–19.

Wold, S. (1991) Validation of QSARs. *Quant Struct–Act Relat* **10**: 191–193.

Xing, L. and Glen, R. C. (2002) Novel methods for the prediction of logP, pKa, and logD. *J Chem Inf Comput Sci* **42**: 796–805.

Yasri, A. and Hartsough, D. (2001) Toward an optimal procedure for variable selection and QSAR model building. *J Chem Inf Comput Sci* **41**: 1218–1227.

Zhao, Y.H., Le, J., Abraham, M.H., Hersey, A. *et al.* (2001) Evaluation of human intestinal absorption data and subsequent derivation of a quantitative structure–activity relationship (QSAR) with the Abraham descriptors. *J Pharm Sci* **90**: 749–784.

Zheng, W. and Tropsha, A. (2000) Novel variable selection quantitative structure–property relationship approach based on the K-nearest-neighbor principle. *J Chem Inf Comput Sci* **40**: 185–194.

Index

Molecular Analysis and Genome Discovery edited by Ralph Rapley and Stuart Harbron
© 2004 John Wiley & Sons, Ltd   ISBN 0 471 49847 5 (cased)   ISBN 0 471 49919 6 (pbk)